Evolution, Games, and God

Evolution, Games, and God

The Principle of Cooperation

Edited by

MARTIN A. NOWAK

SARAH COAKLEY

HARVARD UNIVERSITY PRESS

Cambridge, Massachusetts

London, England

2013

Printed in the United States of America

Chapter 4 was first published as "Five Rules for the Evolution of Cooperation," *Science* 314 (2006): 1560–63, and is reproduced by kind permission of The American Association for the Advancement of Science.

Library of Congress Cataloging-in-Publication Data

Evolution, games, and God : the principle of cooperation /
edited by Martin A. Nowak, Sarah Coakley.
pages cm
Includes bibliographical references and index.
ISBN 978-0-674-04797-6 (hardcover: alk. paper)
Evolution (Biology)—Philosophy. 2. Evolution (Biology)—
Social aspects. 3. Cooperation. 4. Self-sacrifice.
5. Altrusim. I. Nowak, M. A. (Martin A.), editor of
compilation. II. Coakley, Sarah, 1951, editor of compilation.
QH360.5.E965 2013
576.8—dc23

2012041386

To our families,
who continue to teach us
the value of cooperation

Contents

Preface

This book is the product of an ongoing interdisciplinary discussion between Martin A. Nowak and Sarah Coakley on the philosophical, ethical, and theological implications of the evolutionary phenomenon of "cooperation" (and its human variant, "altruism"). From 2005 to 2008 Nowak and Coakley (then Mallinckrodt Professor of Divinity at Harvard Divinity School) co-led a research project at Harvard on "The Theology of Cooperation." This was generously funded by the Templeton Foundation, to which the editors would like to express their profound thanks for its financial support throughout these initial years of research, and also subsequently. The project enabled the editors to assemble a gifted team of visiting professors and postdoctoral researchers in mathematics, biology, evolutionary psychology, ethics, history of science, philosophy, philosophy of science, and theology. This book has subsequently been editorially constructed out of papers and presentations given at a variety of events at Harvard during those years, including a keynote conference at Harvard in May 2007, visiting presentations and discussions, and papers specially commissioned to cover relevant topics. All of these papers have been subsequently revised and updated, and brought into discussion in the various parts of this book so as to create a spiral of interdisciplinary debates at a number of different levels of reflection. The book also reflects important and controversial developments in our fields that have occurred since the end of the initial research project and that are noted *en passant* in our Introduction and elsewhere.

The study of evolutionary cooperation and its cultural implications is a fast-growing field. It seems set to continue to be a topic that invites intense methodological and philosophical scrutiny, and—with that—to suggest new points of debate about the evolutionary origins of ethical and religious systems of belief. Last but not least (as reflected in essays in the last part of this volume) its implications also seem to allow the possibility of a rational rethinking of the relation of evolutionary biology and theology itself. The book is intended for both a general and an academic readership, and for students at a variety of levels: it aims to provide a compact and thought-provoking account of the scholarly advances and disagreements in this intriguing field of study, and to serve as a platform for debate.

The editors would like to thank Michael G. Fisher and his team at Harvard University Press for their patience, support, and professional care in bringing this book to publication. They also especially wish to acknowledge the invaluable work of Drs. Zachary Simpson, Mark McInroy, David Grumett, and Philip McCosker for their superb research and editorial assistance along the way; and Lydia Liu and Michael Wojcik at the Program for Evolutionary Dynamics at Harvard for their attention to some finer details in production.

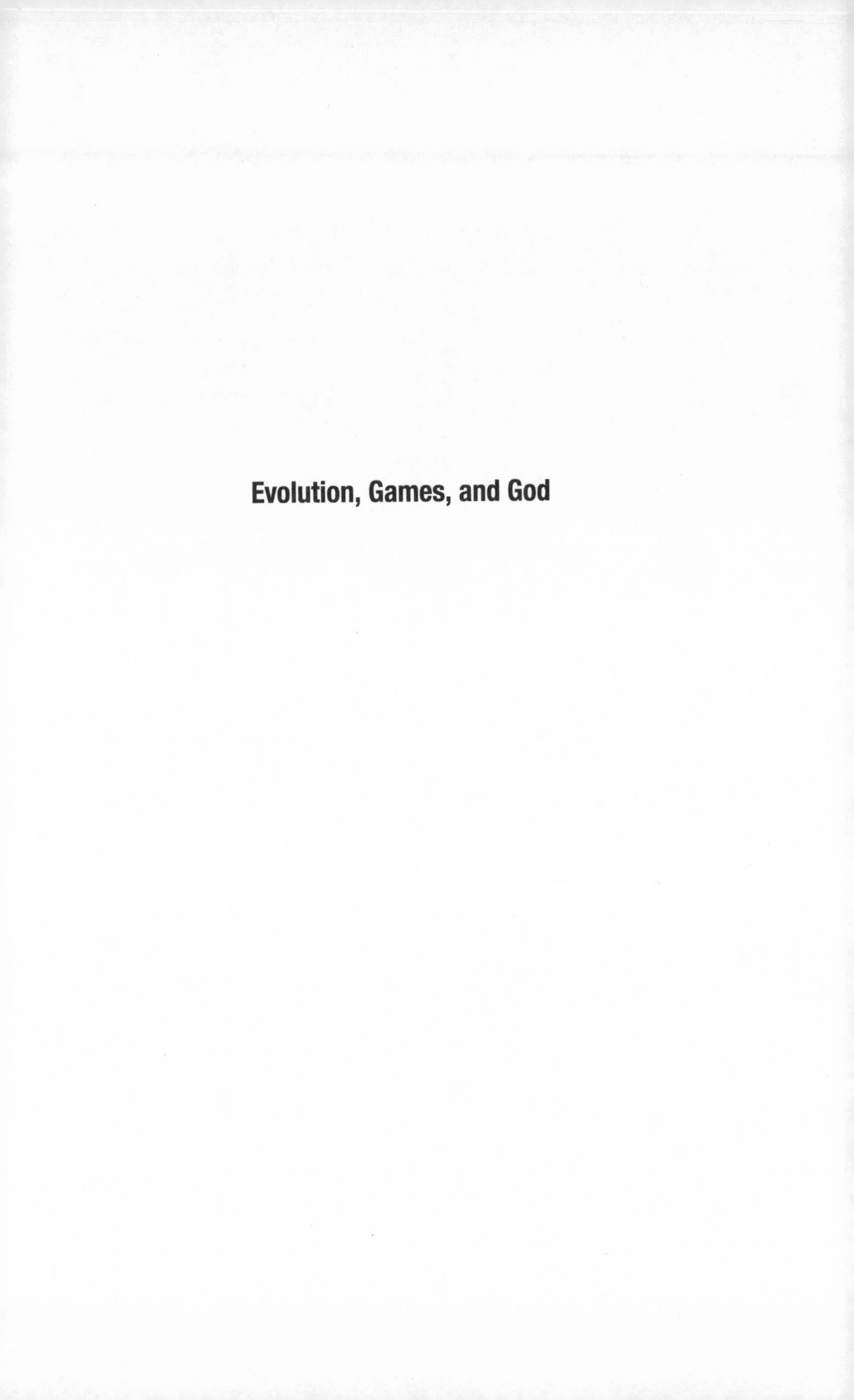

Evolution, Games, and God

Introduction

Why Cooperation Makes a Difference

Sarah Coakley and Martin A. Nowak

This is a book about evolution, and more particularly about the phenomenon known to evolutionary biologists as "cooperation"—in layman's terms, the manifestation of "unselfish" dimensions of evolutionary processes. The aim of the book is to assess the place and significance of cooperation within the dynamics of evolution, and to do so in a radically interdisciplinary way, bringing to bear not only the insights of evolutionary biology (both mathematical and empirical), but also of the social sciences, philosophy, and theology. Our joint undertaking is based on the supposition that recent developments in the understanding of evolutionary cooperation are as important for "cultural evolution" (the evolution of languages, influential ideas, human projects, and elective behaviors) as they are for genetic evolution, and that they thus demand the insights of an array of scholars to ponder and debate the broader implications of this revolution in evolutionary thinking for human behaviors and beliefs.

In attempting to clarify the range of these implications, this book breaks distinctly new ground. We invite the reader to participate in this adventure in ideas. The book is intended both for the general reader and also for students at a variety of levels (undergraduate and graduate): it aims to provide a compact and accessible account of the current scholarly advances and debates in this field of study, and it may fruitfully be used in discussion groups and teaching. The section at the end of this Introduction explains how it may best be used to generate discussion and debate in a teaching context.

As will be demonstrated in this volume, once the impact of cooperation on evolutionary processes is clarified mathematically, it calls into radical question the now well-popularized story of evolution as something propelled entirely by "selfishness." Indeed, it is an interesting question whether future historians of science will look back at the latter part of the twentieth century, specifically, as a period in which the intensification of the idea of evolutionary selfishness (well beyond that hypothesized by Darwin, as will be highlighted in Part I of this book), and allied with a form of genetic determinism of which Darwin knew nothing,[1] was curiously correlated with a Western market boom that has subsequently burst. Yet, ironically, it was in this same time span that many of the breakthroughs in mathematical biology occurred that continue today and are the center of our focus in this book.[2] The new message about the significance of cooperation in evolution is, however, as yet rather little understood in the general populace at large, although—as we believe—its importance can scarcely be underestimated, both culturally and morally. Getting clear about how and when cooperation can flourish as an evolutionary "strategy,"[3] when it can become stable in a species or species subgroup, when it is most likely to be threatened or even obliterated in an evolutionary population, and what the evolutionary implications are, are the first tasks of this book.

As we shall see, one of the most intriguing aspects of this exploration is the discovery that *only* in conditions of heightened (or "stable") cooperation could breakthroughs in our evolutionary past of major significance have occurred in the first place. This principle applies equally to the crucial moments of transition from self-replicating molecules to groups of molecules working together, from individual replicators to chromosomes, from bacteria to eukaryotic cells, from asexual reproduction to sexual reproduction, and so on up to the movement from primate societies to human societies.[4] And this remarkable realization in turn raises speculative questions about whether further evolutionary breakthroughs are possible or imminent, and whether humans, as agents of cultural evolution, may now need to play an intentional and conscious part in the success, or failure, of future such moments of evolutionary challenge. Our current ecological crisis is perhaps the most pressing case in point, although the immediate world economic struggles could also be read in this light.

Although the field of mathematical biology that has charted these dynamics in evolution is still in its relative infancy, it is becoming increasingly clear (and this is another central theme of this volume) that the questions it is opening up press well beyond the merely descriptive and empirical, and lead on to demand normative, *philosophical* analysis. It is no simple matter to move from the discovery of certain kinds of stochastic patternings in evolutionary processes to assessing the implications and *meanings* of such a discovery—but it is inevitable that such issues will arise. Are there, for instance, regularities, even statistically predictable ones, in the dynamics of evolution, and if so, what wisdom and understanding can be gained, ethically and culturally speaking, from comprehending these dynamics?[5] If there are such regularities, could it be that the issue of "teleology" in evolution, long debarred from biological discussion because of its historic theological associations, may be ripe for new analysis?[6] And how then should exponents of *metaphysical* accounts of evolution as a whole, whether secular or theological, respond to the new discoveries made by mathematical biology in this area?[7]

Further, if we can trace in the history of evolution the development of cooperative tendencies that may seem to prefigure important ethical intuitions in the human realm, what does this mean for our own ethical and "metaethical" choices in contemporary debate?[8] What, further, might the presence of such evolutionary cooperative tendencies mean for the disputed relation of ethics and theology? It is the desire to face big questions such as these that has propelled the research group that has produced this volume. While we certainly do not always agree with each other (as will be clear from the pages of this book as the sections unfold: a large spectrum of views is represented among the contributors), we present the reader with a range of debating points to inspire further discussion, teaching, and research.

What Is "Cooperation"?

Before going any further, the main semantic question must first be pressed: what *is* "cooperation"? This is not as simple an issue as it might seem. In ordinary language use, of course, the word means any sort of working together to forward some shared goal. Thus Nowak and Coakley, to take an obvious case in point, have in this ordinary sense "cooperated" in an enjoyable three-year research project together to produce this book. But in evolutionary biology the meaning is more precise than this, implying not just a shared undertaking, but the willingness of the participants to *forego* personal advantages for the sake of it. Even to put it thus, of course, is to continue to smuggle human intentionality and purpose into the definition in a way that is inherently question-begging for the nonhuman world (whereas we may hypothesize "intentions" of a *sort* in the higher animals, it is misleading and nonsensical to speak of bacteria or genes as if they had cooperative "intentions"). For the purposes of this book, then, the generic working definition of "cooperation" in play will be this: *"Cooperation is a form of working together in which one individual pays a cost (in terms of fitness, whether genetic or cultural) and another gains a benefit as a result."*

Now even this definition, as it happens, is not without interesting remaining ambiguity, and this problem continues to bedevil discussion between empirical and mathematical biologists (as will be discussed a little more at the end of this Introduction). A further definitional pitfall is that biologists often fail to distinguish this generic definition of "cooperation" (which clearly can be applied right across the evolutionary spectrum), from a phenomenon which adds to it the special motivational and intentional features of what in ordinary language is now called "altruism."[9] Unfortunately, it has become almost standard in evolutionary biology to apply the term "altruism," or "evolutionary altruism," to family resemblance behaviors or outcomes of a "sacrificial" sort in the entire spectrum of evolutionary development, and thus to blur the crucial distinction between a form of "cooperation" which just *happens,* and "altruism" proper, which is *motivated* by specific goals or affective commitments. Such a blurring then fails precisely to confront one of the most interesting and contested questions of evolutionary progression—that is, when *is* it, exactly, that "cooperation" can properly be called "altruism"? Is this moment precisely coterminous with the emergence of human consciousness and language, or are there already intimations of it in some of the behaviors of the higher mammals?

In order not to suppress or occlude this vital question, we propose here to keep the terms "cooperation" and "altruism" at least nominally distinct.

Thus, for the purposes of this volume, our definition of altruism will be: *"Altruism is a form of (costly) cooperation in which an individual is motivated by good will or love for another (or others)."*

It is unlikely that the distinction between "cooperation" and "altruism" as rehearsed in this volume will have the effect of imposing any new consistency on biologists' untidy semantic usage, or in effecting unanimity over what is essentially a contentious underlying arena of theoretical debate.[10] But perhaps the most important point to cling to as we traverse this semantic minefield is that our own initial, generic definition of "cooperation" ("a form of working together in which one pays a cost in terms of fitness and another gains a benefit"), which is itself admittedly still a little conceptually fuzzy at its edges, can be rendered precise when expressed mathematically in the terms of the Prisoner's Dilemma (PD)[11] or some other "game;" and herein lies the main achievement of mathematical biologists, and the key to their claim to be edging evolutionary theory closer to a precise quantitative understanding than ever heretofore. We are talking, then, about a precision that was not attainable in evolutionary theory until mathematical analyses were utilized, and while this comes with obvious theoretical advantages, there are also accompanying methodological and philosophical dangers that our volume sets out to pinpoint and explore. Key to these debates, needless to say, is the issue of what sort of *explanatory force* mathematical accounts of evolution can rightly claim, especially in comparison to the more conventional study of empirical manifestations and patterns in evolution, with their descriptions in ordinary language.

Let us now explain something of how this much-vaunted mathematical and "game theoretical" precision has been achieved in the field of evolutionary dynamics, and at the same time indicate how the mathematicalization of evolutionary theory has advanced on Darwin's shoulders; these matters will exercise us in Parts I and II of the book.

Darwin, Game Theory, and the Mathematics of "Cooperation"

The genius of Charles Darwin was to disclose and clarify the crucial role of "mutation" and "selection" in the constraining of evolutionary processes. But Darwin himself, especially in his later work, was well aware that there was something else going on in evolution that could not be accounted for solely in terms of these two fundamental principles. In Part I of this book, John Hedley Brooke gives in Chapter 1 an account of Darwin's famous intuition, in *The Descent of Man* (Darwin [1879] 2004, 157–8, my emphasis), that a tribe whose members were "always ready to aid one another, and to sacrifice themselves for the common good would be victorious over most other tribes; *and this would be natural selection*."

Darwin read such a "sacrificial" tendency naturalistically; the notion of an evolved capacity for "sympathy" was crucial to that analysis. But he had no tools for understanding the phenomenon mathematically in the way that "cooperation" (or motivated "altruism" in the human realm) is now analyzed by evolutionary theorists; instead he saw it as a concern (a social instinct, not a conscious calculation) for the welfare of others that was intimately connected with social flourishing and the emergence of a moral sensibility. Even further from his thinking was any kind of utilitarian calculus (as economic game theory now characteristically takes for granted), in which a conscious bid for "success" would be written into the "payoffs" between various strategies of response to others. Darwin's intuitions, nonetheless, were remarkably sound: he had already put his finger on a feature of group dynamics that might act as a countervailing factor against the dominant success of merely individual selfish factors in evolutionary selection. Brooke shows how this insight contributed to a certain "Victorian" optimism in Darwin that human evolution would, in due time, deliver ever more positive moral developments in an unselfish direction; and this presumption Darwin clung to even despite his own progressive loss of faith in a providential God directly ordering the minutiae of evolutionary affairs. Nonetheless, Darwin continued to see patternings and progression in the evolutionary processes he had himself disclosed as a scientist. In that sense he did not deliver any final or fatal blow to the idea of *evolutionary* "teleology," although he went back and forth on the extent to which he was prepared to stress this idea in his work, given its classic connections to "natural theology."[12] In that sense he intriguingly left open some big questions to which this book returns.

In Chapters 2 and 3, Thomas Dixon and Heather D. Curtis add to the historical lessons that may be retrieved from the nineteenth century for our topic. Dixon reminds us that "altruism" was a word originally coined by Auguste Comte (in 1851) in a consciously secular mode, building on Darwin's optimistic sense of moral advance and hypothesizing the emergence of a universal morality *sans* religion. It was a considerable irony, then, as Curtis goes on to outline, that certain liberal churchmen in the United States and Europe thereupon seized on this optimistic and moralistic rendition of evolution to preach a new version of Darwinism fully compatible with Christian thought: "altruism" had been recaptured for the church, and with it a suitably romantic view of gender, in which women (the "angels in the house") were to be allocated the major role in "altruistic" endeavor. This blithely melioristic marriage of Darwinism, liberal Christianity, and romantic gender theory was in due course to be expunged from both memory and influence by the barbarous events of the first World War; and thus what we learn from reminding ourselves of this particular phase of late nineteenth-century Darwinism/Comteianism is the ever-present danger of fanciful or willful "eyes-of-the-beholder" renditions of evolutionary processes that merely express certain regnant cultural obsessions. Not that such difficulties have been expunged from all contemporary mathematicalized accounts of "cooperation" and "altruism," either, as we shall soon discover: *caveat lector.* For Dixon, however, the stern lesson is that evidences of these phenomena in evolution do not, and cannot, *in and of themselves,* dictate any particular moral lessons. Certainly it is naïve and misleading to assume that "cooperation," *in se,* is a good, while its opposite, "defection," is somehow morally culpable. But we shall nevertheless need to return to this issue of whether a natural evolutionary phenomenon like "cooperation" could, according to any particular "metaethic," be seen as *intrinsically* value laden. This is one of the most interesting questions in the burgeoning area of evolutionary ethics, and it has no one ready answer.

Our unfolding story of the study of evolutionary "cooperation" next turns, in Part II, to the twentieth century, and to some of the remarkable developments in the understanding of evolutionary dynamics that have come from the application of game theoretical techniques to it. The detailed story of the invention of game theory, and of its many branches of application in a number of disciplines subsequently, has been told elsewhere, in both technical and more popularized forms.[13] What is of particular note in that story for our purposes here is the way in which its creative application to evolutionary theory heralded the return of confidence in theorizing *social* evolution, especially after the diversion from such a focus created by the recent new excitements in the development of genetic biology.[14] The mathematical study of evolution, however, is intrinsically a study of "populations" and their internal dynamics, and it has made great advances in demonstrating how social dimensions of evolution crucially affect fitness. In this part of the volume we choose to take just three representative "dips" into this research terrain, ones which purposefully indicate some of the most ambitious recent theoretical advances, but which also reveal unfinished business and points of intense methodological debate.

First, in Chapter 4, we present a survey article by Nowak that gathers together the accumulated insights of illustrious forebears in the field of modern evolutionary game theory (including Haldane, Maynard Smith, Hamilton, Trivers, and Axelrod), and adds Nowak's own distinctive summarizing narrative of the *five* particular "mechanisms" that resolve conditions of "social dilemmas" in such a way that cooperation may emerge as favored by natural selection. These mechanisms are "kin selection," "direct reciprocity," "indirect reciprocity," "network reciprocity" (also known as "spatial selection") and "group selection."[15] It is important to understand here that these "mechanisms"[16] are independent of the particular game that is being investigated—such as the PD, for example. They are interaction structures that define who interacts with whom (and how) with what "payoff," and who competes with whom for reproduction (genetic or cultural). Given that in the classic PD "defection" is *prima facie* the best (and for humans, "rational") strategy,[17] the systematic clarification of how cooperation can nevertheless come to dominate and even engender new levels of evolutionary advance (and how, inversely, systematic "defection" in a population actually leads to evolutionary decline), already represents a point of critical reflection against those accounts of the evolutionary spectrum that focus only on "selfishness" as basic. It is precisely the clarification of how these mechanisms can provide the possibility for such a development that constitutes the major advance in evolutionary theory afforded by the mathematical approach. For now we understand much more fully than before how *significant* "cooperation" is in evolutionary processes and in the capacity for species to flourish. Indeed, Nowak is willing in this article to hazard the controversial conclusion that "cooperation" is actually a third "principle of evolution," next to mutation and selection. These three "principles" are, however, arranged in a hierarchy: mutation is first needed to generate (genetic) diversity upon which selection can act; mutation and selection then together give rise to evolutionary processes that sometimes lead to cooperation—if one or more of the five "mechanisms" are operative. For, whenever life discovers a new level of organization (such as the emergence of cells, multicellular organisms, insect societies, or human language), cooperation is involved in one form or another. Mutation and selection alone, without cooperation, may not give rise to complexity.

Nowak's chapter thus provides a succinct account of how evolutionary theory has been rendered more precise in recent decades by the use of mathematical tools, and especially in charting the relation of mutation, selection, and cooperation. But his account also implicitly opens up thorny theoretical issues which it is the task of the rest of this volume to grasp more firmly. Not the least of these is the already-mentioned conundrum about how *true to life* strict mathematical PD dilemmas really are. Do the evidences of empirical biological observation actually confirm the mathematical calculations in all details, and if not, what exactly is the "force" of the mathematical schemas?[18] Moreover, is the intensive focus on the PD found in the current literature distorting, given that a large range of other "games" are also available for investigation?[19] Finally, and perhaps most troublingly, is it intrinsically misleading to subsume nonhuman and human populations together under Nowak's Five Rules without clear differentiation? *Can* mathematics disclose anything about specifically human motivations?

The other two chapters in Part II already alert the reader to some of the unfolding implications of these questions. Christoph Hauert's Chapter 5, which follows Nowak's, chooses to focus on the Snowdrift Game (SG) because it involves a "more relaxed social dilemma" than the PD, and as such is a scenario intrinsically more likely to produce cooperation. It is also a game that appears to have more manifestations in biological life than the strict PD (although in practice it can be difficult to discern a clear difference between the two by empirical observation). The narrative thought experiment of the SG is as follows. Two drivers are both stuck in a snowdrift, and the best strategy depends on the coplayer's decision: if the other driver shovels, I can get away with shirking, but when facing the possibility of a lazy counterpart, it is better to start shoveling myself instead of remaining stuck in the snow. When the SG is run repeatedly, the players can additionally utilize what Hauert calls an "investment" toward their future goal, which also intensifies the cooperative tendency. (Again, here, we see the language of human *motivation* being used to inflect a game initially applied illuminatingly to meerkats or musk ox—an issue that will shortly become a point of focus.) What Hauert argues is that the strict PD (even in its "continuous" form) is relatively uncommon in nature as such, and that an analysis of SG, or other possible variant games, may be more illuminating in explaining the long-term survival and proliferation of cooperative traits.

Similar theoretical complications are at stake in Johan Almenberg and Anna Dreber's account in Chapter 6 of how game theory is utilized by *economists* in ways that often fail to acknowledge the insights of evolutionary *biology* about the relationship of animal and human cooperation. The key issue here is how many of our game strategies are really about cold "rational choice" (as economic game theorists in general assume), and how many are evolutionarily derived or "instinctual": if we have inherited from our evolutionary ancestors a strong primary tendency to cooperate in some circumstances, then willful selfishness may not be so obvious or primary as it seems, nor may it always be to our ultimate benefit. Almenberg and Dreber also probe the questionable assumption of many economists that Homo sapiens should only be interested in short-term *material* "payoffs" (almost always monetary in form), since the very notion of a "utility function" may be systematically ambiguous, or indeed systematically open to question as a metaethical presumption. As it turns out, recent experiments in games in laboratory conditions reported by Almenberg and Dreber show that humans are remarkably willing, seemingly "naturally," to consider options other than merely self-interested ones. Even in the *Public Goods Game* (a form of PD played communally with a strong chance for "free riding," and thus one of the most difficult scenarios in which to produce consistent cooperation: the "rational" option is clearly to contribute nothing and hope others will), 40–60% of people do in fact contribute to the public good. What is going on here, then? Almenberg and Dreber draw on recent results from empirical evolutionary biology to suggest that humans may be *extending,* culturally and morally, a cooperative evolutionary inheritance from our higher mammalian ancestors; as humans, we are overall *more* cooperative than apes, even though there are striking variations in cooperative tendencies even within the ape family itself. In sum, "defection" is by no means the overall default evolutionary pattern, in the PD or variants; in many circumstances the opposite seems to be the case, "against all reason." No wonder, then, that economists are constantly surprised by what they see as "fuzzy thinking" in human game playing!

It is worth pausing at this point to reflect on the lessons of Parts I and II of this volume, because the rest of the book is devoted to exploring, and arguing about, the implications.

Among the major achievements of mathematicalized evolutionary theory to date, as we have seen, are the clarification of a variety of circumstances in which cooperation can take a hold in populations over time, despite the apparently overwhelming unlikelihood of such an outcome, and the demonstration, *vice versa,* that a consistent and thoroughgoing manifestation of "defection" in a population leads to that population's evolutionary decline. However, we have also noted that there are a number of dangers in focusing mathematically on the phenomenon of cooperation across the entire evolutionary spectrum in any simplistic way, and the historical lessons from Part I should give us fresh pause here too and warn us against falling into a new, misleading lurch of the pendulum against the recent cultural obsession with "selfishness." Such might lead to a falsely *optimistic* fantasy about "natural" cooperation, or encourage belief in an ascending moral sensibility arising inexorably from the evolutionary world: we would be back again with Darwin's naïve "Victorian" meliorism.

Moreover, there is another, related, trap: even the literature we have already surveyed shows a tendency to blur the distinction between prehuman and human forms of cooperation, and to attribute intentionality to members of prehuman populations (let us call this the problem of "anthropomorphism"). Yet, paradoxically and conversely, the same literature can often talk carelessly as if even human choices were predetermined by a narrow utilitarian or consequentialist calculus (let us call this the problem of threatened moral "determinism" or "reductionism"). Whatever the remarkable theoretical gains of mathematical evolutionary theory, then, it clearly needs holding to account in the areas in which it tends to make unacknowledged presumptions—those that back onto philosophy of science, metaphysics, and (meta)ethics. And possibly the most immediate and pressing issue is the one that has already tripped us up several times in this Introduction: the problem of how to clarify the "explanatory force" of mathematical accounts of evolution, and in particular the relation of such mathematical accounts to issues of human *motivation.* What purchase can mathematical prediction (albeit stochastic) have on the passage from cooperation to altruism: what specifically human factors are at stake? This takes us on to the central concerns of Parts III and IV of this book.

From "Cooperation" to "Altruism": Social Science Theories and Philosophical Challenges

Part III entertains a number of current speculations from the social sciences about what makes *human* cooperation (= "altruism") distinctive. It is important to be aware that these *are* so far speculations, and as such each introduces extra presumptions into the methodological fray that could themselves be questioned or countered. In Chapter 7, the psychologist Stephen M. Kosslyn suggests that it is our capacity to "borrow"—as it were—each other's minds for the purpose of mutual benefit that marks us off as humans: our brains are already hard-wired for cooperation through the use of "mirror neurons," but only humans can *share* their mental capacities. This then, he thinks, may be the neurophysiological bedrock of human altruism. In Chapter 8, however, Dominic D. P. Johnson avers, in some contrast, that the emergence of specifically human cooperation is not at all straightforward but must be very closely related to the evolution of "religion": according to his "Supernatural Punishment Hypothesis," "gods" are humanly projected to provide rewards and punishments for the performance of ethical demands that can only arise when human language and mental cognition make the intuiting of others' intentions, and thus the attribution of guilt, possible for the first time. For Johnson's theory, prehuman cooperation can *only* be "selfish," and thus for Johnson human intentionality must supervene on this hostile *scenario* in a striking and novel way to achieve altruism. Only the "gods" can do the trick.[20]

In Chapter 9, Maurice Lee, using different methodological tools, takes on the problem of how to explain the felt mental *difficulty* of making truly altruistic decisions: *contra* Kosslyn, he sees the development of other-regarding altruistic behaviors as far from "fun," and wonders what is going on in the brain when humans struggle to achieve such postures of selflessness. Exploring the *opposite* phenomenon from Kosslyn—not the extension of self into others for one's own profit, but the "voluntary withdrawal or disavowal of one's own interests" for the sake of others—he hypothesizes neuroscientifically that the internal, felt conflict between the satisfaction of immediate desires and unselfish love may itself be adaptive. Humans have the notable capacity to think through, and discriminate between, their various desires, and—while this struggle is painful at times—it also gives them great evolutionary advantages. Focusing on the purported "neural structure" of altruistic decisions, Lee points out that conflicts between impulses originating in different parts of the brain can to some extent be tracked via neuroimaging, but the examination of neurobiological processes cannot, as such, solve the *theoretical* mystery of how "altruism" is different from prehuman "cooperation." Lee sharpens his thesis by choosing to push a stronger definition of "altruism" than our volume otherwise proposes: for him it must be "pure, unadulterated, other-directed concern, without any admixture of self-directed motives." However, appealing to Nowak's work, he acknowledges that the widely established existence of prehuman forms of "cooperation" may form a more fertile preparatory ground, even for this strong "altruism," than Johnson's sharp contrast of prehuman and human scenarios assumes.

The essays in Part III of this book thus already reveal striking differences of prior presumption in dealing with "cooperation" and "altruism", and we should be alert to these divergences; none of these tenets is thoroughly argued in such short compass. Although all three authors take it for granted that "selfishness" is *more fundamental* than cooperation or altruism, their ways of accounting for how specifically human altruism comes into being are widely diverse and reflect their particular disciplinary backgrounds. Moreover, in bringing "gods" into his own theory, Johnson assumes, in the spirit of Ludwig Feuerbach or Sigmund Freud, that such "supernatural entities" are human psychological projections (presumably to be outgrown by contemporary sophisticates?), whereas Lee, citing St. Paul and John Calvin, is clearly nuancing his own neurophysiological narrative with underlying suggestions of the theological categories of sin and evil, love and grace. Johnson, however, simply presumes that only punishment (or the threat of it) can bring about human altruism, a matter which is now becoming contentious even in the game theoretical field.[21] It may be clear, then, that although Part III has started to attend to the question of human altruistic motivation using social-psychological and neuroscientific techniques, it has only scratched at the surface of the *philosophical* choices implied by it in the philosophy of mind and in metaethics. The nature of human personhood, intentionality, and freedom is at stake, and the mind/body problem thus cannot be averted, with all its attendant philosophical debates and choices. A deeper analysis of the categories of "explanation" and "cause" too (as pursued explicitly by philosophy of science) will need to be probed if we are to get clearer about the relation of cooperation and altruism at *different* levels of evolutionary progression. It is to these issues that Part IV is devoted.

Jeffrey P. Schloss's contribution, Chapter 10, in fact sharpens the philosophical issues to be faced by adopting the same, strong definition of "altruism" as Lee, and insisting that game theoretical analyses could *never* throw any light on it; in embracing this position he takes it as read that mathematical accounts of evolutionary "cooperation" are, at best, narratives of higher forms of sophisticated "selfishness," and this stance explicitly throws down the big "explanatory" philosophical gauntlet: what *sort* of "explanations" are at play in game theoretical work, and how much of reality can thereby be captured? Is the whole "scientific" story of evolution necessarily one of the forms of "selfishness" (albeit laced with strands of self-interested "cooperation") and "conflict," undergirded by "reductionist" or "utilitarian" norms?

In Chapters 11 and 12 respectively, Justin C. Fisher and Ned Hall take up these philosophical issues with critical verve, and by clearing away a great deal of prevalent philosophical muddle in biological debates about cooperation and altruism, they ironically (for they are both avowedly secular philosophical commentators) keep a space for theistic metaphysics at the discussion table—only providing that its potential explanatory ambitions are rightly *located.*

Fisher sets out to circumscribe, first, the type and scope of "explanation" that a mathematical account of "cooperation" can aspire to. *Prima facie,* he argues, there are at least three reasons why one might be skeptical or anxious about the fulsome explanatory ambitions of game theoretical models of "cooperation" or "altruism": first, in philosophy of mind, if one seeks to maintain a place for some sort of "nonphysical" account of mind; secondly, in the area of human freedom, if—again—one does not choose a "physicalist" or "determinist" option in the range of philosophical accounts of freedom; and third, in the general area of what Fisher, as a philosopher, calls "complexity": human behaviors are rich, complex, and strange, and we should baulk at any mathematical ambition to give full and final explanations of such. But then, having highlighted the problems, Fisher suggests an elegantly simple solution to evade them: provided mathematical accounts do not overstep their proper boundaries and explanatory expectations, "there is a lot to be said for an explanation that picks out a few highly relevant factors, and shows how a pattern of results depends upon those factors." Hence, as long as such mathematical models are aware of their explanatory limitations, and do not succumb to an inappropriate hubris (viz., a desire to explain everything by mathematics alone), there seems no *logical* reason that mathematical accounts of "cooperation" and "altruism" should not do creative business alongside the work of philosophers who hold a variety of views about philosophy of mind, freedom, and the ultimate metaphysics of evolutionary processes. The trouble only comes when evolutionary biologists themselves make unargued *presumptions* about these philosophically disputatious areas, or claim to have "explained" them all away.

Hall follows up, in Chapter 12, with an amusing extended thought experiment (an imaginary debate between a "Sophisticated Christian Theist" [SCT] and a "Reductionist Atheist Physicalist" [RAP]) that is designed to chasten further any falsely simplistic philosophical presumptions accompanying game-theoretical accounts of "cooperation." Rather than simply defending mathematical modeling as providing a *partial* account of reality (as does Fisher), he focuses instead on a central mistake of semantic and methodological blurring that proponents of such modeling often fall prey to. Hall calls this major mistake the "cross-platform" fallacy: it is the idea that evolutionary game theory can explain, *in a uniform fashion,* the development of cooperative behaviors in (say) slime molds, and the development of altruistic behavior in human societies. That really is a "myth," says Hall (in the negative sense of a falsehood), and moreover it is much more of a problem than the rather elastic issue of "reductionism," since those who in some arenas of science or philosophy would count themselves as "physicalists" often, *in fact,* hold quite robust notions of human "intentionality" when put on the spot about freedom of choice. "Reductionism," then, is not a simple or uniform threat to the SCT; even the "RAP-er" may turn out to have a softer, or more flexible, philosophical center here than first imagined. He certainly need not be committed, for instance, to the presumption that all manifestations of human "altruism" are *merely* disguised selfishness. Indeed, he may turn out to have more in common with his "SCT" friend than would initially appear plausible. Provided they both acknowledge the "cross-platform" explanatory danger, they can proceed to an amicable conversation about the relative "depth of applicability" of *particular* mathematical modelings of "co-operation" and "altruism" to *particular* evolutionary populations.

Yet, as Hall admits, what finally will divide them is their vision of "what ultimately grounds" evolution as a whole. They may clear up some initial semantic disagreements, and highlight the lurking philosophical muddles about the type of "explanation" involved in mathematical accounts of "co-operation," but if they are tempted to make any grander judgments about the "meaning" of cooperation and altruism, *tout court,* they will, perforce, find themselves in the realm of metaethics and metaphysics. Even if the RAP-er at this point declares that "there *is* no meaning," this will itself be an (entirely contestable!) hermeneutical and metaphysical judgment. The argument can and must go on, but it will be at a different level of philosophical analysis from before.[22]

From Ethics to Theology: The Implications of Evolutionary "Cooperation" for Metaethical and Metaphysical Choices

By now it will be clear that interdisciplinarity is not just an optional extra in the biological study of "cooperation," but a philosophical necessity. Whether theology may also appropriately enter the discussion here is obviously a much more contentious matter for secular science, but for at least some "metaethical" theories (e.g., the "deontological" theory[23] in its nonsecularized Kantian, or Kierkegaardian, forms), "God" is by no means an optional extra in matters of "altruistic" decision. So if a full range of metaethical theories compatible with cooperation are to be debated, discussions of matters of theology too will be unavoidable. Likewise, if the metaphysical "design" implications of the phenomenon of cooperation are to be carefully weighed, "good old God" may again enter the stage of discussion. If atheistical evolutionary theorists want to debar such a discussion, one may riposte that they should at least be willing to disclose their own fundamental metaphysical presumptions for critical discussion. All's fair in love and war.

Parts V and VI of this book turn, accordingly, to debates about the possible ethical and theological implications of evolutionary "cooperation." Inevitably, some of the sharpest divergences among our contributors are here to be found, but the arrangement of differing views is orchestrated in a way that clarifies fundamental philosophical choices and thus aids further debate.

We recall that in his essay in Part I, Thomas Dixon urged that evolutionary cooperation per se requires no *particular* ethical interpretation or framework. Yet Almenberg and Dreber also underscored, in Part II, the overwhelming prevalence in most game-theoretical analyses (both economic and biological) of "utilitarian"[24] or "consequentialist"[25] metaethical presumptions: "strategies," "payoffs," "utility-functions," and (self-interested) "success" are precisely the name of the game in these discourses. One of the most confusing dimensions of the assessment of game-theoretical renditions of evolution, therefore, is that of untangling the mathematics and the empirical evidences from this oft-accompanying utilitarian dogma. In Part V we thus present four different metaethical alternative renditions for scrutiny. Each attempts to do justice to evolutionary materials; each presents a discerning reading of evolutionary evidences for ethical theory; and each is—whether explicitly or implicitly—highly skeptical about the "utilitarian" alternative.[26]

Marc Hauser's Chapter 13 was written some while before he left his post at Harvard (see Wade 2011), but we have nonetheless chosen to maintain his important theoretical voice in this book. For it represents a distinct, and highly suggestive, account of metaethics in its relation to evolutionary evidences of "cooperation." Hauser argues that humans worldwide are hardwired with evolutionarily derived moral "intuitions" or "principles" that are as basic to us as the capacity for language itself. These intuitions are not, note, specific ethical *directives;* rather (and here Chomsky is appealed to as proposing a parallel theory for language), they are more or less unconscious ethical sensibilities that are even insulated to some extent from local cultural (or indeed religious) influence; they are just evolutionarily "basic." However, when put into action, they generate moral judgments which—as Hauser sees it—are remarkably akin to modern secularized Rawlsianism![27] Still, neither "reason" (as in Kant) nor "emotions" (as in Hume), nor means-end calculations (as in Bentham) are *basic* to ethics, in Hauser's view. More fundamental for him is this inherited intuitional "moral grammar" that supplies the undertow for all particular ethical judgments. And if we cast our eye back over the varying evolutionary contexts of prehuman "cooperation" (Hauser does not, note, commit the "cross- platform" fallacy as highlighted by Hall), we may hypothesize the *gradual* development of this arena of moral intuition from its more immediate "means-end" manifestations to its more "motivational" (deontological) forms. On this rendition, then, "cooperation" and "altruism" are by no means identical but related by family resemblance; and somewhere along the evolutionary line of progression, the sustaining of a cooperative matrix allows the "basic" components of moral intuitions to manifest themselves and to promote, in due course, more sophisticated articulations of altruistic motivation. Like Kosslyn, Hauser thinks that "mirror-neurons," and the capacity for "philosophy of mind" (in the psychologists' sense), are crucial in the transition to "indirect reciprocity" and altruistic motivational strategies in games; but unlike Johnson, Hauser sees no reason to suggest that "gods" played a vital role in primitive ethical development, since for him "basic" ethical sensibilities are *deeper* than later, optional, religious admixtures.

Hauser's ethical theory, then, has the real merit of essaying some sort of account of the transition from "cooperation" to "altruism" proper, but it is precisely the attempt to pinpoint the transition to motivational manifestations in the prehuman realm that is controversial in his account. Friedrich Lohmann's Kantian approach in Chapter 14 is equally, if not more, interested in motivation. Indeed, he makes an initially sharp and critical separation of the non-normative behaviorism of an economic game theorist such as Ken Binmore, and the essentially normative structure of Kant's idea of a moral obligation met by a good will. This first distinction allows Lohmann to make a crucial ethical point that is also often overlooked by game-theoretical biologists: there may well be "cooperation" (even "altruism" in the sense mandated for this book), but it can nonetheless be *falsely* motivated—for example, by vanity or pride, or simply by the wrong goals. It is one of the great values of a Kantian ethic of a universal moral *imperative,* then, that motivation is central to what is weighed in the balance. Perhaps rather surprisingly, Lohmann is willing by the end of his essay (rather in Fisher's earlier spirit of a division of the spoils) to allow that mathematicalized evolutionary accounts of "cooperation" and "altruism" may nonetheless be of *some* use and interest to the Kantian ethicist; the empirical evidences surveyed in business ethics, for instance, may be illuminated to some extent by game theory, but the essentially normative moment in ethics will remain unavoidable and distinct.

Jean Porter's contribution, Chapter 15, is in part a direct response to Hauser's theory, and is written from the perspective of a neo-Aristotelian virtue ethics. Thus it presents a fascinatingly different rendition of the evolutionary evidences from that of Lohmann, and it deserves to be read in critical interaction with him. Whereas Lohmann presumes the classic modern split between "fact" and "value" (or "evidence" and "norms"), which became further intensified after Kant in the neo-Kantian school, Porter returns to Aristotle to suggest a way of overcoming this split from the outset. In so doing she also revives the issue of "teleology" in evolution, and at two importantly distinguishable levels. First, what if evolutionary phenomena, she asks, present to us not merely bare "facts" and "behaviors," over which scientists somewhat arbitrarily impose interpretations (normative or otherwise)? What if, on the contrary, evolution itself discloses to us what Aristotle calls "natural kinds"—beings or groupings already disposed to a normative analysis and which, in turn, are met by our own "natural" capacity to probe the real world with meaningful insight? The very phenomenon of "cooperation," indeed, might suggest precisely such a rendition of evolutionary patterns of reality and thus come front-loaded to educe normative reflection. Secondly, then, there is thus another level of "teleology"—in the human, ethical realm—which may indicate how evolutionary patterns are *met* by appropriate human moral motivations. Here Porter indicates the real genius of "natural law" accounts of ethics, which lies in the analysis of how natural conditions of normativity and actor's intentions purportedly *interrelate*. An Aristotelian/Thomist account of ethical normativity in the human realm thus resides in the idea of what it is to "flourish" humanly—to be and do what is good and "natural" for one, and this may not necessarily equate directly with reproductive success. That is why a teleological account of evolution must, according to such a theory, re-emerge at a second level: not just in tracing the "purposes" of distinct biological entities themselves, but the conditions of human flourishing in relation to them. On this rendition, note, "cooperation" and "altruism" are not in tension (as in Lee's account), or "complementary" (as in Lohmann's), but one builds precisely on the other and fits perfectly with it. Altruism becomes "a reflective pursuit of the values *inherent in* cooperative behavior, reformulated with specific ethical intentions in mind."

Lest this neo-Aristotelian rendition of the ethics of cooperation seem too easy or triumphalist a climax to this section's investigations, our last ethical commentator, Timothy P. Jackson completes the dialectical moves of Part IV in Chapter 16, in which he reacts furiously against *any* positive ethical rendition of evolutionary "cooperation." In particular he rails afresh against the utilitarian metaethic so often applied by game theory to give it moral justification. The reason for Jackson's revulsion is that his own metaethical stance is that of the "agapeistic (love)" command, beloved of Kierkegaard and—so it is claimed—of Jesus behind him, according to which *any* self-interest or calculation (whether genetic/cultural, individual/communal, or short-term/ long-term) must be rigorously excised from consideration if one's motivations are to be genuinely pure. Even the faintest suggestion of "benefit" (and that would also include the Aristotelian/Thomist idea of human "flourishing," just discussed) must be set aside, on this reading, if Jesus's radical sacrificial demands are to be met ethically: to "turn the other cheek," to "love one's enemies," is to court humiliation and even death if need be for the right goals. Several things follow from this agapeistic metaethical commitment, both practical and theoretical. First, as in the earlier chapters by Lee and Schloss, Jackson's required definition of "altruism" is intensified beyond that otherwise used in this book: "altruism" can only be "pure" if it involves no known benefit *whatsoever,* only loss, sacrifice, and self-abnegation for the right motive. Second, on this metaethical theory, "God" is no optional addendum (as modern secular Kantian ethicists have now made Kant's "regulative" God, or as Johnson's psychological theory of the "gods" assumes), but is the only true intentional goal and source of revelatory confirmation—without which the theory founders. Third, if this indeed *is* the correct rendition of "altruism," then evolutionary theorists according to Jackson currently have no way of explaining it except as a "spandrel"—that is, as a sort of "surd" with absolutely no adaptive value.

It is somewhat bracing to end Part V with a theological account of ethics apparently so resistant to "evolutionary" explanation, and one therefore demanding some sort of further response, whether scientific or philosophical. Yet it seems to us that the evolutionary discussion is not yet over here, and that for at least two reasons. First, we have already shown that there are other, competing, theological accounts of the ethics of evolutionary cooperation (whether Kantian or Aristotelian/Thomist), and these too can provide renditions of Jesus's ethic which are more obviously adaptive, and which stand in tension with Jackson's demanding "agapeistic" reading. Thus the debate between these various theological options can at least go on at that level, by peeling back both to the New Testament evidences themselves, and also to the fundamental choices between metaethical principles as outlined here. Secondly, and more probingly, it should be pointed out that Jackson may yet be overly pessimistic about the significance of his own views for evolutionary thinking itself. For once we are in the realm of *cultural* evolution, questions of "fitness" become the more rich and complex and are by no means reducible to mere genetic "success." It follows that it is possible that an ethical approach that currently seems spandrel-like could at some point in the future be embraced by sufficiently large numbers of people so as to effect a modal change in cultural evolutionary options: it could *become* "adaptive" in a sense not yet obvious or clarified. Yet it is difficult to see how one could be an "agapeist" in Jackson's sense without believing in *God,* and thus in a transcendent realm of rewards and punishments, grace and damnation. So, if such a "cultural" belief were to become adaptive, the idea of "God" would be nondispensable. However, it is also not impossible that the trajectories of cultural evolution could, as one option among many, represent a religiously motivated, sacrificial love ethic as the very means of (say) ecological survival. Indeed, as eminent a secular biologist as E. O. Wilson has recently said as much![28] If so, this would represent a new breakthrough in ideational/"cultural" evolutionary possibilities somewhat akin to earlier moments of novel transition but now fueled by specifically ethical and theological motivations. Again, the matter would only become definitively "adaptive" if it could be shown, perhaps retroactively, that without these particular religious motivations (or ones like them) the human race could not have protected itself effectively from ecological disaster.

Ned Hall writes presciently in Part IV that the most contentious issues in evolutionary theory concern the "ultimate grounding" of evolution as a whole. Once this question is finally exposed we are clearly in the realm of the speculative and the metaphysical, and issues of ultimate meaning can neither be ducked nor quickly resolved. This is as true for a reductive, atheistical rendition of the evolutionary whole as it is for other competitors. Empirical or mathematical science, as such, does not deliver answers to big questions such as these, although it could be argued—as we have suggested here—that the mathematicalization of evolutionary processes does sharpen and focus the questions in a new and particular way. In the final section of this book (Part VI) some of the implications of this realization begin to be charted. Here we enter a particularly contentious arena of discussion, but one that we refuse to shirk.

In Chapter 17, Alexander Pruss presses precisely this metaphysical question about evolution as a whole in a pointed and intriguing fashion. Returning to the Aristotelian idea of "natural kinds" already outlined by Porter in this volume, Pruss argues insistently that some facts, even empirical "scientific" facts, are *intrinsically* "normative." An example might be that "altruism" in humans is not only widely manifested but is also considered essentially "appropriate" and "normal" in human life. Exponents of positivistic science may of course vigorously deny this proposal, claiming that empirical evolutionary facts, as such, require *no* such normative analysis, but the debate here must go on, and the Aristotelian vision of "natural" normativity is at least worthy of serious and probing philosophical consideration, but then there is a further complicating wrinkle that has as yet been passed over in this book's unfolding argument. As Pruss puts it, if "altruism" *is* normative for humans, it does not follow that the motivational accompaniments to such altruism are necessarily pure: one could, for instance, be behaviorally "altruistic" in pursuit of a highly dubious goal (think of Nazism), or, more subtly, be consistently self-deluded as to one's own "altruistic" intentions in pursuit of a good goal. Thus to be *properly* "altruistic," says Pruss, one's motivations must not only be actively in play but must also be *objectively* praiseworthy and good, and it is a big step from merely acknowledging the existence of evolutionarily derived "altruism" (motivated sacrificial concern for others) to judging such "altruistic" actions to be *appropriately* directed morally. Science, as such, cannot deliver a judgment on such matters; nor can it explain (at least as yet) the existence of what Pruss calls "supernormal" altruists whose manifestations of sacrificial love far outstrip any "payback" calculation, however long-term or sophisticated. In cases such as these "saints" of altruism, it might well be suggested that only God supplies the motivational tug sufficient to educe such remarkable responses. We are back here to the conundrum left to us by Jackson, but it is here granted a possible solution in the form of a type of "argument to the best explanation."[29] *Excessive* forms of altruism can only be satisfactorily explained—Pruss proposes—if God calls them forth.

Philip Clayton is more cautious in Chapter 18 than Pruss in summoning any argument for God's existence by reference to cooperation or altruism; but what he does tackle is an issue that has remained somewhat moot in earlier parts of the book—that of the precise metaphysical relationship of *different* sorts of "cooperation" and "altruism" at different "levels" in the evolutionary spectrum. Thus Clayton takes Ned Hall's worries about the dangers of the "cross-platform" explanatory fallacy extremely seriously, and he also acknowledges the points raised earlier by Justin Fisher about the importance for intentional human "altruism" of a nonreductive philosophy of mind. So what exactly conjoins these different "levels" of cooperative behavior in the evolutionary spectrum, and what distinguishes them? Clayton's hypothesis is an "emergentist" one—that is, he proposes that evolution itself has the capacity to produce new levels of complexity at certain important junctures of change and transformation (human consciousness being perhaps the most problematic *novum* to explain philosophically). God-of-the-gaps arguments (such as those put forward by exponents of Intelligent Design) are however here sternly eschewed, though it is insisted that, "Recognizing the distinct features of the various levels of emergent complexity is an indispensable condition for the success of theological explanations *which take science seriously*." Clayton finally settles on a claim rather more modest than Pruss's, viz., that one "viable" explanation of emergent complexity could be provided by the hypothesis of divine creative planning. If strong emergence is accepted philosophically, moreover, the picture of a divine plan for *different* and more complex forms of "cooperation" or "altruism" to arise at different levels of evolution becomes the more plausible.

The final two essays in the book tackle some more ramified theological problems that seem to arise if the God hypothesis is invoked to undergird different levels of evolutionary "cooperation." Michael Rota discusses in Chapter 19 the theodicy problem of rampant animal suffering in evolutionary history and notes that it is only partially mitigated by the acknowledgment of the adaptive importance of suffering "cooperation." However, there are two reasons why a good God might tolerate such an apparent excess of suffering by blameless creatures to be written into the evolutionary story.

First, if evolution in general, and "cooperation" in particular, can be perceived as forms of "secondary causation" in Thomas Aquinas's sense, the play between "cooperation" and "defection" might be construed as the generative, but also necessarily suffering, *praeparatio* for the eventual appearance of true human freedom for the good. Secondly, it is a theological mistake to presume that God would make *all* his plans unambiguously and immediately clear through evolutionary processes: since there is a supreme value in the human *choice* for God, this can never be a matter of patency or bludgeoning. In short, God's supreme desire to create rational creatures who choose him for themselves freely maybe is not possible without serious animal suffering en route. Moreover, divine interventions to prevent this might either have jeopardized divine hiddenness and/or involved systematic deception.

In Chapter 20, the final in the volume, Sarah Coakley reminds the reader of some of the major worries that evolution has recently seemed to raise for Christian belief, and asks how a proper understanding of the workings of "cooperation" and "altruism" might change the tone of those worries. In particular, she focuses on the classic Christian doctrine of providence (the idea of ineluctable divine guidance by an atemporal deity), and enquires how it could be compatible with, and even supportive of, a nuanced rendition of the workings of evolution. She highlights three big problems that haunt the believer when considering the doctrine of providence in connection with evolution: the apparent "randomness" of prehuman evolution; the question of the evolutionary transition from prehuman to human and the penchant of much evolutionary theory for reductive physicalist accounts of human "freedom" and the problem of evil. In each case she argues that cooperation *makes a difference* to these problems, provided a) that God's status as sui generis creator ex nihilo is properly conceived theologically; b) that the doctrines of incarnation and Trinity are not left out from the start as unnecessary complications in a generic "theism"; c) that the relation of human freedom to divine providence is carefully explicated; and d) that the physicalist and genetic reductionism of much contemporary evolutionary theory is countered. She ends by insisting that such reflections in support of divine providence are not just matters of willful religious "personal preference," but strongly argued metaphysical positions that, along with others canvassed in this book (including atheistical and physicalist alternatives) have a right to be debated even-handedly at the metaphysical table. Neither mathematics nor empirical biology, as such, can settle these questions; but both mathematical and empirical evolutionary biology constantly court and suggest them.

Concluding Reflections: Where Are We Now?

If this volume has demonstrated nothing else, its unfolding chapters will have provided an initial clarification of the hermeneutical, metaphysical, and metaethical choices to which any discerning evaluation of the significance of evolutionary "cooperation" necessarily leads, but it will also be clear that much unfinished business remains, both empirical and theoretical, as indeed one might expect in a field characterized by such burgeoning scientific creativity.

To take just one salient example of such creativity, there is now an important developing literature on childhood psychological development in empathy, and on "natural" childhood manifestations of "altruism";[30] this new research literature demands close attention and assessment if we are to understand more precisely in human terms the relation of genetic and cultural manifestations of other-regarding behavior. Empirical evidences in laboratory conditions are also now mounting up (as already intimated above) that even adult humans are *initially* more inclined, in social situations, to "cooperate" than to "defect."[31] These findings might have seemed counterintuitive to a slightly earlier generation of investigators, and especially to game-theoretical economists; but now a certain theoretical tide has turned, as this volume has been concerned to chart, and we are required to look again open-mindedly at both the mathematical and the empirical studies of cooperative behaviors and their importance.

Yet while the very definitions of "cooperation" and "altruism" continue to be ambiguous or disputed, it is scarcely surprising that apparently ideological disputations still separate those who utilize mathematical accounts of evolutionary processes for different purposes.[32] There is also the deeper remaining issue of the *extent* to which the mathematical accounts can capture the complexities of the empirical evolutionary evidences in the field and the laboratory, and even more perplexing, as we have seen, is the difficulty of pinpointing whether, and how, mathematical accounts of human "altruism," in various games and guises, can do justice to the phenomenon of human *motivation*. This will surely remain a philosophically and psychologically contested area, the more so when attempts to probe specifically religious motivations are at stake. For if it is simply assumed that "religion" may be explained away in terms of something else, all attempts to clarify its workings will inevitably fall prey to the same reductive principles. Only quite recently have research protocols started to be devised to test genuinely *theological* motivations for "altruistic" human behavior.[33] Such developments at least help to resist the privileged presumption of secular reductionism, which tends to assume that human "religion" arose merely in the context of projective needs among hunter-gatherers and is now scientifically outmoded.

To sum up: it might seem that one of the main morals of this book is that evolutionary biologists working on cooperation need to take a cold *philosophical* bath, but that advice should not be taken punitively nor imply any threatened stultification of their current scientific creativity in this area. On the contrary, the clarification of the points at which their work backs onto philosophy of science, ethics, metaphysics, and theology is itself creative and energizing, as has been the experience of the interdisciplinary team working on this book. One final reflection, indeed, follows from this: it is to note that such interdisciplinary adventures are part and parcel of burgeoning cultural evolution itself—these are new forays in cooperation without which the ingenuity to confront certain global economic and ecological threats may be impaired. Humans who more accurately understand how cooperation works, both biologically and culturally (*qua* "altruism"), can potentially extend the range of its efficacy. Such is the challenge that the editors of this book finally lay before its readers.

How to Use This Book in Teaching and Discussion

The foregoing Introduction will have made it clear that this book is constructed in order to aid further discussion and debate, in an exciting field that is developing and changing very rapidly.[34] Accordingly, there are various "voices" and opinions expressed in this volume, and each section is arranged in such a way as to engender critical conversation and an increased awareness of the implicit methodological, philosophical (and, in Parts V and VI, theological) undertow to current scientific debates about the significance of evolutionary "cooperation." Those using this volume to encourage discussion in the context of teaching should thus first attend carefully to the account of the interdisciplinary "levels" of the book unfolded in this Introduction, and then allow students or discussants to compare and critique the divergent opinions represented in each part of the volume. The theological renditions of "cooperation"/"altruism" in the last parts of the volume may occasion particular debate; they are offered as rational accounts of these evolutionary phenomena when read through a Christian theological lens. A number of secular philosophical renditions of the same phenomena are also on offer in the volume and may fruitfully be brought into critical discussion with the theological alternatives. It is, however, a conviction of the editors, as discussed above, that any "scientific" rendition of these phenomena is bound to run into important interpretative decisions of a metaphysical and metaethical nature. The book is designed to make those decisions clear and to help the reader to make discerning choices between the relevant alternatives presented here.

Notes

1. Richard Dawkins may, of course, be taken as emblematic of this trend. Dawkins 1989, 202–33 provides a lively and accessible introduction to game theoretical accounts of evolution but interprets them as merely complexifying the deviousness of the intentional "selfish gene."
2. For a more varied account of recent developments in game theory than can be attempted in this volume, see the essays in Levin 2009. For a popularized historical account of game theory's application to evolutionary theory and of current mathematical evolutionary dynamics, see Nowak with Highfield 2011.
3. The word "strategy" is used as a shorthand in evolutionary biology equivalent to the term "phenotype" and does not as such indicate any sort of intentionality in advance of the emergence of the higher mammals. However, evolutionary biologists often use it carelessly and metaphorically for nonhuman forms of life, a problem to which we return later in this Introduction.
4. See Maynard Smith and Szathmáry 1997 for close discussion of these "major transitions" in evolution. Within this book, Martin A. Nowak (Chapter 4) and Jeffrey P. Schloss (Chapter 10) also reflect on the biological and philosophical significance of these transitions for our understanding of evolutionary cooperation.

5. It is normally assumed by working evolutionary biologists that biological events are "contingent" and "random," i.e., precisely not subsumable under mathematical laws or predictions. One of the most exciting dimensions of mathematical biology is its challenge to this thinking, a challenge which does not however thereby imply strict determinism. (Note, however, that talk of "stochastic" or statistically probable regularities in evolution is subtly different from talk of "dynamics," which might imply some actual "forces" at work in evolution—clearly a more controversial philosophical position.)
6. "Teleology" is a word that became tainted even in Darwin's time for its association with theological accounts of order and providence, but patterned, norm- or goal-directed features of evolution can occur at several different "levels" of discussion not associated with God: Mayr (2002) therefore prefers to use "teleonomic" language to distinguish it from any theological connotation. For a probing philosophical account of the problems goal-directed language in biology, see Lewens 2004. For an analysis of the relation of these different "levels" of teleonomic/teleological discussion in evolution, including that of a theological rendition, see Coakley 2012, Lecture 5.
7. It is often claimed that science, *qua* science, does not and must not venture into the "metaphysical" (i.e., as here, into speculative theories about the meaning and implications of evolution as a whole). However, it will be explicitly argued in the final sections of this book, and will be strongly implied in the earlier sections, that such metaphysical issues are ultimately unavoidable even for the working scientist, and certainly for the philosopher of biology. True, such discussions are best demarcated as strictly in the arena of "philosophy of science" rather than in empirical science, but it is spurious to debar them from the realm of "science" *tout court*. It is noteworthy that metaphysical theories of a reductionist sort are palpably propounded by contemporary atheistical theorists of evolution, even as they outlaw theological accounts as inherently "unscientific." It is the view of the editors of this book that a *variety* of philosophical and theological responses to evolutionary theory should appropriately be compared, criticized, and adjudicated alongside one another.

   For a discerning recent account of the *philosophical* confusions that are also rampant in contemporary debates between creationist and secular renditions of evolutionary theory, see Cunningham 2010.
8. "Metaethics" is the realm of philosophical debate devoted to establishing the fundamental principles and norms on the basis of which ethical choices are made.
9. This word too has a back history, having first been coined with a rather different meaning by Auguste Comte as a secular, social-scientific alternative to the

Christian language of love. See Thomas Dixon's account of this history in this volume (Chapter 2), and in more detail in Dixon 2008.

10. It should be admitted that Nowak and his team often blur this distinction between "cooperation" and "altruism" in their own published work elsewhere, so common is this semantic tendency in contemporary biological discussion. Thus the discerning reader will notice that such a conflation still occurs in one place in this volume, Chapter 4, which was previously published as Nowak 2006. Sober and Sloan Wilson 1998 are the most important other recent exponents to make a systematic distinction similar to the one applied in this book (they compare "evolutionary altruism" with "psychological altruism," whereas we distinguish "cooperation" and "altruism"). For further discussion of the systematic ambiguity of key terms in this area of research, see the important article Kerr, Godfrey-Smith, and Feldman 2004, which pointedly demonstrates how different definitions of "altruism" (especially divergent accounts of the "benefits" and "costs" of "altruism") lead not only to different research protocols but to correspondingly different conclusions about the *significance* of "altruism" in the evolutionary spectrum.
11. The classic Prisoner's Dilemma thought experiment runs thus (see Nowak 2012b, 36–7): "Imagine that two people have been arrested and are facing jail sentences for having conspired to commit a crime. The prosecutor questions each one privately and lays out the terms of a deal. If one person rats on the other and the other remains silent, the incriminator gets just one year of jail time, whereas the silent person gets slammed with a four-year sentence. If both parties cooperate and do not rat on each other, both get reduced sentences of two years. But if both individuals incriminate each other, they both receive three-year sentences. Because each convict is consulted separately, neither knows whether his or her partner will defect or cooperate. Plotting the possible outcomes on a payoff matrix . . . , one can see that from an individual's standpoint, the best bet is to defect and incriminate one's partner. Yet because both parties will follow the same line of reasoning and choose defection, both will receive the third-best outcome (three-year sentences) instead of the two-year sentences they could get by cooperating with each other." What the PD therefore neatly encodes is a "tight" conundrum in the choice between individual gain and "cooperative" benefits. For discussion of this and other related games' application to evolutionary biology and economics, see Chapters 4–6, below.
12. For recent accounts of different forms of "natural theology" and their theoretical problems, see Fergusson 1998 and Holder 2004. For a new defense of "natural theology" in a redefined sense, see Coakley 2012, Lecture 6.

13. See again Levin 2009 and Nowak with Highfield 2011.
14. See Sober and Sloan Wilson 1998 for this history.
15. The furor that has broken out recently over Nowak's and E. O. Wilson's criticism of Hamilton's inclusive fitness theory (Nowak, Tarnita, and Wilson 2010) does not undermine the general points made in this earlier article (Nowak 2006) about Haldane's original account of "kin selection," despite much misunderstanding on this score: what has been called into question by Nowak and colleagues is not "kin selection" *tout court* but the particular mathematical account of it hypothesized by Hamilton and—ironically—championed for a long time by Wilson himself. For an insightful lay account of this ongoing debate, see Lehrer 2012.
16. In this chapter (orig. Nowak 2006) called "Rules" in the title, although Nowak now generally prefers the language of "mechanisms."
17. See again note 11, above.
18. This is a familiar objection made by empirical evolutionary biologists who work on cooperation in the field to the application of mathematical modeling to empirical reality by mathematical biologists. Tim Clutton-Brock's longstanding study of meerkats in the Kalahari, for instance, has recently refined its understanding of cooperative tendencies right down to individual proclivities and behaviors; see English, Nakagawa, and Clutton-Brock 2010. However, empirical and mathematical biologists arguably need not be in such (false) competition if the mathematicians are aware of the precise force and range of their essentially stochastic "explanation." This issue is discussed at some length in later chapters of this book.
19. In all, sixteen types of game (and their variants) are logically possible, given all the potential interrelations between "cooperation" and "defection."
20. This reductive approach to "religion" advocated by Johnson clashes forcibly, of course, with the theological accounts of God and evolution found elsewhere in this volume (see, e.g., the contributions of Porter, Jackson, Pruss, Rota, Coakley, and—rather differently—Clayton). It should be stressed that Johnson's views are at least as speculative as these more robustly theological accounts in terms of empirical evidences, but much depends here on whether there is an underlying *presumption* in favor of metaphysical "naturalism." For an insightful account of the philosophical implications of such metaphysical naturalism, see Rea 2002.
21. Contrast, e.g., Dreber, Rand, Fudenberg, and Nowak 2008.
22. A recent British *Today* program "debate" (BBC Radio 4, March 17, 2011) between Peter Atkins and Mary Midgley was revealing on this score, Atkins insisting that only a no-meaning option is "scientific," because meaning-ascription cannot arise in the "completely random" realm of evolution, and Midgeley strongly contesting this assertion as a philosopher. Atkins seemed unaware

that a no-meaning position was itself both hermeneutical and implicitly metaphysical, and that the "*completely* random" claim was itself scientifically contestable too.

23. A "deontological" theory in ethics is one that is concerned with duties, obligations, and rights: it focuses on what we *ought* to do and on what grounds (rather than on what sort of *person* we are, as in "virtue" ethics).
24. "Utilitarian" ethics judge the good of an action solely in terms of its overall outcome (whether of fitness, pleasure, happiness, or financial success).
25. "Consequentialism" is the generic term for the view that normative properties depend only on consequences. Utilitarian ethics of various kinds therefore shelter under this overall categorization.
26. This volume therefore presents a range of metaethical alternatives for "evolutionary ethics" resistant to a more commonly held consequentialism; compare, e.g., Alexander 1987, and the critical discussion in Clayton and Schloss 2004.
27. See Hauser 2008, for Hauser's intriguing use of web-based questionnaires to test international sensitivities to (apparently) hard-wired ethical intuitions.
28. See Wilson 2006 for precisely this argument: the affective commitment of religious people may be needed to shift political apathy or resistance where ecological issues are at stake.
29. See Lipton 2004 for the most sophisticated recent philosophical account of "inference to the best explanation," which includes the criterion of "loveliness." John Hare's moral arguments for God, building on Kant, also might seem to tend now in a similar direction, given his recent (more positive) interest in "evolutionary ethics"; see Hare 2012, cp. Hare 1997.
30. See Blake and McAuliffee 2011, Hamann, Warneken, Greenberg, and Tomasello 2011, and Warneken and Tomasello 2012, for some of these recent developments in child psychology and the investigation of "altruism."
31. See Rand, Greene, and Nowak 2012 (in press); also relevant is Rand, Dreber, Ellingsen, Fudenberg, and Nowak 2009.
32. So, e.g., David Queller (with whom the Nowak and Coakley research team interacted actively during the preparation of this volume) distinguishes "cooperation" and "altruism" in a way different from that proposed in this book, and also continues to defend Hamilton's mathematical account of "inclusive fitness"; see Queller 2011.
33. As opposed to testing "religious" motivations on the presumption of a reductive reading of "religion." For an investigation of the impact of specifically theological primes on altruistic giving, see Rand, Dreber, Haque, Kane, Nowak, and Coakley 2012.
34. Recent further work on "structured populations" is particularly significant here; see Tarnita, Ohtsuki, Antal, Fu, and Nowak 2009, and van Veelen, García,

Rand, and Nowak 2012. Also see Nowak 2012a and the whole issue of *Journal of Theoretical Biology* (299: 2012) devoted to exploring the newest developments in mathematical approaches to evolution.

References

Alexander, R. D. 1987. *The Biology of Moral Systems.* Hawthorne, NY: Aldine de Gruyter.

Blake, P. R., and K. McAuliffe. 2011. "'I had so much it didn't seem fair': Eight-year-olds Reject Two Forms of Inequity." *Cognition* 120: 215–24.

Clayton, P., and J. Schloss, eds. 2004. *Evolution and Ethics: Human Morality in Biological and Religious Perspective.* Grand Rapids, MI: Eerdmans.

Coakley, S. 2012. "Sacrifice Regained: Evolution, Cooperation and God." The 2012 Gifford Lectures at Aberdeen University, http://www.abdn.ac.uk/gifford/about/.

Cunningham, C. 2010. *Darwin's Pious Idea: Why the Ultra-Darwinists and Creationists Both Get It Wrong.* Grand Rapids, MI: Eerdmans.

Darwin, C. [1879] 2004. *The Descent of Man, and Selection in Relation to Sex.* London: Penguin.

Dawkins, R. 1989. *The Selfish Gene.* New ed. Oxford: Oxford University Press.

Dixon, T. 2008. *The Invention of Altruism: Making Moral Meanings in Victorian Britain.* Oxford: Oxford University Press.

Dreber, A., D. G. Rand, D. Fudenberg, and M. A. Nowak. 2008. "Winners Don't Punish." *Nature* 452: 348–51.

English, S., S. Nakagawa, and T. H. Clutton-Brock. 2010. "Consistent Individual Differences in Cooperative Behaviors in Meerkats." *Journal of Evolutionary Biology* 23: 1597–1603.

Fergusson, D. A. S. 1998. *The Cosmos and the Creator: An Introduction to the Theology of Creation.* London: SPCK.

Hamann, K., F. Warneken, J. R. Greenberg, and M. Tomasello. 2011. "Collaboration Encourages Equal Sharing in Children but Not in Chimpanzees." *Nature 476:* 328–31.

Hare, J. E. 1997. *The Moral Gap: Kantian Ethics, Human Limits and God's Assistance.* Oxford: Oxford University Press.

———. 2012. "Evolutionary Theory and Theological Ethics." *Studies in Christian Ethics* 25: 244–54

Hauser, M. D. 2008. *Moral Minds: How Nature Designed our Universal Sense of Right and Wrong.* London: Abacus.

Holder, R. D. 2004. *God, the Multiverse, and Everything: Modern Cosmology and the Argument from Design.* Aldershot: Ashgate.

Kerr, B., P. Godfrey-Smith, and M. W. Feldman. 2004. "What is Altruism?" *Trends in Ecology and Evolution* 19: 135–40.

Lehrer, J. 2012. "Kin and Kind." *The New Yorker,* March 5, 36–42.

Levin, S. A., ed. 2009. *Games, Groups, and the Global Good.* London: Springer.

Lewens, T. 2004. *Organisms and Artifacts: Design in Nature and Elsewhere.* Cambridge, MA: The MIT Press.

Lipton, P. 2004. *Inference to the Best Explanation.* 2nd ed. London: Routledge.

Maynard Smith, J., and E. Szathmáry. 1997. *The Major Transitions in Evolution.* Oxford: Oxford University Press.

Mayr, E. 2002. *What Evolution Is.* London: Weidenfeld & Nicolson.

Nowak, M. A. 2006. "Five Rules for the Evolution of Cooperation." *Science* 314: 1560–63.

———. 2012a. "Evolving Cooperation." *Journal of Theoretical Biology* 299: 1–8.

———. 2012b. "Why We Help." *Scientific American* 307: 34–39.

Nowak, Martin A., with Roger Highfield. 2011. *SuperCooperators: Evolution, Altruism and Human Behaviour, or, Why we Need Each Other to Succeed.* New York: Simon & Schuster.

Nowak, M. A., Corina E. Tarnita, and Edward O. Wilson. 2010. "The Evolution of Eusociality." *Nature* 466: 1057–62.

Queller, D. C. 2011. "Expanded Social Fitness and Hamilton's Rule for Kin, Kith, and Kind." *Proceedings of the National Academy of Sciences of the USA* 108: 10792–99.

Rand, D. G., A. Dreber, T. Ellingsen, D. Fudenberg, and M. A. Nowak. 2009. "Positive Interactions Promote Public Cooperation." *Science* 325: 1272–75.

Rand, D. G., A. Dreber, O. S. Haque, R. Kane, M. A. Nowak, and S. Coakley. 2012. "Religious Motivations for Cooperation: An Experimental Investigation Using Explicit Primes." Social Science Research Network, http://papers.ssrn.com/sol3/papers.cfm?abstract_id=2123243.

Rand, D. G., J. D. Greene, and M. A. Nowak. 2012. "Spontaneous Giving and Calculated Greed," *Nature,* in press.

Rea, M. 2002. *World without Design: The Ontological Consequences of Naturalism.* Oxford: Clarendon.

Sober, E., and D. S. Wilson. 1998. *Unto Others: The Evolution and Psychology of Unselfish Behavior.* Cambridge, MA: Harvard University Press, 1998.

Tarnita, Corina E., Hisashi Ohtsuki, Tibor Antal, Feng Fu, and Martin A. Nowak. 2009. "Strategy Selection in Structured Populations." *Journal of Theoretical Biology* 259: 570–81.

van Veelen, M., J. García, D. G. Rand, and M. A. Nowak. 2012. "Direct Reciprocity in Structured Populations." *Proceedings of the National Academy of Sciences of the USA* 109: 9929–34.

Wade, N. 2011. "Scientist under Enquiry Resigns from Harvard." *The New York Times,* July 21, A17.

Warneken, F., and M. Tomasello. 2006. "Altruistic Helping in Human Infants and Young Chimpanzees." *Science 311:* 1301–3.

———. 2012. "Parental Presence and Encouragement do not Influence Helping in Young Children." *Infancy,* in press.

Wilson, E. O. 2006. *The Creation: An Appeal to Save Life on Earth.* New York: Norton.

∴ I ∴

Evolutionary Cooperation in Historical Perspective

1

• • • •

"Ready to Aid One Another"

Darwin on Nature, God, and Cooperation

JOHN HEDLEY BROOKE

> There can be no doubt that a tribe including many members who, from possessing in a high degree the spirit of patriotism, fidelity, obedience, courage, and sympathy, were always ready to aid one another, and to sacrifice themselves for the common good would be victorious over most other tribes; and this would be natural selection.
>
> (Darwin [1879] 2004, 157–58)[1]

In this well-known passage from Darwin's *Descent of Man* lies an invitation to consider his understanding of cooperation and its place in his naturalistic theory of human evolution. How did it feature in his account of the moral sense and its development—a development that Darwin believed had been reinforced by religion? In this introductory chapter my aim is to explore some of the principal reasons why Darwin came to believe that a naturalistic account should be given of what he called our social instincts and their transmission. Central to his analysis was a concept of "sympathy" that allowed him to say that, in the early history of human tribes, a decisive role had been played by an instinctive need for the approbation of others and a strong desire to avoid their disapproval. In order to place Darwin's contribution in perspective, it is, however, necessary to show that there was nothing essentially "natural" about his naturalism. This can best be done—despite its back-to-front appearance—by examining a few theological responses to Darwin's science and by identifying the reasons why he had abandoned a theology of nature that might have motivated him more strongly to see in the creative aspects of evolution the detailed fulfillment of a providential plan.

References to the naturalness, or otherwise, of naturalism immediately raise definitional problems. A distinction is routinely drawn between the search for natural causes as a principle of method in scientific inquiry and naturalism in the much stronger sense of a view of all reality in which there is nothing other than nature, no supernatural being of any kind. As a methodological assumption, naturalism had long proved its fruitfulness, notably by Darwin's mentor, Charles Lyell, in his reconstruction of Earth's physical history in his *Principles of Geology* (1830–1833). Lyell had also expressed the view that the origin of species, while under the control of Providence, might itself prove explicable in terms of natural, or what were often called secondary, causes. Darwin's application of naturalism in this sense to the evolutionary development of human attributes would prove highly controversial. It did not, however, necessitate a naturalistic worldview in the stronger sense of prescribing an ontology in which what could be known of "nature" defined all that is. Not all forms of naturalism have been as extreme as this, since there are both theistic and deistic variants in which a divine being is understood to be working *through* natural causes. In this connection, theological disputes have frequently revolved around the question of whether such a deity need be supposed to have done anything more than establish and possibly sustain the "laws of nature." Darwin would be caught in such debates because his central concept, that of natural selection, was unashamedly naturalistic in that it accorded agency to "nature," not to divine intervention, in the production of new species. What this meant for his contemporaries, however, was a largely cultural matter because the meaning of "nature" was not a given but itself shaped by different presuppositions—whether, for example, it was understood to be an autonomous system of matter in motion, a carefully designed work of art, or a theater of redemption. Because a "naturalistic" explanation did not have to be atheistic, but could also be associated with theism, deism, or agnosticism, it can be simplistic to describe any one of these associations as "natural"! The significance of these distinctions will become clearer as we investigate Darwin's naturalism, which at the time he wrote his *On the Origin of Species* still involved a deity in the design of nature's laws but not in the day-to-day running of the universe.

* * * *

Let us begin at Harvard, where this present book was conceived, but in the Harvard of the mid-nineteenth century when the great Swiss naturalist Louis Agassiz had just been appointed Professor of Geology and Zoology. Why was Agassiz given such a generous and resounding welcome? Part of the answer lies in the interpretation he gave to living forms. His Platonist philosophy of nature struck a chord with the religious ideals of many in the College (Nartonis 2005). Agassiz had no time for theories of evolution that involved material connections between species. Rather, in the fossil record he saw evidence of progressive creation as epoch succeeded epoch. Living things were the instantiation of ideas in the mind of the Creator. As he once put it: "There will be no scientific *evidence* of God's working in nature until naturalists have shown that the whole creation is the *expression of thought* and not the *product of physical agents*" (Roberts 1988, 34).

Agassiz was not alone in that view. In England, Richard Owen also ascribed the common bone structures of the vertebrates to an archetypal idea in the mind of God. Owen had risen to fame through his expertise in anatomy and paleontology, and it was he who coined the word "dinosaur." Owen was willing to see the emergence of new species as the result of natural causes but, at the same time, the whole process was the unfolding of a divine plan. The many different vertebrates looked to him to be instantiations of a common skeletal structure—an archetypal idea in the mind of the Creator. There was a sense in which "creation" was continuous (Owen 1849).

During Agassiz's tenure at Harvard, Darwin published his *On the Origin of Species* (1859). Here was a quite different account of the unity of form. For Darwin it was the consequence of a historical process in which species were related by common descent. The clash was transparent. Whereas Agassiz (Roberts 1988, 34) affirmed that "the intervention of a Creator is displayed in the most striking manner, in every stage of the history of the world," Darwin's mechanism of natural selection required no such intervention. Agassiz communicated his verdict to Asa Gray, who had been Harvard's Professor of Botany since 1842: Darwin's work was "poor—very poor" (Roberts 1988, 35).

A few years later, a scientific meeting was held in Boston at which both Agassiz and the British physicist John Tyndall were present. Tyndall (1879, 182) left a poignant account of a scene that marks the passing of an age:

> Rising from luncheon, we all halted as if by common consent, in front of a window, and continued there a discussion which had been started at table. The maple was in its autumn glory, and the exquisite beauty of the scene outside seemed, in my case, to interpenetrate without disturbance the intellectual action. Earnestly, almost sadly, Agassiz turned, and said to the gentlemen standing round, "I confess that I was not prepared to see [Darwin's] theory received as it has been by the best intellects of our time. Its success is greater than I could have thought possible."

To speak of the passing of an age captures something of the Darwinian impact, but it also misses a vital element. This is the remarkable diversity of the religious response. No interpretation of Darwin's science, whether theistic or atheistic can be singled out as the only "natural" one. Asa Gray, to whom Agassiz confided his poor opinion of Darwin, took a very different view. Gray positively promoted the theory of natural selection, claiming that it had theological advantages (Gray 1963). It underlined the unity of the human races in a way that Agassiz's science did not, and it even helped theologians with their most difficult problem: that of suffering. If competition in a struggle for existence was the motor of evolution, there was perhaps a sense in which the concomitant suffering was a precondition of the very possibility of our existence. Gray even proposed to Darwin that since the cause of the variations on which natural selection worked was, at the time, unknown, there was nothing to say they could not be under the control of Providence.

The Diversity of Reception

Agassiz and Gray represent two poles in the response to Darwin, who had, of course, wondered, and worried, how his theory might be received: "God knows what the public will think," he mused to one correspondent (Darwin 1991, 375). To admit the mutability of species, he once remarked, had been like confessing a murder, so great was the possible stigma. He knew his book was likely to have a polarizing effect, as it often did in public settings. When the politician Benjamin Disraeli suggested that a choice had to be made between apes and angels for the template of human beings, he was depicted in the press as having sprouted large angelic wings (Desmond and Moore 1991, 460–61). The threat to human dignity that so worried Samuel Wilberforce, the Bishop of Oxford, was often captured in cartoons. There were monkeys impatient to have their tails clipped in order to take their true place in society (Brooke 1991, 291). Bruising attacks from some clergymen made Darwin almost say that those who opposed his theory by snarling and baring their teeth merely confirmed their animal origins.

The responses of three women reveal additional problems and other layers of diversity. An elderly Mary Somerville observed with nostalgic regret that the beauty of a bird's plumage and song could no longer be enjoyed as having been designed for our delight. It was their utility to the birds themselves that mattered now (Somerville 1873, 358). For a feminist leader, such as Elizabeth Cady Stanton, Darwinism offered the bright prospect of emancipation. "The real difficulty in woman's case," she wrote, "is that the whole foundation of Christian religion rests on her temptation and man's fall." By accepting the Darwinian theory that "the race has been a gradual growth from the lower to a higher form of life, and the story of the fall is a myth, we can exonerate the snake, emancipate the woman, and reconstruct a more rational religion for the nineteenth century" (Larson 2005, 52). Late in life, and poignantly, Darwin's wife Emma admitted that some aspects of his writing had been painful to her—particularly the view that "*all* morality has grown up by evolution" (Darwin 1958, 93). For so many Victorians, belief in the transcendental significance of moral values could be a way back to an otherwise fractured faith—a route seemingly blocked by a science that had no need of the transcendent. To this question of the moral sense we shall return because, although Darwin's account was naturalistic, it was not relativistic. Nor did it devalue the virtue of cooperation.

There was an even greater variety of religious reaction (Brooke 2003). Geographical parameters played a key role in shaping receptivity to Darwinian ideas, making them seem less natural in some constituencies than others. As David Livingstone (2003, 117–23) has shown, Presbyterians in Princeton reacted very differently from those in Northern Ireland and differently again from those in Scotland. The reasons were often local and related to high-profile public events. In Belfast, the same John Tyndall who recorded the autumnal melancholy of Agassiz, delivered a provocative address in 1874 (1879, 137–203) that associated Darwin's theory with a more forceful naturalism than Darwin's own. Whereas Darwin himself was willing to use theological language when discussing the appearance of the first few living forms (Peckham 2006, 759), Tyndall brooked no compromise. His aggression toward theology in the context of educational priorities sparked an intensity of reaction that had no equivalent in Princeton. It meant that in Belfast Darwin's theory would unequivocally be seen as a vehicle for materialism and atheism.

Many such contrasts could be drawn to indicate the importance of local parameters. On the question of race, for example, geographical location mattered: the reception of Darwin's theory in New Zealand, where it was invoked to justify extermination of the Maori (Stenhouse 1999), was quite different from perceptions in the southern states of America (Stephens 2000). For complex social and political reasons the public spectacle of the "monkey trials" (famously that of the biology teacher John Scopes in Dayton Tennessee in 1925) has been largely confined to North America (Larson 1997). There has been no equivalent in England, where a future archbishop of Canterbury, Frederick Temple, was already speaking in favor of evolution and against a God of the gaps as early as 1860. Because of the prevalent form of naturalism according to which the deity worked through "natural laws" (Kohn 1989; Brooke 2008), the more agnostic forms, such as that to which Darwin eventually tended, require explanation—and all the more so when, as in Darwin's trajectory, there was the loss of an original intention to become an Anglican clergyman.

In Britain, by the close of the nineteenth century, there were, however, Anglican clergy willing to embrace the Darwinian theory for its supposed theological advantages. The Oxford theologian Aubrey Moore declared that under the guise of a foe Darwin had done the work of a friend, protecting Christianity from a deistic travesty in which God was active only when *interfering* in the natural order (Peacocke 1985, 111; England 2001). By explaining naturalistically the origin of new species, Darwin had sharpened the choice between a God active in everything or in nothing. In speaking of Darwin and God we are not therefore dealing with "science versus religion" in any straightforward sense (Moore 1979), tempting though it may be to impose that cliché on the post-Darwinian debates. Contrary to modern creationist rhetoric, Darwin was, in his own words (1887, Volume 1, 304), "never an atheist in the sense of denying the existence of a God." He continued to refer to a Creator, but one who created "by laws" (Brooke 1985, 46). Because Darwin himself anticipated some of the more sophisticated theological moves, his reflections on religion repay closer study. They also help us to understand the metaphysical framework that regulated his explanatory ambitions.

Preliminary Problems

There is, however, a preliminary problem that arises whenever questions are asked about the religious beliefs of scientists from the past. There is no simple answer to the question "What did Darwin believe about God?" There are several reasons why this is so. Most significantly his views changed over time (Brown 1986). Having studied for the Christian ministry during his Cambridge years, he became a deist during the 1850s and increasingly agnostic later in life. Even during one and the same period, it would be difficult to categorize him because he admitted that his beliefs often fluctuated. When referring to himself as an agnostic in May 1879, he would add the caveat "but not always" (1887, Volume 1, 304). At other times he would imply that he deserved to be called a theist (1887, Volume 1, 313). If we try to compress a complex matter into sound bites, we shall certainly get it wrong. Darwin sometimes said that he could not believe that this wonderful universe is the result of chance alone. Such remarks have lent themselves to apologetic exploitation. But to appropriate them in that way misses the nuance that Darwin so often inserted. He could not believe that the universe was the result of chance, but nor could he look at the structures of living organisms and see in them the product of design. As he disarmingly wrote to Asa Gray, he found himself in an "utterly hopeless muddle" (Darwin 1993, 496).

Finally there is the complication that stems from the privacy of belief. Darwin once reproached an enquirer by saying that he could not see why his beliefs should be of concern to anyone but himself (1887, Volume 1, 304). His public remarks were sometimes deliberately calculated to give minimal offence. In contrast to some of his modern disciples he did not believe that religious beliefs, however distasteful, should be confronted head-on. But the consequence is a frustrating degree of ambiguity in Darwin's references to the deity (Kohn 1989). Fortunately, this does not prevent the identification of particular experiences and considerations that weighed heavily with him and paved the way for his mature account of morality, religion, and cooperation.

Formative Experiences

The story of how Darwin's biology was shaped by his experiences on the five-year voyage of *HMS Beagle* has been so brilliantly told (Desmond and Moore 1991; Browne 1995) that I shall not repeat it here. But I would like to abstract a series of encounters that left a permanent mark on his *religious* outlook. One was an encounter with the beauty of a virginal nature. Having been enchanted by the travelogues of Alexander von Humboldt, the young Darwin found the lure of the Brazilian rainforest irresistible. The experience, when it became real rather than vicarious, was to cast a seductive spell on a romantic sensibility. It was during the voyage that Darwin gradually gave up the idea of being a clergyman and one hears the stirrings of a surrogate religion. When he finally arrived in the Brazilian jungle he recorded the experience in almost ecstatic language: "Twiners entwining twiners, tresses like hair, beautiful lepidoptera, Silence, hosannah" (Desmond and Moore 1991, 122).

A second encounter was with what we might call nature in the raw—with various manifestations of a struggle for existence that was to play such a crucial role in his theory of natural selection. The world Darwin encountered in South America was not the happy world of the vicarage garden described by William Paley in his *Natural Theology* (1802). It was, rather, nature red in tooth and claw and writ large, as giant condors menaced their prey. In Argentina Darwin witnessed a human struggle—a colonial struggle in which native Indians were being massacred by the forces of General Rosas. Even Earth itself was turbulent: Arriving in Concepcion, Darwin found the cathedral in ruins, destroyed by a recent earthquake. This prompted him to speculate how the entire condition of England would have been changed had such subterranean forces still been active there. It would have been a tale of famine, pestilence, and death. A struggle to survive in one of the most inhospitable regions on earth was indelibly stamped on his mind when he encountered the "savages" of the Tierra del Fuego.

This particular encounter deserves special consideration. Darwin would eventually ask whether our progenitors had been primitive men like these. His interest in them stemmed in part from an evangelical experiment that he was able to witness. On board ship were three Fuegians who had earlier been taken to England by Robert Fitzroy, captain of the *Beagle,* with a view to their being refined and educated in the gospel. The plan was to return them to their own people in the company of a missionary in the hope they

Figure 1.1 Copper engravings, drawn by R. Fitzroy and engraved by T. Landseer, in King, Fitzroy, and Darwin 1838, 2: 141, 324.

would exert an edifying influence. Darwin was intrigued to see how the experiment would work. In fact it ended in failure, and the missionary had to flee for his life. What was to be made of this? There was clearly an enormous difference between "savage" and civilized humans, but might the veneer of civilization, in general, be thinner than was usually imagined? More crucially perhaps, Darwin found himself asking whether there *is* a clear line of demarcation between humans and animals. His cousin Hensleigh Wedgwood had proposed that there certainly is: only humans have an innate sense of God. After observing the Fuegians (and the natives of Australia) Darwin recorded his doubts. He would do so again, years later, when writing on religion in his *Descent of Man* (1871): "There is ample evidence . . . from men who have long resided with savages, that numerous races have existed, and still exist, who have no idea of one or more gods, and who have no words in their languages to express such an idea" (Darwin [1879] 2004, 116). He had had to come to terms with the fact that there was no universal sense of God.

In an important paper Matthew Day (2008) has reminded us that to speak of the "discovery" of godless savages is simplistic. This is because there was already a prevalent colonial discourse in which to refer to indigenous peoples as godless was a means of denying their full humanity, which in turn provided a convenient license for denying them human rights—even to their own land. Darwin's reflections were, however, colored by his firm belief in the ultimate unity of the human races, one expression of which was his aversion to slavery (Darwin [1879] 2004, xxvii–xxviii). Another of the serious questions raised by his experience of the Fuegians was therefore why some representatives of humanity were religious and others not. The answer he eventually gave was to include reference to cultural as well as biological determinants. What struck him in December 1832, however, was the enormity of the gulf that separated the "savage" from the "civilized." The native Fuegians "seemed the troubled spirits of another world" (Darwin 1985, 307). The gulf was "greater than between a wild and domesticated animal, in as much as in man there is a greater power of improvement" (Darwin 1987, 109).

Darwin would continue to reflect on those differences in the case of both animals and humans. The fundamental thrust of his work would be the affirmation of continuity between the species Homo sapiens and its animal ancestors. In seeking an answer to the question why some humans were religious and others not, he looked to what he considered incipient religious characteristics in the behavior of animals, not least in that of his own dog. As Day (2008) has succinctly put it, Darwin solved the problem of intraspecies variation by appealing to interspecies continuity. When his dog growled at a parasol moved by the breeze, Darwin concluded that "he must, I think, have reasoned to himself in a rapid and unconscious manner, that movement without any apparent cause indicated the presence of some strange living agent, and no stranger had a right to his territory" (Darwin [1879] 2004, 118). There could then be a suggestive continuity between dogs and superstitious "savages" whose world was inhabited by invisible spirits. In both cases a primitive, instinctive notion of causality was evident. In reducing the cognitive gap separating humans and animals, Darwin gained some purchase on how religious beliefs had originated and been sustained. In Day's insightful analysis (2008), "the relationships between savages and the British empire, gods and their followers, and dogs and their masters all share one straightforward trait: they were all established on the asymmetry of power."

We shall return later to Darwin's reflections on religion and to what, in some of its manifestations, he believed had been its ennobling role. But the naturalism that structured his explanatory project deserves further consideration. It can be brought into focus by considering his response to a theological question that actually sprang directly from his theory of natural selection.

A Theological Question

Darwin's study of variation and his conception of natural selection were products of the two years that followed his return to England in October 1836. As with Alfred Russel Wallace later, the reading in September 1838 of Malthus' *Essay on Population* triggered the realization that, in a fiercely competitive struggle for resources, favorable variations would tend to be preserved while unfavorable ones would be destroyed. The transmission and accumulation of variation over many generations would lead to the emergence of new species. Darwin's use of the phrase "natural selection" to capture this process raised a specific theological question. The metaphor of selection was substantiated by reference to the role of human breeders in selecting for the characteristics they wished to enhance in cattle, dogs, or pigeons. If domestic breeders could effect such changes in a short span of time, how much more might not nature achieve given eons of time?

Pigeons in particular gave a boost to Darwin's rhetoric in the *On the Origin of Species.* Unless he knew that all the varieties produced by the pigeon fanciers were derived from the common rock pigeon, even a well-trained ornithologist, Darwin declared, would be inclined to categorize them as separate species. But here is the question: If human intelligence intervenes in the practices of the domestic breeder, might there not, by analogy, be a Selector, with a capital S, working intimately through natural processes? For some of Darwin's contemporaries this seemed a reasonable enough inference and, for them, it took the sting out of the theory. But it was not Darwin's view. To postulate a Selector other than the course of nature was to miss the point, conjuring up the image of God as micromanager of the evolutionary process. This was an image that Darwin steadfastly resisted. In doing so he was obliged to suspend one of the natural (in the sense of seemingly plausible) readings of the analogy that featured so prominently in his rhetoric. He could not deny the part played by human consciousness in the selective role of the breeder: "A man preserves and breeds from an individual with some slight deviation of structure, or takes more care than usual in matching his best animals and thus improves them" (Peckham 2006, 114). But as if to pre-empt the possibility of a theological projection from natural selection, Darwin also deletes the teleology from the production of fancy pigeons: "The man who first selected a pigeon with a slightly larger tail, never dreamed what the descendants of that pigeon would become through long-continued, partly unconscious and partly methodical selection" (Peckham 2006, 113). Despite the fact that breeders did have goals in view in their selective breeding, Darwin finds a way of stressing the *unconsciousness* of their role in relation to the longer-term consequences of their activity. Clearly there was elasticity in the analogy between natural and artificial selection and Darwin himself found it difficult to stabilize a privileged interpretation. Indeed in the preliminary essay he prepared some fifteen years before the publication of his *On the Origin of Species* he had found the invocation of a transcendent being a useful heuristic device to explicate what he meant by natural selection: "Let us now suppose a Being with penetration sufficient to perceive differences in the outer and innermost organization quite imperceptible to man, and with forethought extending over future centuries to watch with unerring care and select for any object the offspring of an organism produced under the foregoing circumstances; I can see no conceivable reason why he could not form a new race . . . adapted to new ends" (Brooke 1985, 55). Nevertheless Darwin could not bring himself to identify this hypothetical being with a providential, continually active God. And so we must ask why.

It seems obvious that his resistance had something to do with the loss of his Christian faith. But then a further question arises. What precisely was the relationship between the gains he made in science and his loss of faith? At this point we often find in the literature one of two extreme positions. Either it is simply assumed that it was his science that destroyed his faith. Or, in complete contrast, it is asserted that it was his loss of faith that made possible his radical science. In a contribution to *Harvard Magazine,* E. O. Wilson (2005, 33) has presented the choice in precisely these stark terms. He writes: "The great naturalist did not abandon Abrahamic and other religious dogmas because of his discovery of evolution by natural selection, as one might reasonably suppose. The reverse occurred. The shedding of blind faith gave him the intellectual fearlessness to explore human evolution wherever logic and evidence took him." But are these the only alternatives? My own view is that a more subtle analysis is necessary. While it is largely true that Darwin's loss of faith was not occasioned by his science, there were, at the very least, indirect connections. Even before their marriage Emma worried that the critical, skeptical mentality necessary for constructive science would corrode his faith: "May not the habit in scientific pursuits of believing nothing till it is proved, influence your mind too much in other things which cannot be proved in the same way" (Darwin 1986, 172). And the second option, preferred by Wilson, cannot be entirely correct because Darwin had not definitively renounced Christianity in the early 1840s when the first substantial draft of his theory was entrusted to Emma for publication in the event of his death (Moore 1989, 195–9).

Another problem with this structuring of alternatives is that some of Darwin's deepest reflections involved both scientific and religious considerations simultaneously. For example, his science highlighted the theological problem of pain and suffering. He once wrote that the existence of so *much* pain and suffering in the world seemed to him one of the strongest arguments against belief in a beneficent deity, but, he continued, it "agrees well with the view that all organic beings have been developed through variation and natural selection" (Darwin 1887, Volume 1, 311). And he would sometimes reflect on features of nature that needed neither sophisticated science not sophisticated theology to read them. These were the gruesome features that deeply offended his aesthetic sensibilities. How could the ichneumon wasp be the product of benevolent design when it laid its eggs in the bodies of caterpillars that were then devoured by the hatching grubs? Was there not something devilish in such a phenomenon? (Darwin 1993, 224).

Darwin's Loss of Faith

I would argue that Darwin's science did to a degree corrode his faith, and for several reasons. The quality of the historical evidence, commonly adduced for the miracles and divinity of Christ, Darwin considered poor compared with the stringent evidential support required of a scientific theory. Moreover, reports of miracles had been rendered increasingly suspect because of advances in scientific understanding to which he was himself contributing. As he put it in his *Autobiography* (1958, 86), "the more we know of the fixed laws of nature the more incredible do miracles become." Considerations drawn from science featured again because, although he was ignorant of the causes of variation, he was convinced that the variations themselves appeared randomly. Many were deleterious and even those that turned out to be advantageous could hardly be said to have been produced with their prospective use in mind. It was on this point that Darwin and Asa Gray eventually parted company. Gray's advice to Darwin was that, until such time as the cause of variation was understood, it would be wise to ascribe it to Providence. This was one reason why Gray felt no dissonance between natural selection and natural theology. He could interpret the variations as having been led in propitious directions. Darwin dissented, arguing that because a builder happened to use a pile of available stones to build a house, it in no way followed that the stones had come into being *in order* that he could build the house. Purposiveness was read out of the story and Gray had to concede that he had no answer—except to say (and there is surely a lesson here for some) that the perception of design is ultimately a matter of faith (Moore 1979, 275–6).

Darwin's science did contribute to his agnosticism in one further respect. It bears on his conviction that the universe as a whole could not be the product of chance. In another of his captivating nuances, he would add that there were reasons why he should not trust even his own convictions. If the human mind was the product of evolution, what guarantees were there that it was equipped to deal with such metaphysical and theological niceties? (Darwin 1958, 93) Having identified these corrosive aspects of his science, I would still want to say that some of the more interesting reasons for his agnosticism had nothing to do with his theory and could have been shared by many of his contemporaries. To that extent I certainly agree with Wilson.

Existential Grounds of Religious Doubt

Darwin was a participant in a well-documented moral revolt against certain Christian teachings, notably the doctrine of eternal damnation for the unredeemed. It was a pressing matter because members of his family were beyond the pale of Christian orthodoxy. His grandfather Erasmus Darwin had been a freethinker; his father took the view that religion was only for women, and his brother Erasmus was an avowed atheist. Emma Darwin later suggested that her husband had reacted against a caricature of Christian doctrine, but there is no doubting the intensity of his reaction. It was in the context of recoil against this "damnable doctrine" that Darwin (1958, 87) let slip his fiercest remarks on Christianity, declaring that he could not see how anyone could even wish it to be true. We have already seen how he was affected by the realization that an innate sense of God was not universal. This militated against arguments for an intelligent deity based on inward convictions and feelings. It was simply not true that "all men of all races had the same inward conviction of the existence of one God" (Darwin 1958, 91). As he had experienced and studied other cultures, he also found it impossible to accept the idea of a unique revelation. The ignorance of the biblical writers was transparent to him and the relationship of New to Old Testament he found incongruous (1958, 85–6).

Furthermore, as James Moore has brilliantly shown, Darwin was deeply affected by the death early in 1851 of his ten-year-old daughter, Annie (Moore 1989). The letters that passed between Charles and Emma as Annie gradually lost her private battle for existence are deeply moving. Even as late as

1851, in distress and under duress, Darwin still invoked the name of God in his most intimate correspondence (Darwin 1989, 13–24). But the loss of so innocent a child (and his favorite) left a terrible scar. Was this not the problem of pain and suffering, justice even, experienced with piercing intimacy? In one of Darwin's letters to Asa Gray we find yet another consideration that could scarcely be counted as scientific. Darwin (1993, 275) in July 1860 asked Gray whether, if a man standing under a tree were struck by lightning, he really believed that the accident had happened designedly. Many people, Darwin supposed, did believe this, but he could not. When, on a summer night, a swallow caught a gnat, did Gray really believe that it had been predetermined that that particular swallow should swallow that particular gnat at that particular moment? For Darwin, the particularities, the contingencies, the accidents of both human and nonhuman life—the absence of any intelligible pattern—made belief in a caring Providence extremely difficult, if not impossible.

Nature, God, and Cooperation

Enough has now been said of Darwin's spiritual and intellectual biography to explain why he would seek a naturalistic account of human morality and its development—naturalistic in the sense of appealing to natural instincts and rejecting the intervention of a supernatural Being. It was, nevertheless, an account that still allowed the possibility that the evolutionary process rested on laws that had been prescribed by a prescient deity. Indeed, in successive editions of *On the Origin of Species,* a Creator who had "impressed laws on matter" and even breathed life into the first living things was deliberately retained (Peckham 2006, 757–9). Darwin genuinely hoped there could still be common ground between his deistic metaphysics and the natural theology of Christian theists who were so prone to criticize him (Brooke 2008). He even glimpsed a possible means of rapprochement in a view expressed by his cousin Hensleigh Wedgwood whom Darwin described as a "very strong theist." Darwin asked him whether he thought that each time a fly was snapped up by a swallow, its death was designed—the same question he put to Gray. Wedgwood replied that he did not believe so. It was rather that God ordered general laws and left the result to "what may be so far called chance" (Darwin 1993, 350). This was the formula that Darwin tried on Gray when divulging his own take on evolution and design: "I am inclined to look at everything as resulting from designed laws, with the details, whether good or bad, left to the working out of what we may call chance" (1993, 224); not, he added, that he was at all satisfied by it.

The secularizing force of Darwin's brand of naturalism, particularly as applied to mental evolution and the study of what came to be called the emotions, has been discussed in depth by Thomas Dixon (2001; 2003). The question to which I return here concerns the interplay between his understanding of religion, morality, and that cooperation within a tribe which, in *The Descent of Man,* he supposed would have survival value. Those tribes whose members were ready to aid each other stood the better chance of survival. Contemporary advocates of group selection, such as David Sloan Wilson (2002), have no difficulty in achieving an integrated vision. A spirituality in which there is concern for others can be "understood from a purely evolutionary and naturalistic perspective, as a strategy designed to generate societal benefits, often at an individual cost" (Wilson 2008, 42). Was Darwin's understanding similar? How did he address the issue of cooperation?

To ask whether Darwin understood cooperation in the sense commended in this book is not a straightforward question. There are discontinuities as well as continuities between Darwin's understanding of biological and cultural evolution and that of contemporary analysts—not least because a genetic calculus of fitness was some way in the future. Nor should we forget that when Mendelian genetics first featured in evolutionary theory in the early years of the twentieth century, it initially tended to displace natural selection by promoting mutation as an almost sufficient mechanism for change—a mechanism that would have alienated Darwin, who consistently favored the gradual accumulation of variation. Mathematical models constructed around the dilemmas of game theory would certainly have intrigued him but, when describing selfless behavior, Darwin's first port of call was the role of deep-seated social instincts in inducing an *unhesitating,* rather than a calculating, response. And because Darwin found himself balancing many different forces at work in human evolution, conceding in his *Descent of Man* that he had probably given too much weight to natural selection in the first edition of *On the Origin of Species,* his admiration for those attempting to quantify the various parameters with precision would probably have been tempered with caution. Darwin himself never ceased to emphasize the massive number of contingencies in the production of evolutionary change, complicating the picture even while balancing the respective contributions of natural and sexual selection. As he charmingly put it ([1879] 2004, 263), "the power to charm the female has sometimes been more important than the power to conquer other males in battle."

Rather than pursue counterfactual questions, it is, however, perfectly possible to recognize in *The Descent of Man* an account of morality in which the theme of cooperation is conspicuous. A concern for the welfare of others constitutes for Darwin the highest form of morality. Cooperation is an expression of a social instinct that he chooses to illustrate with an example of self-sacrifice. In this respect he does conform to a concept of cooperation defined as a "form of working together in which one individual pays a cost . . . and another gains a benefit." Darwin's example ([1879] 2004, 145) also challenges a purely utilitarian ethics:

> Under circumstances of extreme peril, as during a fire, when a man endeavours to save a fellow-creature without a moment's hesitation, he can hardly feel pleasure; and still less has he time to reflect on the dissatisfaction which he might subsequently experience if he did not make the attempt. Should he afterwards reflect over his own conduct, he would feel that there lies within him an impulsive power widely different from a search after pleasure or happiness; and this seems to be the deeply planted social instinct.

Darwin was in part responding to the reproach that he had laid the foundation of the noblest part of our nature in the "base principle of selfishness." On the contrary, he points out, his argument hinges on a willingness to risk one's life on the principle that, at an early period in human history, the expressed wishes of the community will have influenced to a large extent the conduct of each member. A very particular feature of this relationship between the individual and the community is given prominence when Darwin charts the evolution of conscience. This is the fundamental feeling of sympathy that would compel a person to have regard for the approbation and disapprobation of his fellows. Obedience to the wishes of the community would be strengthened by habit so that an act of theft, for example, would induce feelings of dissatisfaction with oneself. And it was the fundamental feeling of sympathy that would have been intensified by natural selection, since "those communities, which included the greatest number of the most sympathetic members, would flourish best, and rear the greatest number of offspring" ([1879] 2004, 163). In statements such as these one sees very clearly an interdependence between his science and ethical values that does have modern equivalents. It is, however, worth re-emphasizing that the ultimate axiom on which his explanation rests concerns the drive for approbation, cooperative behavior being catalyzed by it.

In discussing moral improvement Darwin assumed that virtuous tendencies could be inherited. Advocating a "principle of the transmission of moral tendencies" ([1879] 2004, 148), he regarded a disinterested love for all living creatures as the most noble and distinctive attribute of humankind. Because he believed so firmly in the inheritance of moral feelings, he took issue with John Stuart Mill who had described them as acquired rather than innate. In a courteous but censorious footnote, Darwin opined that "the ignoring of all transmitted mental qualities will . . . be hereafter judged as a most serious blemish in the works of Mr. Mill" ([1879] 2004, 121).

How had the feeling of sympathy been strengthened in primitive societies? Darwin's answer was through the habit of performing benevolent actions. Whence the primal motivation for these? In a few lines that do perhaps presage later discussions of reciprocal altruism, Darwin recognized the part played by a "low motive;" for "each man would soon learn that if he aided his fellow-men, he would commonly receive aid in return" ([1879] 2004, 156). This was emphatically not, however, the stimulus Darwin wished to accentuate. He immediately referred to "another and much more powerful stimulus to the development of the social virtues." This was afforded by the "praise and blame of our fellow-men."[2] One reason perhaps why Darwin gave so much weight to this characteristic was that it re-established continuity with the animal kingdom. His protoreligious dog made another appearance: "It appears that even dogs appreciate encouragement, praise, and blame." Because the foundation-stone of morality was that one should "do unto others as ye would they should do unto you," Darwin considered it "hardly possible to exaggerate the importance during rude times of the love of praise and the dread of blame." In this respect, self-interest had played a crucial role in the emergence of cooperative beings. Hence we find Darwin's summary of a complex matter: "Ultimately our moral sense or conscience becomes a highly complex sentiment—originating in the social instincts, largely guided by the approbation of our fellow-men, ruled by reason, self-interest, and in later times by deep religious feelings, and confirmed by instruction and habit" ([1879] 2004, 157).

Darwin's reference here to "deep religious feelings" shows that he could not exclude them from his story. Reverence or fear of gods or spirits had intensified feelings of acceptance or rejection, making possible higher degrees of remorse and repentance. Religious beliefs themselves he traced back to three human capacities that had been crucial for survival: a basic concept of causality, a capacity for reason, and a curiosity about the world. One might observe that these could equally be seen as the preconditions of the possibility of a science of nature and it is perhaps not surprising that, for the comprehension of religion, he added emotional parameters. In *The Expression of the Emotions in Man and Animals* ([1872] 1965, 217), he wrote of religious devotion as, in some degree, "related to affection, though mainly consisting of reverence, often combined with fear."

Reading Darwin's account of religion and its role in reinforcing a moral sense deeply rooted in a social instinct of cooperation, it would be easy to conclude that he had explained it away. Certainly he had no compunction in describing many religious beliefs as absurd. There are those who would like to find in Darwin a blanket denunciation, but, unlike some of his modern disciples, he showed discrimination. There was always what Darwin called the "higher" question—namely, "whether there exists a Creator and Ruler of the universe." This question, he noted, "has been answered in the affirmative by some of the highest intellects that have existed" ([1879] 2004, 116). If the laws governing human development, which had made religious beliefs possible, were ultimately derived from such a ruler, then the higher question was not necessarily vacuous. Darwin repeatedly described belief in a universal and beneficent God as "ennobling" and springing from a long and elevating culture ([1879] 2004, 116, 151, and 682). Darwin's account of the moral sense was often judged to be relativistic and his own wife was deeply troubled by it. But though it could be disturbing, Darwin did not intend it to be seen as relativizing. He explicitly wished to privilege the golden rule as the foundation and highest expression of morality: "As ye would that men should do to you, do ye to them likewise." His avowed object had not been to explain this principle away, but to show how it had been engendered naturally ([1879] 2004, 151).

One consequence of his evolutionary account was that it was possible to see a trajectory of moral improvement. Whereas many have seen grounds for pessimism in Darwin's dethronement of the human, his vision was full of optimism: "There is no cause to fear that the social instincts will grow weaker, and we may expect that virtuous habits will grow stronger, becoming perhaps fixed by inheritance. In this case the struggle between our higher and lower impulses will be less severe, and virtue will be triumphant" ([1879] 2004, 150).

For all his prescience, Darwin remains a Victorian.

Notes

1. References to Darwin's *The Descent of Man* are to the second edition, published in 1874 but corrected in the 1879 printing, on which Adrian Desmond and James Moore have based their recent (2004) edition.
2. Or, for modern game theorists, "reputation."

References

Brooke, J. H. 1985. "The Relations between Darwin's Science and His Religion." In *Darwinism and Divinity,* ed. John R. Durant. Oxford: Blackwell, 40–75.

———. 1991. *Science and Religion: Some Historical Perspectives.* Cambridge: Cambridge University Press.

———. 2003. "Darwin and Victorian Christianity." In *The Cambridge Companion to Darwin,* ed. Jonathan Hodge and Gregory Radick. Cambridge: Cambridge University Press, 192–213.

———. 2008. " 'Laws Impressed on Matter by the Deity?': The *Origin* and the Question of Religion." In *The Cambridge Companion to the Origin of Species,* ed. Robert J. Richards and Michael Ruse. Cambridge: Cambridge University Press.

Brown, F. B. 1986. "The Evolution of Darwin's Theism." *Journal of the History of Biology* 19: 1–45.

Browne, J. 1995. *Charles Darwin: Voyaging.* London: Pimlico.

Darwin, C. 1859. *On the Origin of Species By Means of Natural Selection.* London: Murray.

———. [1872] 1965. *The Expression of the Emotions in Man and Animals,* with an introduction by Konrad Lorenz. Chicago: University of Chicago Press.

———. [1879] 2004. *The Descent of Man, and Selection in Relation to Sex,* with an introduction by James Moore and Adrian Desmond. London: Penguin.

———. 1958. *The Autobiography of Charles Darwin 1809–1882, With Original Omissions Restored*, ed. Nora Barlow. London: Collins.

———. 1985. *The Correspondence of Charles Darwin, Volume 1 1821–1836*, ed. Frederick Burkhardt and Sydney Smith. Cambridge: Cambridge University Press.

———. 1986. *The Correspondence of Charles Darwin, Volume 2 1837–1843*, ed. Frederick Burkhardt and Sydney Smith. Cambridge: Cambridge University Press.

———. 1987. "Diary of the Voyage of the H.M.S. Beagle." In *The Works of Charles Darwin, Volume One*, ed. Paul Barrett and R. B. Freeman. New York: New York University Press.

———. 1989. *The Correspondence of Charles Darwin, Volume 5 1851–1855*, ed. Frederick Burkhardt and Sydney Smith. Cambridge: Cambridge University Press.

———. 1991. *The Correspondence of Charles Darwin, Volume 7 1858–1859*, ed. Frederick Burkhardt and Sydney Smith. Cambridge: Cambridge University Press.

———. 1993. *The Correspondence of Charles Darwin, Volume 8 1860*, ed. Frederick Burkhardt and Sydney Smith. Cambridge: Cambridge University Press.

Darwin, F. 1887. *The Life and Letters of Charles Darwin*. 3 vols. London: Murray.

Day, M. 2008. "Superstitious Dogs and Godless Savages: Charles Darwin, Imperial Ethnography and the Problem of Human Uniqueness." *Journal of the History of Ideas* 69: 49–70.

Desmond, A., and J. Moore. 1991. *Darwin*. London: Michael Joseph.

Dixon, T. 2001. "The Psychology of the Emotions in Britain and America in the Nineteenth Century: The Role of Religious and Antireligious Commitments." *Osiris* 16: 288–320.

———. 2003. *From Passions to Emotions: The Creation of a Secular Psychological Category*. Cambridge: Cambridge University Press.

England, R. 2001. "Natural Selection, Teleology, and the Logos: From Darwin to the Oxford Neo-Darwinists, 1859–1909." *Osiris* 16: 270–85.

Gray, A. 1963. *Darwiniana: Essays and Reviews Pertaining to Darwinism*, ed. A. Hunter Dupree. Cambridge, MA: Harvard University Press.

King, P. P., R. Fitzroy, and C. Darwin. 1838. *Narrative of the Surveying Voyages of His Majesty's Ships Adventure and Beagle, between the Years 1826 and 1836, Describing their Examination of the Southern Shores of South America, and the Beagle's Circumnavigation of the Globe*. 3 vols. London: Colburn.

Kohn, D. 1989. "Darwin's Ambiguity: The Secularization of Biological Meaning." *British Journal for the History of Science* 22: 215–39.

Larson, E. J. 1997. *Summer for the Gods: The Scopes Trial and America's Continuing Debate over Science and Religion*. Cambridge, MA: Harvard University Press.

———. 2005. "Evolutionary Dissent." *Science and Spirit*, March/April: 48–52.

Livingstone, D. 2003. *Putting Science in Its Place: Geographies of Scientific Knowledge*. Chicago : University of Chicago Press.

Moore, J. R. 1979. *The Post-Darwinian Controversies.* Cambridge: Cambridge University Press.

———. 1989. "Of Love and Death: Why Darwin 'Gave Up Christianity.'" In *History, Humanity and Evolution,* ed. James R. Moore. Cambridge: Cambridge University Press, 195–229.

Nartonis, D. K. 2005. "Louis Agassiz and the Platonist Story of Creation at Harvard, 1795–1846." *Journal of the History of Ideas* 66: 437–49.

Owen, R. 1849. *On the Nature of Limbs.* London: van Voorst.

Peacocke, Arthur. 1985. "Biological Evolution and Christian Theology—Yesterday and Today." In *Darwinism and Divinity,* ed. John R. Durant. Oxford: Blackwell, 101–30.

Peckham, M. 2006. *The Origin of Species: A Variorum Text.* Philadelphia: University of Pennsylvania Press.

Roberts, J. H. 1988. *Darwinism and the Divine in America.* Madison, WI: University of Wisconsin Press.

Somerville, M. 1873. *Personal Recollections from Early Life to Old Age of Mary Somerville with Selections from her Correspondence.* London: Roberts Brothers Publishing.

Stenhouse, J. 1999. "Darwinism in New Zealand, 1859–1900." In *Disseminating Darwinism: The Role of Place, Race, Religion, and Gender,* ed. Ronald L. Numbers and John Stenhouse. New York: Cambridge University Press.

Stephens, L. D. 2000. *Science, Race, and Religion in the American South: John Bachman and the Charleston Circle of Naturalists, 1815–1895.* Chapel Hill, NC: University of North Carolina Press.

Tyndall, J. 1879. *Fragments of Science.* Vol. 2. London: Longmans, Green and Co.

Wilson, D. S. 2002. *Darwin's Cathedral: Evolution, Religion, and the Nature of Society.* Chicago: University of Chicago Press.

———. 2008. "Our Superorganism: Evolutionary Biology Gazes on Religion and Spirituality." *Science and Spirit* 19: 39–42.

Wilson, E. O. 2005. "Intelligent Evolution." *Harvard Magazine* November/December: 29–33.

2

Altruism

Morals from History

Thomas Dixon

This book represents the continuation of a project that began in earnest in the nineteenth century: the attempt to understand human morality in the light of modern science, with particular reference to theories of evolution, and especially to their implications for traditional religious and ethical teachings. The present chapter returns to the invention of "altruism" in the nineteenth century to ask what lessons can be learned from the pioneers of evolutionary ethics and their critics about the prospects and perils of bringing together science, God, and morality.

Today "altruism" is a term that conjures up two different sets of ideas. On the one hand, it stands for heroic human selflessness and is associated with celebrated and saintly individuals such as Oskar Schindler and Mother Teresa. On the other hand, it is a term with technical scientific meanings—used by neuroscientists, evolutionary theorists, and sociologists. In fact, the term has had simultaneously religious and scientific connotations from its earliest uses one hundred and fifty years ago—although the religion and the science in question have changed quite considerably.[1]

It was in 1851 that the French sociologist Auguste Comte first used the term *altruisme* in his *System of Positive Polity* (1851–1854) as the name for a group of other-regarding instincts, which he located physically toward the front of the human brain, as summarized in tabular form by Comte in Figure 2.1 (Comte 1875–1877, Volume 1, 73, 558–59, 592–93; Volume 2, 172). It was also a key term in his sociological plans for the reorganization of society under the influence of his atheistic "Religion of Humanity," which aimed to see egoism subordinated to altruism as human civilization evolved. In the following decades, "altruism" became the watchword first of scientific atheists sympathetic with Comte's "Religion of Humanity," and latterly of philanthropists and socialists of various kinds. It was initially resisted by clergymen as an unnecessary scientific neologism but was later appropriated by some, most notably the Scottish evangelical Henry Drummond, as nothing less than a synonym for Christian love.

A specific problem that has hampered discussions of science and ethics, both in the nineteenth century and in more recent years, has been the

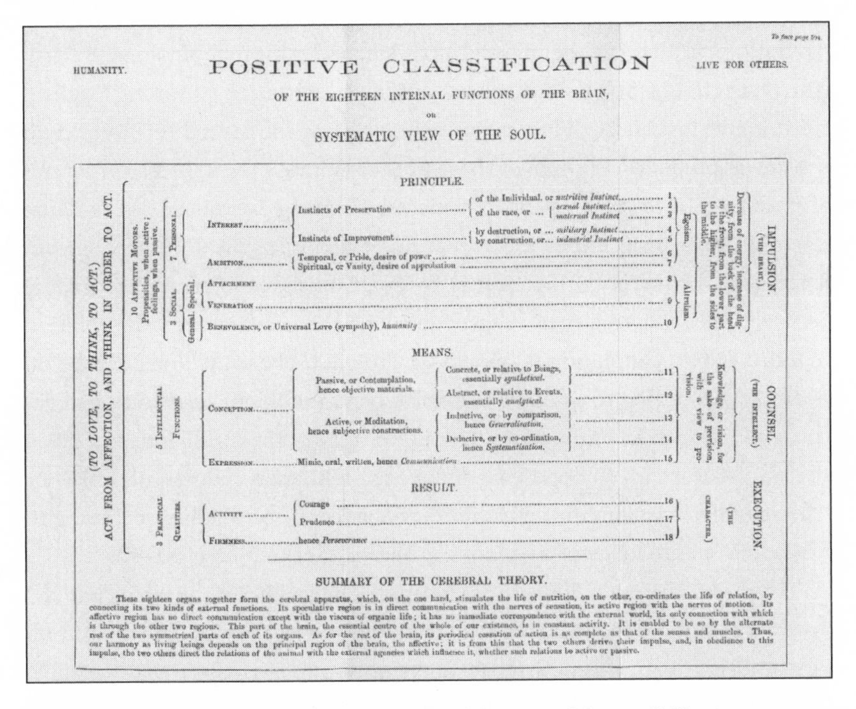

To face page 594.

HUMANITY. POSITIVE CLASSIFICATION LIVE FOR OTHERS.

OF THE EIGHTEEN INTERNAL FUNCTIONS OF THE BRAIN,

OR

SYSTEMATIC VIEW OF THE SOUL.

ACT FROM AFFECTION, AND THINK IN ORDER TO ACT.
(TO LOVE, TO THINK, TO ACT.)

PRINCIPLE.

10 Affective Motors. Propensities, when active; feelings, when passive.

7 Personal.
- Interest
  - Instincts of Preservation: of the Individual, or *nutritive Instinct* ... 1; of the race, or ... *sexual Instinct* ... 2, *maternal Instinct* ... 3
  - Instincts of Improvement: by destruction, or ... *military Instinct* ... 4; by construction, or ... *industrial Instinct* ... 5
- Ambition: Temporal, or Pride, desire of power ... 6; Spiritual, or Vanity, desire of approbation ... 7

Egoism.

3 Social.
- Special: Attachment ... 8; Veneration ... 9
- General: Benevolence, or Universal Love (sympathy), *humanity* ... 10

Altruism.

Decrease of energy, increase of dignity, from the back of the head to the front, from the lower part to the higher, from the sides to the middle.

IMPULSION. (THE HEART.)

MEANS.

5 Intellectual Functions.
- Conception
  - Passive, or Contemplation, hence objective materials: Concrete, or relative to Beings, essentially *synthetical*. ... 11; Abstract, or relative to Events, essentially *analytical*. ... 12
  - Active, or Meditation, hence subjective constructions: Inductive, or by comparison, hence *Generalisation*. ... 13; Deductive, or by co-ordination, hence *Systematisation*. ... 14
- Expression ... Mimic, oral, written, hence *Communication* ... 15

Knowledge, or vision, for the sake of prevision, with a view to provision.

COUNSEL. (THE INTELLECT.)

RESULT.

3 Practical Qualities.
- Activity: Courage ... 16; Prudence ... 17
- Firmness ... hence *Perseverance* ... 18

EXECUTION. (THE CHARACTER.)

SUMMARY OF THE CEREBRAL THEORY.

These eighteen organs together form the cerebral apparatus, which, on the one hand, stimulates the life of nutrition, on the other, co-ordinates the life of relation, by connecting its two kinds of external functions. Its speculative region is in direct communication with the nerves of sensation, its active region with the nerves of motion. Its affective region has no direct communication except with the viscera of organic life; it has no immediate correspondence with the external world, its only connection with which is through the other two regions. This part of the brain, the essential centre of the whole of our existence, is in constant activity. It is enabled to be so by the alternate rest of the two symmetrical parts of each of its organs. As for the rest of the brain, its periodical cessation of action is as complete as that of the senses and muscles. Thus, our harmony as living beings depends on the principal region of the brain, the affective; it is from this that the two others derive their impulse, and, in obedience to this impulse, the two others direct the relations of the animal with the external agencies which influence it, whether such relations be active or passive.

Figure 2.1 Table illustrating Comte's cerebral theory. Of the ten "affective motors," seven were associated with "egoism" and three—attachment, veneration, and benevolence—with "altruism." From Comte 1875–77, Vol. 1, facing 594.

multivalence of this key word "altruism." There are three different clusters of concepts that have been associated with the term. Some, including the term's originator, Auguste Comte, have used "altruism" to refer to selfless or other-regarding instincts or intentions. Others have defined "altruism" in terms of actions rather than intentions. This second approach was spread through the influence of the English evolutionist Herbert Spencer's writings from the 1870s onward. It has also become the standard use in the biological sciences today, which normally define behaviors as altruistic if they increase another's chances of reproductive success at the expense of the actor, regardless of intentions (Okasha 2005; Foster 2008;). Finally, "altruism" is sometimes invoked, in opposition to "individualism," as the name of any ethical principle that asserts an identity between moral goodness and the good of others. From the 1850s onward, "altruism" was frequently used in this third sense, as a term for a wide range of humanistic and socialistic ideologies (and, later, even for Christianity too).

These three different sets of "altruism" concepts can be described as psychological altruism, behavioral altruism, and ethical or ideological altruism, respectively (Sober and Wilson 1998, 6–8; Dixon 2005). Among English-speaking writers, it was Herbert Spencer who had the most direct impact on the development of all three of these uses of Comte's term. In *The Principles of Psychology* (1870–1872) Spencer proposed adopting "altruism" as the name for other-regarding impulses, in place of existing terms such as "benevolence" and "beneficence" (Spencer 1870–1872, Volume 2, 607n.). In *The Study of Sociology,* the following year, Spencer developed an account of social evolution in terms of two competing religions—the religion of enmity and the religion of amity. According to Spencer, primitive pagan religions of enmity had encouraged egoism and militarism. A reaction against these religions had given rise to a diametrically opposed religion, the Christian religion of amity, or "the religion of unqualified altruism," as Spencer also called it, according to which one was to love one's enemies as oneself (Spencer 1873, 182–83).

Then in *The Data of Ethics* (1879), Spencer's definition of the term changed again, this time to focus on physical behavior rather than on either psychology or ideology. In this new sense, Spencer wrote, altruism was discernible from the very dawn of life, in the lowest and simplest creatures, and especially in the evolution of the parental instincts, which ultimately evolved into social sympathy. He stipulated that "altruism" was to mean "all action which, in the normal course of things, benefits others instead of benefiting self." This was to include all "acts by which offspring are preserved and the species maintained," in nonhuman as well as human species; "acts of automatic altruism" were to be included along with those with some conscious motivation. Reproductive fission of the simplest single-celled organisms, such as an infusorium or a protozoon, was also to qualify as an act of "physical altruism" (Spencer 1879, 201–2). Spencer taught that the evolution of life was a process in which egoism and altruism were both primordial and evolved simultaneously, with altruism gradually coming to predominate in more civilized human societies. These evolutionary ideas were subsequently enthusiastically taken up by Patrick Geddes, Henry Drummond, and others (Geddes and Thomson 1889; Drummond 1894; see also Heather Curtis's chapter in this book).

The definitions of "cooperation" and "altruism" recommended by the editors of this book draw on two of the three clusters of concepts I have just outlined, namely the behavioral and the psychological, and other contributors to the book extend these to suggest a more ideological or ethical commitment to cooperation and altruism as defining features of moral goodness. The editors suggest that "cooperation" should be used to mean any action in which an individual pays a cost and another gains a benefit. Cooperation, for them, is defined in terms of outcomes, not intentions. "Altruism," for their purposes, refers to a subset of cooperative behaviors, namely those "in which an individual is motivated by goodwill or love for another." This definition combines behavioral and psychological aspects together. In that respect, it preserves both the Comtean and Spencerian heritage. It is also worth noting that, depending very much upon one's view of what sort of creatures can be motivated by "goodwill" or "love for another," it might seem, at least to some, to be a definition that can apply only to humans. Again, this is consistent with many nineteenth-century authors, including Comte and Spencer, who thought that altruism only reached its full development in the human race.

The Natural Selection of Sympathy

Having got this far without mentioning his name, it will be as well now to discuss the role of Charles Darwin in the history of ideas about altruism, since his theory of evolution by natural selection still casts such a long shadow over debates about cooperation and conflict in the natural world. Darwin has often been remembered as someone who showed nature to be a blood-soaked arena of vicious competition in which the only way to succeed was the ruthless pursuit of self-interest. On this view, Darwin discovered that, from the point of view of evolution, the survival of the fittest amounted to the survival of the most selfish. As John Hedley Brooke's chapter in this volume has already established, however, such a suggestion fails to capture Darwin's true views on the evolution of morality.

Charles Darwin, like anyone who had made a serious study of animal behavior, including the French natural historian Charles Georges Leroy (1802), the inventor of "altruism" Auguste Comte (1875–1877, Volume 1, 486–517), and many others before them, saw that the natural world was full of examples of creatures behaving in cooperative (in the lay sense) and sympathetic ways toward each other. Darwin documented, in *The Descent of Man* (1871), the existence of loving and cooperative behaviors among a wide range of birds and mammals, including dogs, elephants, baboons, and pelicans. It was clear to Darwin that it would benefit each individual creature to help others if such help would be reciprocated. Darwin thought it also apparent that sympathetic and cooperative tribes and groups would flourish in comparison with communities made up of more purely selfish individuals. Once equipped with reason and with language, human communities would have been able to reinforce existing tendencies toward loving and collaborative behavior through explicit approval and disapproval, and through the formulation of moral codes. Darwin saw every reason to predict that this evolution of human cooperation would, after the nineteenth century, continue to even higher levels. "Looking to future generations," he predicted in the *Descent,* "there is no cause to fear that the social instincts will grow weaker, and we may expect that virtuous habits will grow stronger, becoming perhaps fixed by inheritance. In this case the struggle between our higher and lower impulses will be less severe, and virtue will be triumphant" (Darwin 1882, 125).

Darwin returned to his ideas about the social instincts in his *Autobiography,* written toward the end of his life. In a section dealing with his views on religion and morality, Darwin recalled how he had pondered the question of belief in a Creator. Darwin found it hard to believe that the universe was the product of chance or necessity, but equally hard to imagine that it was the work of a divine craftsman, and, in any case, Darwin asked himself, "Can the mind of man, which has, as I fully believe, been developed from a mind as low as that possessed by the lowest animal, be trusted when it draws such grand conclusions?" Darwin concluded that "the mystery of the beginning of all things is insoluble to us, and I for one must be content to remain an Agnostic." Turning to morality, Darwin thought it quite possible to live a moral life while lacking religious faith. A person in that position, Darwin wrote, "can have for his rule of life, as far as I can see, only to follow those impulses and instincts which are the strongest or which seem to him the best ones." Of course following one's strongest instincts and following those that seem best are quite different ethical rules. Darwin was aware of this but never really resolved the problem.

In the same passage in the *Autobiography,* Darwin wrote that the unbeliever following his social instincts acts as a dog does but, unlike the dog, can reflect rationally on which impulses, if followed, would lead to the greatest approbation from others. Receiving the approval of others and love from those with whom one lives, Darwin wrote, was "undoubtedly the highest pleasure on this earth." As in the passage in the *Descent of Man,* when he had predicted ever-increasing levels of virtue in future societies, Darwin here also wrote optimistically that this process of following social instincts would lead to a stage when it became intolerable for anyone "to obey his sensuous passions rather than his higher impulses, which when rendered habitual may be almost called instincts." Having considered inherited instincts and the desire for social approval as two possible spurs to moral action, Darwin finally also mentioned occasions when a person may act against both of these motives in order to satisfy "his innermost guide or conscience" (Darwin 1958, 93–95).

As these passages reveal, Darwin had quite complex ideas about the evolution of human motivation and morality. His writings revealed a basic tension between the view that good actions were the products of evolved instincts and the alternative suggestion that humans, unlike other animals, acted on the basis of such principles as conscience and rational reflection and were thus able to follow their best rather than their strongest instincts. Broadly speaking, Darwin held that human qualities of sympathy and co-operation, which he considered to be self-evidently virtuous, had evolved from animal origins and would continue to develop to still higher levels in future civilizations, even in the absence of religion. So he was not a prophet of individualism, nor did he read the book of nature as a story of the triumph of self-interest. Rather he hoped and believed that loving sympathy would triumph over selfish passions in the long run and would then become fixed by heredity.

A second common misconception about Darwin is that he thought there was a "problem of altruism" that posed a threat to his theory of evolution by natural selection. In this context Darwin's concerns about how to explain the existence and behavior of neuter (or sterile) social insects is sometimes mentioned. But the problem Darwin struggled with here had nothing to do with "altruism" (indeed, he never used the word) but concerned the difficulty of explaining how sterile castes of insects could differ so drastically, both anatomically and behaviorally, from their parents and their siblings. The problem that Darwin thought might be "actually fatal to my whole theory" was to explain these radical dissimilarities within families of insects, not to explain the existence of altruism (Cronin 1991, 299; Darwin 2006, 245–50). Darwin believed he had a solution: "The difficulty, though appearing insuperable, is lessened, or, as I believe, disappears, when it is remembered that selection may be applied to the family, as well as to the individual, and may thus gain the desired end" (Darwin 2006, 245). Darwin's suggestion here that natural section could act at the level of the "family" is the closest that he came to articulating anything like the modern theory of "kin selection." He did not distinguish carefully, even in this case, however, between "family" and "tribe" or "community," so it is a moot point whether he intended to prioritize the social group or the family as the primary unit of selection. Elsewhere he clearly envisaged natural selection operating between communities rather than families, favoring those that happened to produce valuable variations such as sterile worker castes or, for that matter, high levels of cooperation and sympathy.[2]

We should, as John Hedley Brooke has also warned in his chapter above, resist the temptation to read twentieth- or twenty-first-century biological concerns back into a work composed in the middle of the nineteenth century. "Altruism" has been recognized as a potential problem for modern neo-Darwinism, especially since W. D. Hamilton proposed his solution in terms of "kin selection" in the 1960s. The neo-Darwinian insistence that evolutionary explanations must connect all traits with direct advantages to individuals or their genes (rather than to groups or species as a whole) brought the question of "altruism" much more sharply into focus. As Helena Cronin put it in the early 1990s: "Only once it was solved was the problem clearly seen. Only with gene-centred hindsight were Darwinians able to formulate sharply what really should have been considered altruistic, and why. . . . For a century, altruistic behaviour was hardly discussed at all; until recently most Darwinians did not even appreciate that altruism posed a problem" (Cronin 1991, 265). But Charles Darwin, of course, was not a Hamiltonian, nor a neo-Darwinian. He favored explanations that referred to the heritability of acquired physical and moral characteristics, and believed that natural selection could act at the level of groups and tribes as well as individuals.[3]

Charles Darwin, then, was neither a prophet of selfish individualism nor a neo-Darwinian theorist trying to find a solution to what was then a nonexistent "problem of altruism." Darwin himself never used the new language of "altruism," and instead thought about how natural selection, acting at the level of tribes and communities, in combination with the inheritance of acquired characteristics, could account for the fact that human beings had such strong feelings of love and sympathy for each other—feelings so strong that they could even come to override the urgings of passions such as anger and desire.

As Brooke's chapter in this volume has already shown, Darwin thought that his natural historical account of the evolution of the human conscience had shown that utilitarians and political economists had been wrong to try to explain all human behavior as the product of self-interest. On his new evolutionary view, Darwin wrote in *The Descent of Man,* "the reproach is removed of laying the foundation of the noblest part of our nature in the base principle of selfishness; unless the satisfaction which every animal feels, when it follows its proper instincts, and the dissatisfaction felt when prevented, be called selfish" (Darwin 1882, 121). This sentiment captures perfectly the characteristically Darwinian combination of Victorian ethos with biological bathos. Darwin's *The Descent of Man* similarly included the statement that the highest of human actions were those in which a person would unhesitatingly risk his life to save another. In such cases, Darwin wrote, the selfless hero was acting "in the same manner as does probably a bee or ant, when it blindly follows its instincts" (Darwin 1882, 120).

Life and Her Children

Darwin was, of course, a pioneer in many respects. But he can also be placed in a longer tradition of those who tried to use natural history and science to understand human morality. I have already mentioned Georges Leroy and Auguste Comte as contributors to this tradition. Another was the Congregationalist divine and hymn-writer Isaac Watts, whose *Divine and Moral Songs for Children* (1715) included tributes to the industry of the bee and the providence of the ant (Watts 1866, 65–66, 103–4). The anatomy and instincts of insects provided inspiration for works of natural theology too. William Gould's *An Account of English Ants* (1747) and William Kirby and William Spence's *Introduction to Entomology* (1815–1826) were both works that considered ants from the points of view of natural history and natural theology, as examples of the power, wisdom, and goodness of their Creator. Kirby and Spence's book had been studied by Darwin as he prepared to write *On the Origin of Species,* and also by the leading popularizer of the Darwinian account of the evolution of morality in the late nineteenth century, Arabella Buckley. Buckley recognized that highly organized communities of social insects could be used as models of human societies, albeit imperfect ones, and she admired the qualities of determination and obedience in the "industrious law-abiding bee" (Buckley 1891, 100).[4]

Arabella Buckley had worked as personal secretary to one of Charles Darwin's mentors, the geologist Sir Charles Lyell, before turning her hand to literary work. She became a very successful writer of books about science, especially for children. Her titles included *The Fairy-Land of Science* (1879) and its sequel *Through Magic Glasses* (1890). Her works about the moral messages to be read in the book of nature included *Life and Her Children: Glimpses of Animal Life from the Amoeba to the Insects* (1880) and *Winners in Life's Race; or, The Great Backboned Family* (1882). Nature taught that qualities of mutual help and sympathy, as Buckley wrote, "are among the most powerful weapons, as they are also certainly the most noble incentives, which can be employed in fighting the battle of life" (Buckley 1880, 301). Her conclusion was that "one of the laws of life which is as strong, if not stronger, than the law of force and selfishness, *is that of mutual help and dependence*" (Buckley 1882, 351). Although Buckley saw great value in acts of altruism, she drew an important distinction between mere self-sacrifice, which she detected in abundance in the highly organized societies of insects, and the higher moral qualities of love, mutual help, and sympathy. Such affectionate bonds were to be observed in "that higher devotion of mother to child, and friend to friend, which ends in a tender love for every living being" (Buckley, 1880, 301).

This distinction between blind self-sacrifice and sympathetic affection was made by several Victorian commentators on Darwin, and on the writings of his friend the entomologist Sir John Lubbock. These commentators pointed out that there was a difference between mechanically following an instinct in the manner of a worker ant, on the one hand, and experiencing feelings of love and sympathy for a particular individual, on the other. Buckley suggested that while social insects might be devoted to the good of their community, they were not devoted to each other. In other words, there was more to the well-lived life than laboring for the good of a faceless collective. Such a life must involve bonds of affection and sympathy between distinct individuals (Buckley 1880, 300–1; Buckley 1891, 67).

Buckley's works belonged to an alternative Darwinian tradition that flourished from the 1870s onward, in which love and cooperation were emphasized instead of individualism and competition (Dixon 2008a, 157–58). This tradition included a range of philosophical, religious, and political reinterpretations of the natural word. We have already seen that Herbert Spencer and Charles Darwin both contributed to this tradition in England; in Germany, Ludwig Büchner was an important figure, and in France, Alfred Espinas (Espinas 1877; Büchner 1879, 1880). Later, the Russian anarchist Peter Kropotkin's *Mutual Aid* (1902), which focused on cooperation in both animal and human societies, suggested that life's message to her children was: "Don't compete!—competition is always injurious to the species, and you have plenty of resources to avoid it! . . . Therefore combine—practice mutual aid! That is the surest means of giving to each and to all the greatest safety, the best guarantee of existence and progress, bodily, intellectual, and moral" (Kropotkin 1939, 73).

Another significant work in this tradition was Patrick Geddes and J. Arthur Thomson's *The Evolution of Sex* (1889), which included the graphic representation of the relationship between egoism and altruism shown in Figure 2.2. This essentially Spencerian work suggested that egoism evolved from self-preserving instincts, and altruism ultimately from sexual and reproductive instincts. Geddes and Thomson produced a strongly gendered version of the

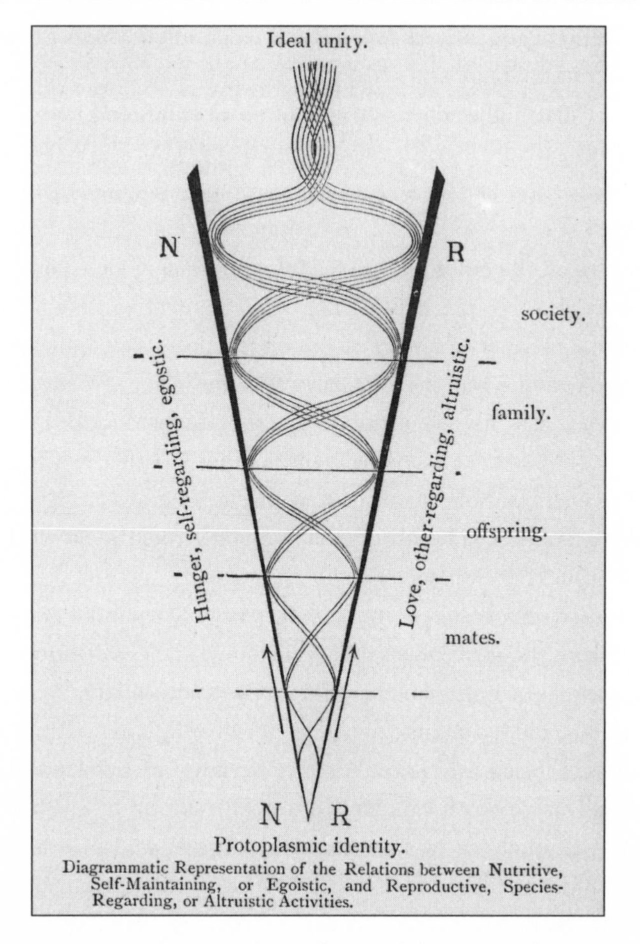

Figure 2.2 Diagram representing the evolution of altruistic and egoistic activities in the animal world, from Geddes and Thomson 1889, 280.

evolution of altruism, in which maternal self-sacrifice was especially emphasized. Examples of maternal care among crayfish, cuttlefish, bees, spiders, and ants were given. Geddes and Thomson agreed with Buckley, however, that this sort of devotion reached its highest form among mammals, such as the female opossum, pictured in Figure 2.3, and ultimately in humans (Geddes and Thomson 1889, 270–74, 279–81, 312–14).

This version of the story of evolution reached a still wider audience through the Lowell Lectures delivered in Boston in 1893 by the Scottish writer

Figure 2.3 An illustration of the altruistic maternal activities of an opossum from Geddes and Thomson 1889, 312.

Henry Drummond, and published as a very successful book, *The Ascent of Man,* the following year. Drummond, pictured in Figure 2.4, was an evangelical theologian and a popularizer of science whose work is explored further in Heather Curtis's chapter in this book. Drummond saw evolution as an epic love story, starting with the protoplasmic fission of the earliest single-celled organisms and ending with the emergence of the human mother—the peak of altruistic love. One reviewer wrote that Drummond's sentiments on the evolution of maternal love "are more than philosophy; they are poetry, soaring almost into rhapsody" (Smith 1895, 236). Drummond surveyed the whole history of animal life from the amoeba to the invertebrates to fish, amphibians, reptiles, and birds and "then—What? The Mammalia, THE MOTHERS. There the series stops. Nature has never made anything since." Drummond even wondered whether the "one motive of organic Nature was to make Mothers." The evolution of parental care was, Drummond said, the way that nature had chosen "to teach the youth of the world the Fifth Commandment"—that is, to honor their fathers and mothers. Drummond described the struggle for life as the domain in which fathers prevailed and the "struggle for the life of others" as the quintessentially maternal preserve. In so doing, he reinforced the Victorian domestic ideology of separate masculine and feminine spheres of influence—the mother devoted to nurturing

Figure 2.4 Portrait of Henry Drummond (1851–1897). Date of photograph unknown; reproduced from frontispiece of Smith 1899.

her family at home, the father to making his way in the public square (Drummond 1899, 22–23, 282–85, 342–43, 355).

Killing the "Angel in the House"

The nineteenth-century passion for altruism was expressed in medical works as well as evolutionary and religious ones. According to many nineteenth-century physicians, overweening egoism was a sign of mental illness. The leading British psychiatrist Henry Maudsley, for instance, wrote that whoever was destitute of proper moral feeling was

> to that extent a defective being; he marks the beginning of race-degeneracy; and if propitious influences do not chance to check

> or to neutralize the morbid tendency, his children will exhibit a further degree of degeneracy and be actual morbid varieties. . . . Any course of life then which persistently ignores the altruistic relations of an individual as a social unit, which is in truth a systematic negation of the moral law of human progress, deteriorates his higher nature, and so initiates a degeneracy which may issue in actual mental derangement in his posterity. (Maudsley 1879, 102–3; quoted in Skultans 1975, 197)

Max Nordau, the author of the best-selling medicomoral tract, *Degeneration* (1892), drawing on scientific authorities such as the Italian criminologist Cesare Lombroso, also denounced modern-day "egomania" as a symptom of psychopathology. His examples of degenerate egomaniacs included Oscar Wilde and Friedrich Nietzsche. The truly healthy individual, Nordau explained, would develop beyond the stage of egoism to one of wholesome altruism. This altruism was defined in terms of the underlying motives and intentions—it was not just a physical action but a genuine concern for another based on "sympathy or curiosity": "Not till he attains to altruism is man in a condition to maintain himself in society and in nature" (Nordau 1993, 244, 252–53).

It is not hard to see why this sort of authoritarian anti-individualistic writing could be experienced as repressive rather than uplifting. Women suffered especially at the hands of these repressive ideologies of altruism. We have seen how writers such as Patrick Geddes and Henry Drummond used the story of evolution to locate the roots of morality in biological reproduction and to celebrate its apogee in modern female self-sacrifice. This was, from a contemporary perspective, a very back-handed compliment to women, and one that seemed designed to give the authority of nature to their confinement within the domestic sphere.

Virginia Woolf's modernist individualism was premised on a rejection of just this kind of Victorian sexism. Lecturing on the subject of "Professions for Women" in 1931, Woolf recalled how, at the beginning of her career as a writer, she had been haunted by a Victorian phantom, namely the heroine of a famous poem, "The Angel in the House." "You who come of a younger and happier generation may not have heard of her," Woolf acknowledged to her audience. "The Angel in the House," she explained, was "immensely sympathetic. She was immensely charming. She was utterly unselfish. She excelled in the difficult arts of family life. She sacrificed herself daily. If there was chicken, she took the leg; if there was a draught she sat in it—in short she was so constituted that she never had a mind or a wish of her own, but preferred to sympathize always with the minds and wishes of others." Woolf recalled that only once she had killed this phantom had she been able to find the confidence and the authorial voice required to embark on her chosen profession. She remembered the moment with satisfaction: "I took up the inkpot and flung it at her. She died hard. Her fictitious nature was of great assistance to her. It is far harder to kill a phantom than a reality" (Woolf 1993, 102–3).

The Paradox of Christ

If Woolf reminds us that an ideology of self-sacrifice can be constraining, especially for women, other criticisms of Victorian "altruism" teach us the lesson that such an ideology was by no means identical with Christian ethics. Initially coined by the atheist Auguste Comte, the term was originally considered a watchword of atheism, before later being appropriated by some as a synonym for Christian love. Christian thinkers were, in fact, prominent as both advocates and critics of the new ideal of "altruism." Christian critics of altruism saw it as an exaggerated and one-sided ideal, and one that distorted ethics by neglecting love of God and love of self. One Anglican clergyman preached about the contrast between the Christian teaching "Love thy neighbor as thyself" and the unfortunate new altruistic teaching, "Love thy neighbor and not thyself" (Mozley 1886, 65). Many others authors also objected to the reduction of Christian ethics to mere altruism.

To take just two examples, Henry Sidgwick and Oscar Wilde both argued that the Christian injunctions to forgive one's enemies and to give to the poor should be followed primarily for one's own sake rather than as ways to improve either one's enemies or the poor. Sidgwick thought that when Jesus told the young man to sell his possessions and give the proceeds to the poor this instruction was "surely given, not primarily for the sake of the poor, but for the sake of the young man himself: it was a test, not of philanthropy, but of faith" (Sidgwick 1904, 18–20). In the spiritual autobiography he wrote while in prison for committing acts of gross indecency, *De Profundis,* Oscar Wilde insisted that Christ had been the "the supreme individualist . . . the first individualist in history." "People have tried to make him out an ordinary philanthropist," Wilde complained, "or ranked him as an altruist with the unscientific and sentimental. But he was really neither one nor the other" (Wilde 2005, 176).

The perspectives of three different Christian preachers at the end of the nineteenth century encapsulate the fraught relationship they perceived between their faith and the fashionable ideal of altruism. In Manchester in 1899 the Lancastrian theological lecturer Robert Mackintosh informed his students with disapproval: "Innumerable writers, Christian as well as non-Christian, have come to employ the term 'Altruism' as a synonym for goodness" (Mackintosh 1899, 45). A couple of months after Queen Victoria's death, in March 1901, however, the archdeacon of London was preaching on "Christian altruism," which he took to be the best description of the text of John 13:34, "A new commandment I give unto you, That ye love one another; as I have loved you, that ye also love one another." Tellingly, the archdeacon went on to instance the sympathy and philanthropy of the new queen (Edward VII's wife, Alexandra) as an example of Christian altruism, and to express the hope that her "gracious charities and works of benevolence" would "continue to give her an increasing happiness, and to crown the years to come with a golden halo of the people's gratitude and Heaven's benediction."[5] The happiness of the individual, both in this world and the next, still seemed to be a very prominent and legitimate concern, even when speaking of "Christian altruism." The Methodist preacher Hugh Price Hughes perhaps expressed the Christian hope for self-realization through self-sacrifice best: "Happy is the man who understands the characteristic and oft-repeated paradox of Christ, 'He that findeth his life shall lose it; and he that loseth his life for my sake shall find it'" (Hughes 1894, 41–42).

That "altruism" is synonymous either with goodness in general or with Christian love in particular is taken for granted by many writers still today (for example, Collins 2006, 27–28; Dugatkin 2006, ix; Montague and Chiu 2007, 137). But it should not be. Self-sacrifice is not the same thing as goodness, nor is it all there is to Christian ethics. If just one moral were to be taken from the history of controversies about "altruism" it should be this.

Biology and Ethics

What lessons, finally, can be learned today about the problems involved in nineteenth-century attempts to produce ethical and religious arguments informed by natural science? There are several morals, even warnings, that readers of the present volume might like to keep in mind as they proceed through subsequent chapters. All of these arise from an examination of nineteenth-century controversies, but are hopefully still pertinent to contemporary debates.[6]

The first and most obvious issue is what we could call the general problem of ethical naturalism. Of what use, if any, is nature to the moralist? Scientists might be able show that a certain tendency or behavior is "natural." So what? How do we leap from there to any ethical conclusion? Presumably selfish and violent behaviors, as much as cooperative and loving ones, have an evolutionary history and some genetic basis. Even if we can establish that a certain instinct is natural, that leaves open the ethical question of whether we should follow or resist it. Darwin, Buckley, and other nineteenth-century students of nature were well aware that there were aggressors, slave-makers, thieves, and parasites in the natural world, as well as cooperators and altruists. The bare existence of an evolutionary origin for an instinct does nothing to answer questions about its moral value. Richard Dawkins was well aware of this point when he wrote, in *The Selfish Gene* (1976), about the ethical imperative to resist nature. Dawkins thought that those hoping to build a more cooperative society could expect "little help from biological nature." Instead, he exhorted his readers: "Let us try to *teach* generosity and altruism, because we are born selfish." "We, alone on earth," Dawkins wrote, "can rebel against the tyranny of the selfish replicators" (Dawkins 1989, 3, 200–1). But someone with a different moral philosophy and a different interpretation of evolutionary science might equally say: "Let us try to *teach* self-realization and individualism, because we are born altruistic."

So it would seem that nature can provide, at the most, suggestive moral parables—natural models for human emulation—rather than substantive ethical rules. The next problem is to determine which natural behaviors exactly it is that we wish to emulate. Is it "altruism," or "cooperation," or something else that we think a particularly admirable natural propensity? Even among those who have wished to pursue some fairly strong form of ethical naturalism, history shows that they have not agreed about exactly which aspect of natural life it is they wish to see human individuals and societies emulate.

As I mentioned at the outset, even those who are agreed in advocating "altruism" do not always agree whether it is an inward motive, an outward act, or a more general ideology that they are recommending. If "altruism" is defined in terms of emotional motives such as love and goodwill, it is hard for biological scientists to study it. Yet if it is defined purely in terms of outcomes rather than motives, it seems to be of relatively little ethical interest. In his *Data of Ethics,* Herbert Spencer suggested, in offering a physical definition of altruism, that even mere "loss of physical substance" might qualify as a kind of altruism. Friedrich Nietzsche retorted that in that case even urination should be counted among the altruistic virtues (Spencer 1879, 201–2; Small 2005, 174–75). It often seems that the evolutionary roots of human morality are themselves amoral. The problem was also evident in Drummond's *Ascent of Man.* He acknowledged that the automatic processes that took place within simple single-celled organisms could not be described in moral terms. Yet he still suggested that in the process of cellular fission "there lies a prophecy, a suggestion of the day of Altruism" (Drummond 1899, 282–83), and although he acknowledged that, in purely biological terms, the struggle for life and the "struggle for the life of others" had "no true moral content," he could not resist going on to add: "yet the one marks the beginning of Egoism, the other of Altruism" (Drummond 1899, 284–85).

A similar problem arises for contributors to the present volume. "Cooperation" is taken here to refer to any action with some cost to the agent, and that benefits another individual (for example, by increasing that individual's reproductive fitness). Under such a definition, the deaths of slime mould cells when they are turned into a stalk for the dispersal of spores; the release of digestive enzymes by individual microbes that break down food for the benefit of genetically related cells; the donation of regurgitated blood among vampire bats; and the human menopause can all be included as examples of cooperation (see Foster 2008; Foster et al. 2007). Such natural actions are of no obvious ethical interest. What human moral lessons can be learned from microbial digestion or blood-drinking? And, again, even if they did somehow correspond to a kind of cooperation also operative among humans, why should we feel inclined to follow rather than resist this tendency?

What then of "altruism"—that subset of cooperative actions taken to be motivated "by goodwill or love for another"? Such actions may or may not be considered a paradigm of moral goodness. That question will be answered with reference to social and ethical considerations, not biological ones about cooperation and reproductive fitness. The problem here is the same one encountered by Herbert Spencer and Henry Drummond in the nineteenth century. The interest of taking an evolutionary approach to ethics derives from the hope that one might discover the biological roots of human moral ideas and altruistic actions. However, it is difficult to see how such explanations should change our ethical views about rightness and goodness, if at all, since the putative evolutionary roots of human goodness seem very remote from the fully developed version.

Even those behaviors in nature that seem to come closer to human virtue than do urination or regurgitation may not be much more attractive as moral models. The sorts of attributes that have been emphasized in these contexts have been instinctive self-sacrifice and cooperation among microbes, insects, birds, and monkeys. Yet there is more to the well-lived life than sacrificing one's own interests. So perhaps it is not blind "altruism" that should be emulated, but qualities that involve a greater affective element. Darwin wrote about "love" and "sympathy," for example, and Peter Kropotkin admired the reciprocal "mutual aid" he observed in nature. Arabella Buckley, as we have seen, insisted that her moral paradigm was not self-sacrifice but a reflective and affectionate bond between two individuals.

Nineteenth-century writers worried that evolutionary ethicists who championed "altruism" had forgotten about the importance of self-love, individualism, the value of human affection, and the duty sometimes to resist rather than to follow our natural instincts. Although both the scientific and religious contexts have changed considerably over the last century, their worries may still be worth remembering.

Notes

1. The material in this chapter is drawn from the research undertaken for Dixon 2008a, which documents in detail the early history of debates about "altruism" in the second half of the nineteenth century in the context of Christian belief, evolutionary science, and new political and philosophical movements.
2. For a more detailed explanation of Darwin's arguments here, see Richards 1987, 142–52; Cronin 1991, 298–308; Gayon 1998, 70–73; Dixon 2008a, 142–48.
3. For a reassertion of the plausibility of group selection as a Darwinian explanation, see Sober and Wilson 1998; D. S. Wilson and Sober 2002.

4. On Buckley, see Gates 1997, 1998, 2004. On the cultural and moral meanings of insects through history, see Sleigh 2001; Sleigh 2003, esp. 58–86; Lustig 2004; Preston 2006; Sleigh 2007. On the natural theology of instincts, and the role it played in shaping Darwin's theories, see Richards 1987, 127–56.
5. As reported in *The Times,* "Ecclesiastical Intelligence," 11 March 1901, 8.
6. Some of these ideas are developed more fully in Dixon 2008a, 360–74; see also Dixon 2008b, 114–26.

References

The writings of Charles Darwin, including his autobiography, are available online at *The Complete Works of Charles Darwin Online:* http://darwin-online.org.uk/. Many of Darwin"s letters are available online at *The Darwin Correspondence Project:* http://www.darwinproject.ac.uk/.

Browne, J. 2002. *Charles Darwin: The Power of Place.* London: Jonathan Cape.

Büchner, L. 1879. *Liebe und Liebes-Leben in der Thierwelt.* Berlin: U. Hofmann.

———. 1880. *Mind in Animals,* trans. A. Besant. London: Freethought Publishing Company.

Buckley, A. B. 1880. *Life and her Children: Glimpses of Animal Life, from the Amoeba to the Insects.* London: Edward Stanford.

———. 1882. *Winners in Life's Race; or, The Great Backboned Family.* London: Edward Stanford.

———. 1891. *Moral Teachings of Science.* London: Edward Stanford.

Collins, F. S. 2006. *The Language of God: A Scientist Presents Evidence for Belief.* New York: Free Press.

Comte, A. 1875–1877. *System of Positive Polity, or Treatise on Sociology, Instituting the Religion of Humanity.* 4 vols. Trans. E. S. Beesly, J. H. Bridges, F. Harrison, R. Congreave, and H. D. Hutton. London: Longmans, Green & Co.

Cronin, H. 1991. *The Ant and the Peacock: Altruism and Sexual Selection from Darwin to Today.* Cambridge: Cambridge University Press.

Crook, P. 1994. *Darwinism, War and History: The Debate over the Biology of War from the "Origin of Species" to the First World War.* Cambridge: Cambridge University Press.

Darwin, C. [1859] 2006. *The Origin of Species: A Variorum Text,* ed. Morse Peckham. Philadelphia: University of Pennsylvania Press.

———. 1882. *The Descent of Man, and Selection in Relation to Sex.* 2nd ed., revised and augmented. London: Murray.

———. 1958. *The Autobiography of Charles Darwin,* ed. Nora Barlow. London: Collins.

Dawkins, R. [1976] 1989. *The Selfish Gene.* New edition. Oxford: Oxford University Press.

Dixon, T. 2005. "Altruism." In *New Dictionary of the History of Ideas, ed.* M. C. Horowitz. 6 vols. New York: Scribner's. Vol. 1, 49–53.

———. 2008a. *The Invention of Altruism: Making Moral Meanings in Victorian Britain.* Oxford: Oxford University Press for the British Academy.

———. 2008b. *Science and Religion: A Very Short Introduction.* Oxford and New York: Oxford University Press.

Drummond, H. 1899. *The Lowell Lectures on the Ascent of Man.* London: Hodder and Stoughton.

Dugatkin, L. A. 2006. *The Altruism Equation: Seven Scientists Search for the Origins of Goodness.* Princeton: Princeton University Press.

Espinas, A. 1877. *Des Sociétés Animales.* Paris: G. Baillière.

Foster, K. R. 2008. "Behavioral Ecology: Altruism." In *Encyclopedia of Ecology,* ed. S. E. Jørgensen. 8 vols. Amsterdam: Elsevier, 154–59.

Foster, K. R., K. Parkinson, and C. R. Thompson. 2007. "What can microbial genetics teach sociobiology?" *Trends in Genetics* 23: 74–80.

Gates, B. T. 1997. "Revisioning Darwinism with Sympathy: Arabella Buckley." In *Natural Eloquence: Women Reinscribe Science,* ed. B. T. Gates and A. B. Shteir. Madison: University of Wisconsin Press, 164–76.

———. 1998. *Kindred nature: Victorian and Edwardian Women Embrace the Living World.* Chicago and London: University of Chicago Press.

———. 2004. "Arabella Buckley." In *The Dictionary of Nineteenth-Century British Scientists,* ed. B. Lightman. 4 vols. Bristol: Thoemmes Continuum. Vol. 1, 337–39.

Gayon, Jean. 1998. *Darwinism's Struggle for Survival: Heredity and the Hypothesis of Natural Selection,* trans. M. Cobb. Cambridge: Cambridge University Press.

Geddes, P., and J. A. Thomson. 1889. *The Evolution of Sex.* London: Walter Scott.

Hughes, H. P. 1894. *Essential Christianity: A Series of Explanatory Sermons.* London: Ibister and Co.

Kropotkin, P. [1902] 1939. *Mutual Aid: A Factor of Evolution.* Harmondsworth: Penguin.

Leroy, C. G. 1802. *Lettres Philosophiques sur l'"Intelligence et la Perfectibilité des Animaux.* Paris: Valade.

Lustig, A. 2004. "Ants and the Nature of Nature in Auguste Forel, Erich Wasmann, and William Morton Wheeler." In *The Moral Authority of Nature, ed.* L. Daston and F. Vidal. Chicago: University of Chicago Press, 282–307.

Mackintosh, R. 1899. *From Comte to Benjamin Kidd: The Appeal to Biology or Evolution for Human Guidance.* London and New York: Macmillan.

Maudsley, H. 1879. *The Pathology of Mind.* London: Macmillan.

Montague, P. Read, and Pearl H. Chiu. 2007. "For Goodness' Sake." *Nature Neuroscience* 10(2): 137–38.

Mozley, J. B. [1876] 1886. *Sermons Preached Before the University of Oxford and on Various Occasions.* 6th ed. London: Rivingtons.

Nordau, M. [1895] 1993. *Degeneration,* trans. and ed. George L. Mosse. Lincoln: University of Nebraska Press.

Okasha, S. 2005. "Biological Altruism." In *The Stanford Encyclopedia of Philosophy,* ed. E. N. Zalta. Summer 2005 ed., http://plato.stanford.edu/archives/sum2005/entries/altruism-biological/. Last accessed December 16, 2009.

Preston, C. 2006. *Bee.* London: Reaktion Books.

Richards, R. J. 1987. *Darwin and the Emergence of Evolutionary Theories of Mind and Behavior.* Chicago: University of Chicago Press.

Sidgwick, H. 1904. *Miscellaneous Essays and Addresses.* London: MacMillan and Co.

Skultans, V., ed. 1975. *Madness and Morals: Ideas on Insanity in the Nineteenth Century.* London: Routledge and Kegan Paul.

Sleigh, C. 2001. "Empire of the Ants: H. G. Wells and Tropical Entomology." *Science as Culture,* 10: 33–71.

———. 2003. *Ant.* London: Reaktion Books.

———. 2007. *Six Legs Better: A Cultural History of Myrmecology.* Baltimore and London: Johns Hopkins University Press.

Small, R. 2005. *Nietzsche and Rèe: A Star Friendship.* Oxford: Clarendon Press.

Smith, G. 1895. "Guesses at the Riddle of Existence." *North American Review* 161: 230–48.

Smith, G. A. 1899. *The Life of Henry Drummond.* London: Hodder and Stoughton.

Sober, E., and D. S. Wilson. 1998. *Unto Others: The Evolution and Psychology of Unselfish Behavior.* Cambridge, MA, and London: Harvard University Press.

Spencer, H. [1855]1870–1872. *Principles of Psychology.* 2 vols. 2nd ed. London: Williams and Norgate.

———. 1873. *The Study of Sociology.* London: H. S. King.

———. 1879. *The Data of Ethics.* London: Williams and Norgate.

Tankersley, D., C. J. Stowe, and S. A. Huettel. 2007. "Altruism is Associated with an Increased Neural Response to Agency." *Nature Neuroscience* 10 (2): 150–51.

Watts, I. 1866. *Divine and Moral Songs for Children.* New York: Hurd and Houghton. First published as *Divine Songs* in 1715.

Wilde, O. [1905] 2005. *The Complete Works of Oscar Wilde.* Vol. 2. *De Profundis: "Epistola in Carcere et Vinculis,"* ed. Ian Small. Oxford: Oxford University Press.

Wilson, D. S., and E. Sober. 2002. "The Fall and Rise and Fall and Rise and Fall and Rise of Altruism in Biology." In *Altruism and Altruistic Love: Science, Philosophy and Religion in Dialogue,* ed. S. G. Post, L. G. Underwood, J. P. Schloss, and W. B. Hurlbut. Oxford and New York: Oxford University Press, 182–91.

Woolf, V. 1993. *The Crowded Dance of Modern Life. Selected Essays.* Vol. 2, ed. Rachel Bowlby. London: Penguin.

3

Evolution and "Cooperation" in Late Nineteenth- and Early Twentieth-Century America

Science, Theology, and the Social Gospel

Heather D. Curtis

In 1893 Henry Drummond, a Scottish naturalist and evangelical minister, delivered a series of addresses entitled *The Ascent of Man* at the Lowell Institute in Boston. The title of Drummond's lectures evoked Charles Darwin's 1871 text, *The Descent of Man,* a work in which Darwin had (among other things) offered a "natural history" of the "moral sense." Focusing on the phenomenon of mutual aid in the animal world, Darwin argued that social instincts such as sympathy and love could be adaptive for both humans and other species, and that natural selection favored cooperative strategies at the communal or tribal level. Although Drummond's title implied opposition to Darwin's naturalistic account of ethics, Drummond was actually engaged in a very similar project. He too sought to "find a sanction for morality" in the "biological fact" of cooperation. While acknowledging that Darwin had recognized the importance of sociability in his later work, Drummond complained that *The Origin of Species* had emphasized "the Struggle for Life" as the "governing factor in development" to such an extent that subsequent investigators had overlooked a "*second* factor" that had, in his view, played an equally, if not more "prominent part" in the evolutionary process. That "missing factor," Drummond insisted, was "the Struggle for the Life of Others" or "Altruism" (1894, 48, 51, 220, 12–3).[1]

Drummond's Lowell lectures provoked a range of reactions from reviewers on both sides of the Atlantic. While critics derided his ideas as pseudoscientific, philosophically derivative, or insufficiently theological, supporters commended his efforts to reassess the relationship between evolution and ethics from an emphatically Christian perspective.[2] Among his most enthusiastic respondents were a number of American Protestant pastors and seminary professors who were also grappling with the implications of Darwinian theory for Christian doctrine and morality. For theologians seeking to harmonize the discoveries of science with the dogmas of Christianity, Drummond's assertion that "co-operation and sympathy" were as "engrained in the world-order" as struggle and selfishness, and his corresponding claim that evolution was "not a tale of battle" but a "Love-story" or even a "revelation—the phenomenal expression of the Divine" provided welcome confirmation of their own convictions about the relationships between God, nature, and the development of human ethical capacities (1894, 238, 29, 218, 339). If cooperation and altruism were embedded within the evolutionary process, as Drummond insisted that they were, interpreting the natural world as an essentially benevolent arena in and through which an immanent God made known the moral imperatives of "love, service, and sacrifice" gained credibility on both scientific and theological grounds (Abbott 1922, 454–55).

This essay explores the centrality of what is now called "cooperation" in late nineteenth- and early twentieth-century attempts to integrate the findings of science and the precepts of Christian faith. Examining how several American Protestants assessed the significance of evolutionary theories of "altruism" for Christian belief and practice shows that theological responses to Darwin were far more complex and often more constructive than some accounts of the science–religion encounter have suggested. For Drummond and his American admirers, reflecting upon the phenomenon of "altruism" afforded insights about God's involvement in the evolutionary process that, in their view, redressed the deficiencies of theological models that emphasized God's distance and detachment from the natural world. Rather than undermining Christian beliefs about God and the moral life, in other words, evolutionary theory offered resources for reconfiguring and enhancing theology and ethics.[3]

Analyzing enthusiasm for evolutionary altruism among a certain class of late nineteenth- and early twentieth-century Protestant leaders also exposes the underlying ambitions and assumptions that motivated their efforts to reconcile scientific accounts of human morality with Christian theories of virtue. During this period, many American Protestants were seeking new ways to combat social crises such as poverty, crime, racism, and class conflict. For participants in the social gospel movement, evolutionary theories that elevated cooperation over competition provided powerful ammunition for critiquing laissez-faire capitalism and for promoting an alternative economic order based upon the "natural" ideals of equity and justice. Science, in this view, did not sanction selfishness but, rather, endorsed Christian values of sympathy, solidarity, and self-sacrifice.

Because supporters of the social gospel movement subscribed to an optimistic ideology of progress, they also assumed that altruism would eventually vanquish egoism, that sympathy would subdue strife, that cooperation would overcome contestation. Evolution, from this perspective, was a teleological process, continually and inevitably advancing humanity toward a state of scientific, social, and ethical perfection. As events of the twentieth century undermined confidence in melioristic visions of inexorable improvement, Protestant theologians in America and elsewhere largely abandoned efforts to probe connections between evolution, cooperation, and God. The recent revival of interest in exploring the significance of evolutionary altruism for Christian theism and ethical theory thus represents a new phase in a much longer, but frequently overlooked, tradition of inquiry. By placing current investigations within a wider historical context, I aim to encourage reflection upon the opportunities and challenges involved in constructing theologies of evolution and cooperation.

Cooperation, not "Warfare" or "No Contest" between Science and Theology

Contrary to much popular perception, the relationship between science and religion has always involved a fair amount of cooperation, in the lay sense. In the United States, especially, the "warfare" metaphors that have often dominated descriptions of the science–religion dialogue have obscured a much more complicated history of interactions, many of which have been significantly more collaborative than images of contest and combat imply. Even in the late nineteenth century, when the "battle" between evolution and Christianity is supposed to have been particularly fierce, many American theologians were, in fact, actively striving to reconcile scientific developments and Christian doctrines.[4]

The American ministers and seminary professors who embraced Drummond's *The Ascent of Man* and applauded his emphasis on altruism were deeply committed to this broader project of integrating evolutionary theory and Christian theology. As participants in Protestant modernism, a movement that sought to adapt Christian beliefs and practices to contemporary culture, these theologians worried that conflict between science and religion would ultimately undermine the credibility of Christianity. Rather than adopting a hostile posture toward Darwinism, or claiming that theology and evolutionary biology were separate domains with distinctive methodologies and different subjects of inquiry, liberal clergymen such as James T. Bixby, Washington Gladden, and George Harris insisted that scientific investigation and theological reflection were complementary rather than competing endeavors. Lyman Abbott, the pastor of Plymouth Congregational Church in Brooklyn, New York, during the 1890s and a principal spokesperson for Protestant modernism argued that a "method which sets theological theories against scientifically ascertained facts is fatal to . . . theology and injurious to the spirit of religion." By contrast, a "method which frankly recognizes the facts of life, and appreciates the spirit of the scientists, whose patient and assiduous endeavor has brought those facts to light, will commend the spirit of religion to the new generation, and will benefit—not impair—theology as a science, by compelling its reconstruction" (1897, 37).[5]

Cooperation Reveals "a Purpose of Benevolence" in Nature and a "God Whose Name Is Love"

Abbott made these comments in a book entitled *The Theology of an Evolutionist,* published in 1897. In this work, he drew heavily upon Drummond's Lowell lectures in an effort to "apply the fundamental principles of evolution to the problems of religious life and thought." Abbott was especially impressed with Drummond's account of the role that cooperation played in the "history of life" (1897, vii). By focusing on the phenomena of mutual aid, self-sacrifice, and altruism in nature, Abbott maintained, Drummond made it possible to reimagine God's relationship to the evolutionary process.

Drummond had been particularly critical of the implicit deism he observed in many traditional theological responses to evolution. "There are reverent minds who ceaselessly scan the fields of Nature and the books of Science in search of gaps—gaps which they will fill up with God," Drummond complained (coining a now-famous phrase). In Drummond's view, a God-of-the-gaps approach was problematic for several reasons. First, as science advanced, theology would be forced to retreat. "If God is only to be left to the gaps in our knowledge, where shall we be when these gaps are filled up?" Drummond asked. Second, and perhaps even more troubling for Drummond, was the tendency to view God as separate from, or outside, developments in the natural world. "Those who yield to the temptation to reserve a point here and there for special divine interposition are apt to forget that this virtually excludes God from the rest of the process," he declared. "If God appears periodically, He disappears periodically. If He comes upon the scene at special crises, He is absent from the scene in the intervals." The discoveries of science, Drummond concluded, had "discredited" this "older view" of the deity, and helped Christians to conceive of an ever-present God who is actively and intimately involved in the on-going creation of the world. "The idea of an immanent God, which is the God of Evolution," he insisted, "is infinitely grander than the occasional wonder-worker, who is the God of an old theology" (1894, 333–34).

Abbott and other American Protestant modernists shared Drummond's misgivings about theologies that envisioned "a God external to nature" who interfered "now and then to repair the machinery." "An absentee God . . . is not the Christian idea of God," declared George Harris, a professor of theology at Andover Seminary and president of Amherst College, in his 1896 book *Moral Evolution* (1896, 416). Like Drummond, theologians such as Harris and Abbott suggested that reading nature as inherently cooperative, rather than entirely "red in tooth and claw," resulted in a new appreciation of God's "perpetual presence" in the evolutionary process (Abbott 1897, 138). "All nature and all life is one great theophany," Abbott asserted (1897, 9). Therefore, he explained, "the evolutionist does not believe that God created protoplasm and left protoplasm to create everything else. Evolution is God's way of doing things. . . . It is literally true that in Him we live and move and have our being" (1897, 76). God, in this view, ought not to be construed as "a Great First Cause which made and wound up the universe a long while ago and set it going and interferes with it occasionally as a clockmaker might with his clock" (1922, 454). Rather, Abbott avowed, "The Christian evolutionist believes that God is the one universal and always present Cause; that there are no secondary causes, and that God's method of manifesting His eternal presence is the method of growth, not of manufacture, by a power dwelling within nature and working outward, not by a power dwelling without and working upon nature" (1897, 138). As the Unitarian philosopher John Fiske put it, "God is the ever-present life of the world; it is through him that all things exist from moment to moment, and the natural sequence of events is a perpetual revelation of the divine wisdom and goodness" (quoted in Harris 1896, 417).

Cooperation Promotes the Evolution of Ethics and Foreshadows Christian Morality

The notions that nature was infused with divinity and that the unfolding of natural processes revealed God's loving purposes for the world fit well with Protestant modernist assumptions about the connection between evolution and ethics. Drummond's declaration that "the principle of cooperation" was not a "late arrival" in the "history of humanity" but rather "one of the earliest devices hit upon in the course of Evolution" suggested that other-regarding behavior was deeply rooted within the natural order (1894, 155, 216). "The significance of this interpretation of nature cannot be exaggerated," wrote the prominent liberal pastor Washington Gladden in a review of *The Ascent of Man*. Drummond's assertion that the "Struggle for the Life of Others" was "destined from the first to replace the Struggle for Life," meant that nature provided, in Gladden's words, "a good foundation for morality." "Surely," Gladden surmised, "our ethical doctrine must be in harmony with the nature of things" (1894, 242).

The Unitarian minister James Bixby, chairman of the Liberal Ministers' Association and a professor at Meadville Theological School, reached a similar conclusion after reflecting upon the necessity of "mutual help," "disinterested sympathy and cooperation" for the survival and development of species (1895, 452). "Natural science," Bixby contended, "shows that it is not individual self-seeking, but social cooperation, that is the more effective factor in evolution" (1893, 214). Although a "superficial acquaintance with the facts of evolution" might suggest that "struggle, selfishness, and cruelty" were its "prominent features," Bixby argued, "a deeper and keener study shows that from the outset of life there have been . . . instincts of solidarity and sympathy involved that irresistibly carry the individual beyond the circle of his own interests." Because "nature, instead of frowning upon and repressing this altruistic tendency, has constantly favored and sanctioned it," Bixby concluded, evolution actually promotes the emergence of ethical behavior and even Christian virtue. "In the simplest cell which in obedience to the expansive tendency of life, splits into two," he proclaimed, "the philosophic eye beholds the germ of the moral law and the promise of the Beatitudes" (1895, 444–58).

For Bixby and his modernist colleagues, this attempt to establish "a sanction for morality" in the evolutionary process stemmed in part from a desire to challenge other prominent arguments about the relationship between biology and ethics, especially versions of social Darwinism that were being put forward by various philosophers and social theorists during the latter decades of the nineteenth century. Arguing against T. H. Huxley's gladiatorial view of nature and Friedrich Nietzsche's claim that virtue is a cultural construction, Bixby insisted that an ethic of sympathy, self-sacrifice, and "altruistic giving" was woven into the fabric of creation from the very start and had steadily been gaining ground over selfishness, egoism, and competition. "Morality," he wrote, "is no invention of priests, statesmen, or philosophers. It is . . . the age-long victory and product of that Divine Life in the Universe that has ever moved onward from chaos to cosmos, from carnal to spiritual." The evolution of ethics, in this view, involved "a patient ascent through successive planes of wider and more intimate cooperation, fusing individual in families, families in tribes, tribes in nations, and nations in the universal family of God's children, in which Jew and Greek, male and female, black and white, must have their equal right and place before the tribunal of Christian equity and sympathy" (1895, 458).[6]

Evolution, according to this interpretation, provided a rationale for a Christian ethics of equality and service to others, rather than authorizing competitive individualism or endorsing the belief that "might makes right"

(Abbott 1922, 454–55). Participants in the American social gospel movement that sought to redress the crises of poverty and urbanization brought about through industrialization and laissez-faire capitalism seized upon the idea that natural selection actually favored cooperative strategies over egoistic behavior. "Some of us have long been urging that the principle of cooperation must somehow be brought into industrial society, to mitigate the strife arising from competition," Washington Gladden observed in 1894. "The answer of the multitude has been that the scheme was impracticable; that natural law, which is the law of competition, was the only regulative principle in economics." In his glowing assessment of Drummond's work, Gladden rejoiced that "a student of evolution" had provided solid evidence that unselfishness, rather than selfishness, was the "eternal law" of nature. "If such is the law even of the lower nature," Gladden mused, "it seems not altogether quixotic to believe that there must be room in human nature and in industrial society for a large infusion of altruistic purpose, and for a liberal application of the cooperative principle" (1894, 242).[7]

For Gladden and like-minded theologians, evolutionary theories that emphasized "good-will and helpfulness" were compelling because they made possible a re-envisioning of the social order along more cooperative lines. During the 1890s, a number of prominent Protestant modernists predicted that the evolution of cooperation would eventually usher in a just and harmonious society. "In the course of human association and progress of civilization, this law of cooperation slowly yet surely gains extension and power, while, on the other hand, the method of conflict becomes secondary in importance, and circumscribed in its operation," wrote Andover Seminary professor Newman Smyth in his 1892 text *Christian Ethics*. "The history of civilization indicates already an immense gain of the action of the principle of cooperation over the action of the principle of competition" (1892, 247). Bixby's forecast for the future of human society was equally sanguine. "The social life and the sympathetic forces gain steadily upon the isolated and selfish life," he asserted, "so that the union of man with man, and of humanity with all the rest of creation, steadily increases. The shining ideals of beauty, truth, and virtue draw us upward and onward, toward the goal of an ever-enlarging perfection ahead of us" (1893, 219). Gladden was perhaps the most confident of all about the trajectory of moral and social evolution. Because the golden rule—"Thou shalt love thy neighbor as thyself"—was "incorporated into the nature of man at his creation," Gladden reasoned, human beings would eventually learn to conform to its demands. That "this law . . . will be perfectly obeyed in the perfect society of the future is now recognized as a scientific certainty," he declared (1887, 232).

The Death and Resurrection of Theologies of Evolution and Cooperation in the Twentieth Century

From a twenty-first-century perspective, the optimism that Protestant modernists such as Bixby, Smyth, and Gladden expressed about the inevitable outcome of the evolutionary process seems distressingly naïve. Indeed, by the end of World War I, many critics had become disillusioned with the meliorism that characterized so much of liberal Protestant theology. The ideology of progress had fallen on hard times, and ensuing events—the Great Depression of the 1930s, the rise of the Third Reich and the Holocaust, the invention and deployment of the atomic bomb (to name just a few of the most grievous episodes in twentieth-century history)—would do nothing to improve its fortunes. In the years between the world wars, neoorthodox theologians and Protestant fundamentalists alike fervently condemned liberal assumptions about God's immanence, the perfectibility of human nature, and the certainty of social improvement. Unlike their liberal predecessors, Protestant "realists" were far more interested in describing the vast gulf between divine holiness and human sinfulness than in closing any supposed gap between God and the evolutionary process.[8]

Increased specialization in biology and the advent of neo-Darwinism in the 1920s also contributed to a decline in speculation about the significance of the evolutionary phenomenon of cooperation for Christian theology and ethics. As scientific discourse became more technical, interdisciplinary dialogue grew progressively more challenging. Keeping up with developments in evolutionary theory was increasingly difficult for nonbiologists, especially since fewer scientists wrote for lay audiences during this period. This growing isolation between theology and evolutionary biology, combined with the dominance of neo-Darwinian theory, helps explain why the tradition of theological reflection upon cooperation in the natural world dissipated over the course of the twentieth century. Although a number of American ecologists continued to emphasize the importance of mutual aid for biological evolution and even considered the implications of their theories for political and social life, for example, their insights do not seem to have generated a great deal of interest among theologians.[9]

Only in the last decade or so have American theologians reengaged in efforts to understand how the evolution of cooperation might intersect with Christian ethics and metaphysics. Drawing upon the remarkable resurgence of scientific research on the adaptive functions of cooperative behavior since the 1960s, a number of Christian thinkers have recently begun to reflect upon the relationship between biological accounts of altruism and theological interpretations of morality and divinity. Current attempts to explore the implications of evolutionary theories of cooperation for Christian theology, including efforts represented in this book, share several common features with late nineteenth- and early twentieth-century endeavors.[10]

Like their late nineteenth-century forerunners, many contemporary theologians interested in evolutionary altruism contest the assumption that science and religion are fundamentally hostile or competing forces. They also resist the idea that because evolutionary biology and Christian theology are discrete disciplines that explore different questions using dissimilar tools, their conclusions have no bearing outside their isolated fields. Instead, participants in this growing conversation hope that the results of their investigations will show that biological experimentation and theological reflection can be complementary and perhaps even mutually constructive.

Focusing on the evolutionary phenomenon of cooperation has also prompted some twenty-first-century interlocutors to question theological accounts that rely on a God-of-the-gaps approach to describe the relationship between divine providence and natural developments. Just as late nineteenth-century liberals rejected deism as a deficient model for construing God's relation to the natural world, so too have some current thinkers expressed discontent with the notion that God is completely "distanced" from the evolutionary process. Although there are important differences between modernist appeals to God's immanence and more recent evocations of God's incarnational and trinitarian character, both methods suggest that emphasizing evolutionary altruism enables Christian theologians to avoid the inadequacies of alternative perspectives that place God wholly outside the natural order.

Finally, contemporary conversations about the evolution of cooperation often involve inquiries about the origins of human morality. While Christian thinkers in the early twenty-first century largely eschew "progressive" and melioristic presumptions about the teleology of evolution, many do evince a keen interest in the possibility that there may be some sort of causal connection between the emergence of altruism and the development of human ethical faculties. Like their modernist predecessors, some current theologians seem hopeful that the phenomenon of cooperation might provide evidence that moral capacities in general, and (more optimistically) Christian ethics in particular, derive in part from biological or natural processes.

Whether or not this sensibility is linked with a larger political or social agenda, as it was for many theologians of the social gospel movement who endorsed a "cooperative" evolutionary ethics, is a crucial question that contemporary investigators would do well to confront. What, in other words, are the underlying and perhaps unconscious assumptions that motivate and shape current efforts to integrate evolutionary theory and Christian theology? More broadly, why are so many scientists, theologians, philosophers, and ethicists in the contemporary United States (and beyond) interested in the evolution of cooperation? With what social, economic, political, or theological visions are these research projects inadvertently (or even consciously) entangled? Highlighting both the resonances and disjunctions between contemporary conversations and earlier attempts to reconcile the evolutionary theory of cooperation with Christian theism will, I hope, encourage serious and sustained consideration of these larger issues.

Notes

1. For this interpretation of both Darwin and Drummond, I am indebted to Thomas Dixon (2008).
2. For a fuller discussion of responses to Drummond's work, see Dixon (2008, Chapter 7).
3. As Thomas Dixon has shown in Chapter 2 of this volume, the term "altruism" has encompassed a multiplicity of meanings since its invention by August Comte in the mid-nineteenth century. Nineteenth-century figures such as Henry Drummond used "altruism" and "cooperation" in a variety ways, frequently blurring definitional boundaries and failing to distinguish between physiological and ideological connotations. The theologians whom I discuss in this chapter (including Drummond) were also imprecise in their usage of these terms and often conflated what the editors of this volume define as "cooperation" and "altruism." They rarely, if ever, parsed the moral or ethical implications of "altru-

ism" from strictly "scientific" descriptions of "cooperative" behavior. Readers should be aware that the figures under discussion in this chapter did not ascribe to the "tight" definitions of altruism and cooperation proposed by the editors of this book.

4. The literature on the historical relationship between science and religion or theology is voluminous. I am especially indebted to John Hedley Brooke's discussions of this subject (1991; also see Brooke and Cantor, 2000). For a recent addition to the literature that focuses primarily on Protestant theology, see Bowler (2007). Helpful studies of the reception of Darwinism in the American context include Moore (1979), Roberts (1988), Numbers (1998) Livingstone (2001), and Ryan (2002).
5. For a fuller history of Protestant Modernism, see Williams (1970) and Hutchison (1976). Biographical information on liberal clergymen seeking to integrate evolutionary theory and Christian theology may be found in Ryan (2002).
6. On Social Darwinism see Hofstadter (1959), Bannister (1979), Degler (1991), Bowler (1993), and Hawkins (1997).
7. For background on the social gospel movement, see Handy (1966), Hutchison (1976), Hopkins (1982), and Phillips (1996).
8. A fuller account of this history is given in Hutchison (1976, Chapters 6–8).
9. For a helpful discussion of the effects of scientific specialization on the science–religion dialogue in the early twentieth century, see Laurent (2002, viii). On American ecology and cooperation, see Mitman (1992).
10. Other theologians who have demonstrated an interest in the evolution of cooperation include Grant (2001), Post (2002), and Clayton and Schloss (2004).

References

Abbott, L. 1897. *The Theology of an Evolutionist*. Boston: Houghton Mifflin.

———. 1922 (March). "In the Workshop of God." *Outlook* 22: 454–55.

Bannister, R. C. 1979. *Social Darwinism: Science and Myth in Anglo-American Social Thought*. Philadelphia: Temple University Press.

Bixby, J. T. 1893. "Morality on a Scientific Basis." *The Andover Review*, 19(110), March/April: 208–20.

———. 1895, September. "The Sanction for Morality in Nature and Evolution." *The New World: A Quarterly Review of Religion and Ethics* 4(15): 444–58.

Bowler, P. J. 1993. *Biology and Social Thought*. Berkeley, CA: Office for History of Science and Technology, University of California at Berkeley.

———. 2007. *Monkey Trials and Gorilla Sermons: Evolution and Christianity from Darwin to Intelligent Design*. Cambridge, MA: Harvard University Press.

Brooke, J. H. 1991. *Science and Religion: Some Historical Perspectives*. New York: Cambridge University Press.

Brooke, J. H., and G. Cantor. 2000. *Reconstructing Nature: The Engagement of Science and Religion*. New York: Oxford University Press.

Clayton, P., and J. Schloss. 2004. *Evolution and Ethics: Human Morality in Biological and Religious Perspective*. Grand Rapids, MI: W. B. Eerdmans.

Darwin, C. 1871. *The Descent of Man, and Selection in Relation to Sex*. 2 vols. London: Murray.

Degler, C. 1991. *In Search of Human Nature: The Decline and Revival of Darwin in American Social Thought*. New York: Oxford University Press.

Dixon, T. 2008. *The Invention of Altruism: Making Moral Meanings in Victorian Britain*. Oxford: Oxford University Press for the British Academy.

Drummond, H. 1894. *The Lowell Lectures on the Ascent of Man*. London: Hodder and Stoughton.

Dugatkin, L. A. 2006. *The Altruism Equation: Seven Scientists Search for the Origins of Goodness*. Princeton, NJ: Princeton University Press.

Fiske, J. 1885. *The Idea of God as Affected by Modern Knowledge*. Boston: Houghton Mifflin.

Gladden, W. 1887. *Applied Christianity: Moral Aspects of Social Questions*. Boston: Houghton Mifflin.

———. 1894. "The New Evolution." *McClure's Magazine*, 3(3) August: 242.

Grant, C. 2001. *Altruism and Christian Ethics*. New York: Cambridge University Press.

Handy, R. T., ed. 1966. *The Social Gospel in America, 1870–1920*. New York: Oxford University Press.

Harris, G. 1896. *Moral Evolution*. Boston: Houghton Mifflin.

Hawkins, M. 1997. *Social Darwinism in European and American Thought, 1860–1945*. New York: Cambridge University Press.

Hofstadter, R. 1959. *Social Darwinism in American Thought*. New York: G. Braziller.

Hopkins, C. H. 1982. *The Rise of the Social Gospel in American Protestantism, 1865–1915*. New York: AMS Press.

Hutchison, W. R. 1976. *The Modernist Impulse in American Protestantism*. Cambridge, MA: Harvard University Press.

Laurent, M. 2002. Introduction to *Darwinism and Theology in America: 1850–1930*. Vol. 1, *The Benevolent Hand*, ed. Frank X. Ryan. Bristol, UK: Thoemmes.

Livingstone, D. N. 2001. *Darwin's Forgotten Defenders: the Encounter between Evangelical Theology and Evolutionary Thought*. Vancouver: Regent College Publishing.

Mitman, G. 1992. *The State of Nature: Ecology, Community, and American Social Thought*. Chicago: University of Chicago Press.

Moore, J. R. 1979. *The Post-Darwinian Controversies: A Study of the Protestant Struggle to Come to Terms with Darwin in Great Britain and America, 1870–1900*. New York: Cambridge University Press.

Numbers, Ronald L. 1998. *Darwinism Comes to America.* Cambridge: Harvard University Press.

Phillips, Paul T. 1996. *A Kingdom on Earth: Anglo-American Social Christianity, 1880–1940.* University Park, PA: Pennsylvania State University Press.

Post, S. G., ed. 2002. *Altruism and Altruistic Love: Science, Philosophy and Religion in Dialogue.* New York: Oxford University Press.

Richards, R. J. 1987. *Darwin and the Emergence of Evolutionary Theories of Mind and Behavior.* Chicago, IL: University of Chicago Press.

Roberts, J. H. 1988. *Darwinism and the Divine in America: Protestant Intellectuals and Organic Evolution, 1859–1900.* Madison, WI: University of Wisconsin Press.

Ryan, F. X. 2002. *Darwinism and Theology in America, 1850–1930,* 4 vols. Bristol, UK: Thoemmes.

Smyth, Newman. 1892. *Christian Ethics.* New York: Charles Scribner's Sons.

Sober, E., and D. S. Wilson. 1998. *Unto Others: The Evolution and Psychology of Unselfish Behavior.* Cambridge, MA: Harvard University Press.

Williams, Daniel Day. 1970. *The Andover Liberals: A Study in American Theology.* New York: Octagon Books.

II

Mathematics, Game Theory, and Evolutionary Biology

The Evolutionary Phenomenon of Cooperation

4

Five Rules for the Evolution of Cooperation

Martin A. Nowak

Evolution is based on a fierce competition between individuals and should therefore only reward selfish behavior. Every gene, every cell, and every organism should be designed to promote its own evolutionary success at the expense of its competitors. Yet we observe cooperation on many levels of biological organization. Genes cooperate in genomes. Chromosomes cooperate in eukaryotic cells. Cells cooperate in multicellular organisms. There are many examples for cooperation among animals. Humans are the champions of cooperation: from hunter gatherer societies to nation states, cooperation is the decisive organizing principle of human society. No other life form on Earth is engaged in the same complex games of cooperation and defection. The question of how natural selection can lead to cooperative behavior has fascinated evolutionary biologists for several decades.

A cooperator is someone who pays a cost, c, for another individual to receive a benefit, b. A defector has no cost and does not deal out benefits. Cost and benefit are measured in terms of fitness. Reproduction can be genetic or cultural. In any mixed population, defectors have a higher average fitness than cooperators (Figure 4.1). Therefore, selection acts to increase the relative abundance of defectors. After some time, cooperators vanish from the

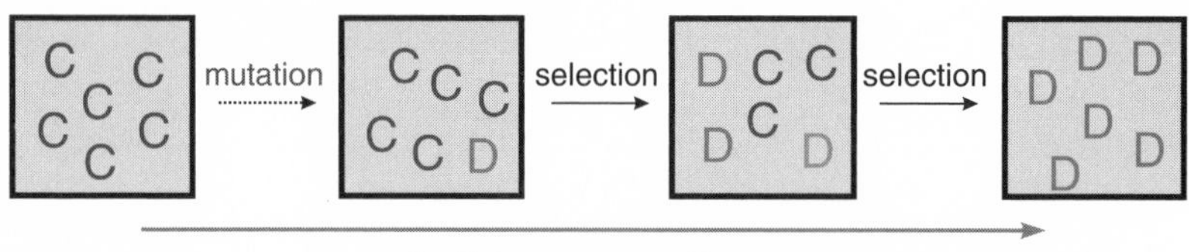

Figure 4.1 Without any mechanism for the evolution of cooperation, natural selection favors defectors. In a mixed population defectors, *D*, have a higher payoff (= fitness) than cooperators, *C*. Therefore, natural selection continuously reduces the abundance, *i*, of cooperators until they are extinct. The average fitness of the population also declines under natural selection. The total population size is given by *N*. There are *i* cooperators and $N-i$ defectors. The fitness of cooperators and defectors is respectively given by $f_C = [b(i-1)/(N-1)] - c$ and $f_D = bi/(N-1)$. The average fitness of the population is given by $\bar{f} = (b-c)i/N$. From "Five Rules for the Evolution of Cooperation" by Martin A. Nowak, *Science* 314: 1560–63 (2006). Reprinted with permission from AAAS.

population. Remarkably, however, a population of only cooperators has the highest average fitness, while a population of only defectors has the lowest. Thus, natural selection constantly reduces the average fitness of the population. Fisher's fundamental theorem, which states that average fitness increases under constant selection, does not apply here because selection is frequency-dependent: the fitness of individuals depends on the frequency (= relative abundance) of cooperators in the population. We see that natural selection in well-mixed populations needs help for establishing cooperation.

Kin Selection

When J. B. S. Haldane remarked, "I will jump into the river to save two brothers or eight cousins," he anticipated what became later known as Hamilton's rule (Hamilton 1964). This ingenious idea is that natural selection can favor cooperation if the donor and the recipient of an altruistic act are genetic relatives. More precisely, Hamilton's rule states that the coefficient of relatedness, *r*, must exceed the cost-to-benefit ratio of the altruistic act:

$$r > c/b \qquad (1)$$

Relatedness is defined as the probability of sharing a gene. The probability that two brothers share the same gene by descent is ½, while the same probability for cousins is ⅛. Hamilton's theory became widely known as "kin selection" or "inclusive fitness" (Grafen 1985; Taylor 1992; Queller 1992; Frank 1998; West, Pen, and Griffin 2002; Foster, Wenseleers, and Ratnieks 2006). When evaluating the fitness of the behavior induced by a certain gene it is important to include the behavior's effect on kin who might carry the same gene. Therefore, the "extended phenotype" of cooperative behavior is the consequence of "selfish genes" (Wilson 1975; Dawkins 1976).

Direct Reciprocity

It is unsatisfactory to have a theory that can only explain cooperation among relatives. We also observe cooperation between unrelated individuals or even between members of different species. Such considerations led Trivers (1971) to propose another mechanism for the evolution of cooperation: direct reciprocity. There are repeated encounters between the same two individuals. In every round, each player has a choice between cooperation and defection. If I cooperate now, you may cooperate later. Hence, it might pay off to cooperate. This game theoretical framework is known as the repeated Prisoner's Dilemma, but what is a good strategy for playing this game? In two computer tournaments, Axelrod (1984) discovered that the "winning strategy" was the simplest of all, tit for tat. This strategy always starts with cooperation, then does whatever the other player has done in the previous round: a cooperation for a cooperation, a defection for a defection. This simple concept captured the fascination of all enthusiasts of the repeated Prisoner's Dilemma. Many empirical and theoretical studies were inspired by Axelrod's ground breaking work (Axelrod and Hamilton 1981; Milinski 1987; Dugatkin 1997).

But soon an Achilles heel of the world champion was revealed: if there are erroneous moves, caused by "trembling hands" or "fuzzy minds," the performance of tit for tat declines (Selten and Hammerstein 1984; Fudenberg and Maskin 1990). Tit for tat cannot correct mistakes, because an accidental defection leads to a long sequence of retaliation. At first, tit for tat was replaced by generous tit for tat (Nowak and Sigmund 1992), a strategy that cooperates whenever you cooperate but sometimes cooperates although you have defected (with probability 1—c/b). Natural selection can promote forgiveness.

Subsequently, tit for tat was replaced by win-stay, lose-shift, which is the even simpler idea of repeating your previous move whenever you are doing well but changing otherwise (Nowak and Sigmund 1993). By various measures of success, win-stay, lose-shift is more robust than either tit for tat or generous tit for tat (also Fudenberg and Maskin 1990). Tit for tat is an efficient catalyst of cooperation in a society where nearly everybody is a defector, but once cooperation is established win-stay, lose-shift is better able to maintain it.

The number of possible strategies for the repeated Prisoner's Dilemma is unlimited, but a simple, general rule can be shown without any difficulty. Direct reciprocity can only lead to the evolution of cooperation if the probability, w, of another encounter between the same two individuals exceeds the cost-to-benefit ratio of the altruistic act:

$$w > c/b \tag{2}$$

Indirect Reciprocity

Direct reciprocity is a powerful mechanism for the evolution of cooperation but leaves out certain aspects that are particularly important for humans. Direct reciprocity relies on repeated encounters between the same two individuals, and both individuals must be able to provide help, which is less costly for the donor than beneficial for the recipient. But often the interactions among humans are asymmetric and fleeting. One person is in a position to help another, but there is no possibility for a direct reciprocation. We help strangers who are in need. We donate to charities that do not donate to us. Direct reciprocity is like a barter economy based on the immediate exchange of goods, while indirect reciprocity resembles the invention of money. The money that fuels the engines of indirect reciprocity is reputation—helping someone establish a good reputation that will be rewarded by others. When deciding how to act we take into account the possible consequences for our reputation. We feel strongly about events that affect us directly but also take a keen interest in the affairs of others, as demonstrated by the contents of gossip.

In the standard framework of indirect reciprocity, there are randomly chosen, pairwise encounters, where the same two individuals need not meet again. One individual acts as donor, the other as recipient. The donor can decide whether or not to cooperate. The interaction is observed by a subset of the population who might inform others. Reputation allows evolution of cooperation by indirect reciprocity (Nowak and Sigmund 1998). Natural selection favors strategies that base the decision to help on the reputation of the recipient. Theoretical and empirical studies of indirect reciprocity show that people who are more helpful are more likely to receive help (Wedekind and Milinski 2000; Leimar and Hammerstein 2001; Milinski, Semmann, and Krambeck 2002; Fishman 2003; Hauser et al. 2003; Brandt and Sigmund 2004; Ohtsuki and Iwasa 2004; Panachanathan and Boyd 2004; Nowak and Sigmund 2005).

Although simple forms of indirect reciprocity can be found in animals (Bshary and Grutter 2006), only humans seem to engage in the full complexity of the game. Indirect reciprocity has substantial cognitive demands. Not only do we have to remember our own interactions but also monitor the ever-changing social network of the group. Language is needed to gain the information and spread the gossip associated with indirect reciprocity. Presumably, the selection for indirect reciprocity and human language has played a decisive role in the evolution of human intelligence (Nowak and Sigmund 2005). Indirect reciprocity also leads to the evolution of morality (Alexander 1987) and social norms (Brandt and Sigmund 2004; Ohtsuki and Iwasa 2004).

The calculations of indirect reciprocity are complicated and only a tiny fraction of this universe has been uncovered, but again a simple rule has emerged (Nowak and Sigmund 1998). Indirect reciprocity can only promote cooperation if the probability, q, to know someone's reputation exceeds the cost-to-benefit ratio of the altruistic act:

$$q > c/b \qquad (3)$$

Network Reciprocity

The argument for natural selection of defection (Figure 4.1) is based on a well-mixed population, where everybody interacts equally likely with everybody else. This approximation is used by all standard approaches to evolutionary game dynamics (Maynard Smith 1982; Hofbauer and Sigmund 1998, 2003; Nowak and Sigmund 2004). But real populations are not well mixed. Spatial structures or social networks imply that some individuals interact more often than others. One approach of capturing this effect is evolutionary graph theory (Liebermann, Hauert, and Nowak 2005), which allows us to study how spatial structure affects evolutionary and ecological dynamics (Durrett and Levin 1994; Hassell, Comins and May, 1994; Hauert and Doebeli 2004; May 2006).

The individuals of a population occupy the vertices of a graph. The edges determine who interacts with whom. Let us consider plain cooperators and defectors without any strategic complexity. A cooperator pays a cost, c, for each neighbor to receive a benefit, b. Defectors have no costs, and their neighbors receive no benefits. In this setting, cooperators can prevail by forming network clusters, where they help each other. The resulting "network reciprocity" is a generalization of "spatial reciprocity" (Nowak and May 1992).

Games on graphs are easy to study by computer simulation but difficult to analyse mathematically because of the enormous number of possible configurations that can arise. Nevertheless, a surprisingly simple rule determines if network reciprocity can favor cooperation (Ohtsuki et al. 2006). The benefit-to-cost ratio must exceed the average number of neighbors, k, per individual:

$$b/c > k \tag{4}$$

Group Selection

Selection does not only act on individuals but also on groups. A group of cooperators might be more successful than a group of defectors. There have been many theoretical and empirical studies of group selection with some controversy, and most recently there is a renaissance of such ideas under the heading of "multilevel selection" (Williams and Williams 1957; Wilson 1975; Taylor and Wilson 1988; Rogers 1990; Keller 1999; Michod 1999; Paulsson 2002; Rainey and Rainey 2003; Wilson and Hölldobler 2005).

A simple model of group selection works as follows (Traulsen and Nowak 2006). A population is subdivided into groups. Cooperators help others in their own group. Defectors do not help. Individuals reproduce proportional to their payoff. Offspring are added to the same group. If a group reaches a certain size, it can split into two. In this case, another group becomes extinct in order to constrain the total population size. Note that only individuals reproduce, but selection emerges on two levels. There is competition between groups, because some groups grow faster and split more often. In particular, pure cooperator groups grow faster than pure defector groups, while in any mixed group defectors reproduce faster than cooperators. Therefore, selection on the lower level (within groups) favors defectors, while selection on the higher level (between groups) favors cooperators. This model is based on "group fecundity selection," which means groups of cooperators have a higher rate of splitting in two. We can also imagine a model based on "group viability selection," where groups of cooperators are less likely to go extinct.

In the mathematically convenient limit of weak selection and rare group splitting, we obtain a simple result (ibid.): if n is the maximum group size and m the number of groups, then group selection allows evolution of cooperation provided

$$b/c > 1 + n/m \qquad (5)$$

Evolutionary Success

Before proceeding to a comparative analysis of the five mechanisms, let me introduce some measures of evolutionary success. Suppose a game between two strategies, cooperators C and defectors D, is given by the payoff matrix

$$\begin{array}{ccc} & C & D \\ C & \alpha & \beta \\ D & \gamma & \delta \end{array}$$

The entries denote the payoff for the row player. Without any mechanism for the evolution of cooperation, defectors dominate cooperators, which means $\alpha < \gamma$ and $\beta < \delta$. A mechanism for the evolution of cooperation can change these inequalities.

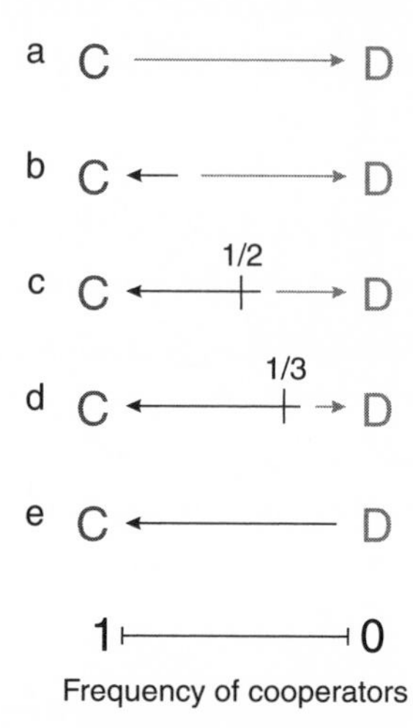

Figure 4.2 Evolutionary dynamics of cooperators and defectors. The arrows indicate selection favoring defectors and cooperators, respectively. (a) Without any mechanism for the evolution of cooperation, defectors dominate. A mechanism for the evolution of cooperation can allow cooperators to be the evolutionarily stable strategy (ESS), risk dominant (RD) or advantageous (AD) in comparison with defectors. (b) Cooperators are ESS if they can resist invasion by defectors. (c) Cooperators are RD if the basin of attraction of defectors is less than ½. (d) Cooperators are AD if the basin of attraction of defectors is less than ⅓. In this case, the fixation probability of a single cooperator in a finite population of defectors is greater than the inverse of the population size (for weak selection). (e) Some mechanisms allow cooperators to dominate defectors. From "Five Rules for the Evolution of Cooperation" by Martin A. Nowak, *Science* 314: 1560–63 (2006). Reprinted with permission from AAAS.

(i) If $\alpha > \gamma$, then cooperation is an evolutionarily stable strategy (ESS). An infinitely large population of cooperators cannot be invaded by defectors under deterministic selection dynamics (Hofbauer and Sigmund 1998).

(ii) If $\alpha + \beta > \gamma + \delta$, then cooperators are risk-dominant (RD). If both strategies are ESS, then the risk-dominant strategy has the bigger basin of attraction.

(iii) If $\alpha + 2\beta > \gamma + 2\delta$, then cooperators are advantageous (AD). This concept is important for stochastic game dynamics in finite populations. Here, the crucial quantity is the fixation probability of a strategy, defined as the probability that the lineage arising from a single mutant of that strategy will take over the entire population consisting of the other strategy. An advantageous strategy has a fixation probability greater than the inverse of the population size, $1/N$. The condition can also be expressed as a ⅓ rule: if the fitness of the invading strategy at a frequency of ⅓ is greater than the fitness of the resident, then the fixation probability of the invader is greater than $1/N$. This condition holds in the limit of weak selection (Nowak et al. 2004).

A mechanism for the evolution of cooperation can ensure that cooperators become ESS, RD or AD (Figure 4.2). Some mechanisms even allow cooperators to dominate defectors, which means $\alpha > \gamma$ and $\beta > \delta$.

Comparative Analysis

We have encountered five mechanisms for the evolution of cooperation (Figure 4.3). Although the mathematical formalisms underlying the five mechanisms are very different, at the center of each theory is a simple rule. I will now present a coherent mathematical framework that allows the derivation of all five rules. The crucial idea is that each mechanism can be presented as a game between two strategies given by a 2×2 payoff matrix (Figure 4.4). From this matrix, we can derive the relevant condition for evolution of cooperation.

For kin selection, I use the approach of inclusive fitness proposed by Maynard Smith. The relatedness between two players is r. Therefore, your payoff multiplied by r is added to mine. A second method leads to a different matrix but the same result.[1] For direct reciprocity, the cooperators use tit for tat while the defectors use "always-defect." The expected number of rounds is $1/(1-w)$. Two tit-for-tat players cooperate all the time. Tit for tat versus always-defect cooperates only in the first move and then defects. For indirect reciprocity, the probability to know someone's reputation is given by q. A cooperator helps unless the reputation of the other person indicates a defector. A defector never helps. For network reciprocity, it can be shown that the expected frequency of cooperators is described by a standard replicator equation using a transformed payoff matrix (Ohtsuki and Nowak 2006). For group selection, the payoff matrices of the two games—within and between groups—can be added up.[2]

For kin selection, the calculation shows that Hamilton's rule, $r > c/b$, is the decisive criterion for all three measures of evolutionary success: ESS, RD, and AD. Similarly for network reciprocity and group selection, we obtain the same condition for all three evaluations, namely $b/c > k$ and $b/c > 1 + n/m$, respectively. The reason is the following: if these conditions hold, then cooperators dominate defectors. For direct and indirect reciprocity, we find that the ESS conditions lead to $w > c/b$ and $q > c/b$, respectively. Slightly more stringent conditions have to hold for cooperation to be risk-dominant (RD) or advantageous (AD).

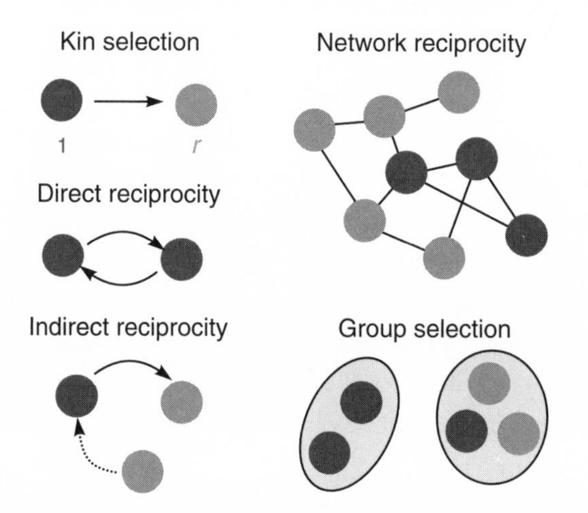

Figure 4.3 Five mechanisms for the evolution of cooperation. Kin selection operates when the donor and the recipient of an altruistic act are genetic relatives. Direct reciprocity requires repeated encounters between the same two individuals. Indirect reciprocity is based on reputation: a helpful individual is more likely to receive help. Network reciprocity means that clusters of cooperators outcompete defectors. Group selection is the idea that competition is not only between individuals but also between groups. From "Five Rules for the Evolution of Cooperation" by Martin A. Nowak, *Science* 314: 1560–63 (2006). Reprinted with permission from AAAS.

	Payoff matrix	Cooperation is ... ESS	RD	AD	
Kin selection	$\begin{array}{c\|cc} & C & D \\ \hline C & (b-c)(1+r) & (br-c) \\ D & (b-rc) & 0 \end{array}$	$\frac{b}{c} > \frac{1}{r}$	$\frac{b}{c} > \frac{1}{r}$	$\frac{b}{c} > \frac{1}{r}$	r...genetic relatedness
Direct reciprocity	$\begin{array}{c\|cc} & C & D \\ \hline C & (b-c)/(1-w) & -c \\ D & b & 0 \end{array}$	$\frac{b}{c} > \frac{1}{w}$	$\frac{b}{c} > \frac{2-w}{w}$	$\frac{b}{c} > \frac{3-2w}{w}$	w...probability of next round
Indirect reciprocity	$\begin{array}{c\|cc} & C & D \\ \hline C & b-c & -c(1-q) \\ D & b(1-q) & 0 \end{array}$	$\frac{b}{c} > \frac{1}{q}$	$\frac{b}{c} > \frac{2-q}{q}$	$\frac{b}{c} > \frac{3-2q}{q}$	q...social acquaintanceship
Network reciprocity	$\begin{array}{c\|cc} & C & D \\ \hline C & b-c & H-c \\ D & b-H & 0 \end{array}$	$\frac{b}{c} > k$	$\frac{b}{c} > k$	$\frac{b}{c} > k$	k...number of neighbors; $H = \frac{(b-c)k-2c}{(k+1)(k-2)}$
Group selection	$\begin{array}{c\|cc} & C & D \\ \hline C & (b-c)(m+n) & (b-c)m-cn \\ D & (bn) & 0 \end{array}$	$\frac{b}{c} > 1 + \frac{n}{m}$	$\frac{b}{c} > 1 + \frac{n}{m}$	$\frac{b}{c} > 1 + \frac{n}{m}$	n...group size; m...number of groups

Figure 4.4 Each mechanism can be described by a simple 2×2 payoff matrix that specifies the interaction between cooperators and defectors. From these matrices we can directly derive the necessary conditions for evolution of cooperation. The parameters c and b denote, respectively, the cost for the donor and the benefit for the recipient. All conditions can be expressed as the benefit-to-cost ratio exceeding a critical value. The concepts of evolutionarily stable (ESS), risk-dominant (RD) and advantageous strategies (AD) are defined in the text. From "Five Rules for the Evolution of Cooperation" by Martin A. Nowak, *Science* 314: 1560–63 (2006). Reprinted with permission from AAAS.

Conclusion

I have discussed five mechanisms for the evolution of cooperation: kin selection, direct reciprocity, indirect reciprocity, network reciprocity, and group selection. Each mechanism can be described by a characteristic 2×2 payoff matrix, from which we can directly derive the fundamental rules that specify if cooperation can evolve (Figure 4.4). Each rule can be expressed as the benefit-to-cost ratio of the altruistic act being greater than some critical value. The payoff matrices can be imported into standard frameworks of evolutionary game dynamics. For example, we can study replicator equations for games on graphs (Ohtsuki and Nowak 2006), for group selection and for kin selection. This creates interesting new possibilities for the theory of evolutionary dynamics (Nowak 2006).

I have not discussed all potential mechanisms for the evolution of cooperation. An interesting possibility is offered by "green-beard" models where cooperators recognize each other via arbitrary labels (Riolo, Cohen, and Axelrof 2001; Traulsen and Schuster 2003; Jansen and van Baalen 2006). Another option to obtain cooperation is making the game voluntary rather than obligatory: if players can choose between cooperation, defection or not playing at all, some level of cooperation usually prevails in dynamic oscillations (Hauert et al. 2002). Punishment is an important factor that can promote cooperation (Clutton-Brock and Parker 1995; Fehr and Gaetcher 2002; Fehr and Fischbacher 2003; Camerer and Fehr 2006; Guërk, Irlenbusch, and Rockenbach 2006). It is unclear, however, if punishment alone constitutes a mechanism for the evolution of cooperation. All evolutionary models of punishment so far are based on underlying mechanisms such as indirect reciprocity (Sigmund, Hauert, and Nowak 2001), group selection (Boyd et al. 2003; Bowles and Gintis 2004), or network reciprocity (Nakamaru and Iwasa 2005). Punishment can enhance the level of cooperation that is achieved in such models.

Kin selection has led to mathematical theories (based on the Price equation) that are more general than just analyzing interactions between genetic relatives (Queller 1992; Frank 1998). The interacting individuals can have any form of phenotypic correlation. Therefore, kin selection theory also provides an approach to compare different mechanisms for the evolution of cooperation (Fletcher and Zwick 2006; Lehmann and Keller 2006). I have not taken this approach in the present chapter.

The two fundamental principles of evolution are mutation and natural selection. But evolution is constructive because of cooperation. New levels of organization evolve when the competing units on the lower level begin to cooperate. Cooperation allows specialization and thereby promotes biological diversity. Cooperation is the secret behind the open-endedness of the evolutionary process. Perhaps the most remarkable aspect of evolution is its ability to generate cooperation in a competitive world. Thus, we might add "natural cooperation" as a third fundamental principle of evolution besides mutation and natural selection.[3]

Notes

1. Supporting material for this chapter was published in *Science* and is online at http://www.sciencemag.org/content/suppl/2006/12/05/314.5805.1560.DC1/Nowak.SOM.pdf.
2. The details of all these arguments and their limitations are given online in the *Science* link in Note 1.
3. Support from the John Templeton Foundation and the NSF/NIH joint program in mathematical biology (NIH grant 1R01GM078986–01) is gratefully acknowledged. The Program for Evolutionary Dynamics at Harvard University is sponsored by Jeffrey Epstein.

References

Alexander, R. D. 1987. *The Biology of Moral Systems.* New York: Aldine de Gruyter.

Axelrod, R. 1984. *The Evolution of Cooperation.* New York: Basic Books.

Axelrod, R., and W. D. Hamilton. 1981. "The Evolution of Cooperation." *Science* 211: 1390–96.

Bowles, S., and H. Gintis. 2004. "The Evolution of Strong Reciprocity: Cooperation in Heterogeneous Populations." *Theoretical Population Biology* 65: 17–28.

Boyd, R., H. Gintis, S. Bowles, and P. J. Richerson. 2003. "The Evolution of Altruistic Punishment." *Proceedings of the National Academy of Sciences of the USA* 100: 3531–35.

Brandt, H., and K. Sigmund. 2004. "The Logic of Reprobation: Assessment and Action Rules for Indirect Reciprocation." *Journal of Theoretical Biology* 231: 475–86.

Bshary, R., and A. S. Grutter. 2006. "Image Scoring Causes Cooperation in a Cleaning Mutualism." *Nature* 441: 975–78.

Camerer, C. F., and E. Fehr. 2006. "When does 'Economic Man' Dominate Social Behavior?" *Science* 311: 47–52.

Clutton-Brock, T. H., and G. A. Parker. 1995. "Punishment in Animal Societies." *Nature* 373: 209–16.

Dawkins, R. 1976. *The Selfish Gene.* Oxford: Oxford University Press.

Dugatkin, L. A. 1997. *Cooperation among Animals.* Oxford: Oxford University Press.

Durrett, R., and S. A. Levin. 1994. "The Importance of being Discrete (and Spatial)." *Theoretical Population Biology* 46: 363–94.

Fehr, E., and S. Gaechter. 2002. "Altruistic Punishment in Humans." *Nature* 415: 137–40.

Fehr, E., and U. Fischbacher. 2003. "The Nature of Human Altruism." *Nature* 425: 785–92.

Fishman, M. A. 2003. "Indirect Reciprocity among Imperfect Individuals." *Journal of Theoretical Biology* 225: 285–92.

Fletcher, J. A., and M. Zwick. 2006. "Unifying the Theories of Inclusive Fitness and Reciprocal Altruism." *The American Naturalist* 168: 252–62.

Foster, K. R., T. Wenseleers, and F. L. W. Ratnieks. 2006. "Kin Selection is the Key to Altruism." *Trends in Ecology and Evolution* 21: 57–60.

Frank, S. A. 1998. *Foundations of Social Evolution.* Princeton, NJ: Princeton University Press.

Fudenberg, D., and E. Maskin. 1990. "Evolution and Cooperation in Noisy Repeated Games." *American Economic Review* 80: 274–79.

Grafen, A. 1985. "A Geometric View of Relatedness." In *Oxford Surveys in Evolutionary Biology, vol. 2,* ed. R. Dawkins and M. Ridley. Oxford: Oxford University Press, 28–89.

Gürerk, Ö., B. Irlenbusch, and B. Rockenbach. 2006. "The Competitive Advantage of Sanctioning." *Science* 312: 108–111.

Hamilton, W. D. 1964. "The Genetical Evolution of Social Behaviour." *Journal of Theoretical Biology* 7: 1–16.

Hassell, M. P., H. N. Comins, and R. M. May. 1994. "Species Coexistence and Self-Organizing Spatial Dynamics." *Nature* 370: 290–92.

Hauert, C., and M. Doebeli. 2004. "Spatial Structure often Inhibits the Evolution of Cooperation in the Snowdrift Game." *Nature* 428: 643–46.

Hauert, C., S. De Monte, J. Hofbauer, and K. Sigmund. 2002. "Volunteering as Red Queen Mechanism for Cooperation in Public Goods Games." *Science* 296: 1129–32.

Hauser, M. D., M. K. Chen, F. Chen, and E. Chuang. 2003. "Give unto Others: Genetically Unrelated Cotton-Top Tamarin Monkeys Preferentially Give Food to Those Who Altruistically Give Food Back." *Proceedings of the Royal Society of London B* 270: 2363–70.

Hofbauer, J., and K. Sigmund. 1998. *Evolutionary Games and Population Dynamics.* Cambridge: Cambridge University Press.

———. 2003. "Evolutionary Game Dynamics." *Bulletin of the American Mathematical Society* 40: 479–519.

Jansen, V. A., and M. van Baalen. 2006. "Altruism through Beard Chromodynamics." *Nature* 440, 663–66.

Keller, L., ed. 1999. *Levels of Selection in Evolution.* Princeton, NJ: Princeton University Press.

Lehmann, L., and L. Keller. 2006. "The Evolution of Cooperation and Altruism: A General Framework and Classification of Models." *Journal of Evolutionary Biology* 19: 1365–76.

Leimar, O., and P. Hammerstein. 2001. "Evolution of Cooperation through Indirect Reciprocity." *Proceedings of the Royal Society of London B* 268: 745–53.

Lieberman, E., C. Hauert, and M. A. Nowak. 2005. "Evolutionary Dynamics on Graphs." *Nature* 433: 312–16.

May, R. M. 2006. "Network Structure and the Biology of Populations." *Trends in Ecology and Evolution* 21: 394–99.

Maynard Smith, J. 1982. *Evolution and the Theory of Games.* Cambridge: Cambridge University Press.

Michod, R. E. 1999. *Darwinian Dynamics.* Princeton, NJ: Princeton University Press.

Milinski, M. 1987. "Tit for Tat in Sticklebacks and the Evolution of Cooperation." *Nature* 325: 434–35.

Milinski, M., D. Semmann, and H. J. Krambeck. 2002. "Reputation Helps Solve the 'Tragedy of the Commons.'" *Nature* 415: 424–26.

Nakamaru, M., and Y. Iwasa. 2005. "The Evolution of Altruism by Costly Punishment in Lattice-Structured Populations: Score-Dependent Viability versus Score-Dependent Fertility." *Evolutionary Ecology Research* 7: 853–70.

Nowak, M. A. 2006. *Evolutionary Dynamics: Exploring the Equations of Life.* Cambridge, MA: Harvard University Press.

Nowak, M. A., and R. M. May. 1992. "Evolutionary Games and Spatial Chaos." *Nature* 359: 826–29.

Nowak, M. A., A. Sasaki, C. Taylor, and D. Fudenberg. 2004. "Emergence of Cooperation and Evolutionary Stability in Finite Populations." *Nature* 428: 646–50.

Nowak, M. A., and K. Sigmund. 1992. "Tit for Tat in Heterogeneous Populations." *Nature* 355: 250–53.

———. 1993. "A Strategy of Win-Stay, Lose-Shift that Outperforms Tit-for-Tat in Prisoner's Dilemma." *Nature* 36456–58.

———. 1998. "Evolution of Indirect Reciprocity by Image Scoring." *Nature* 393: 573–77.

———. 2004. "Evolutionary Dynamics of Biological Games." *Science* 303: 793–99.

———. 2005. "Evolution of Indirect Reciprocity." *Nature* 437: 1291–98.

Ohtsuki, H., C. Hauert, E. Lieberman, and M. A. Nowak. 2006. "A Simple Rule for the Evolution of Cooperation on Graphs and Social Networks." *Nature* 441: 502–5.

Ohtsuki, H., and Y. Iwasa. 2004. "How Should We Define Goodness? Reputation Dynamics in Indirect Reciprocity." *Journal of Theoretical Biology* 231: 107–20.

Ohtsuki, H., and M. A. Nowak. 2006. "The Replicator Equation on Graphs." *Journal of Theoretical Biology* 243: 86–97.

Panchanathan, K., and R. Boyd. 2004. "Indirect Reciprocity Can Stabilize Cooperation without the Second-Order Free Rider Problem." *Nature* 432: 499–502.

Paulsson, J. 2002. "Multileveled Selection on Plasmid Replication." *Genetics* 161: 1373–84.

Queller, D. C. 1992. "Quantitative Genetics, Inclusive Fitness and Group Selection." *American Naturalist* 139: 540–58.

Rainey, P. B., and K. Rainey. 2003. "Evolution of Cooperation and Conflict in Experimental Bacterial Populations." *Nature* 425: 72–74.

Riolo, R. L., M. D. Cohen, and R. Axelrod. 2001. "Evolution of Cooperation without Reciprocity." *Nature* 414: 441–43.

Rogers, A. R. 1990. "Group Selection by Selective Emigration: The Effects of Migration and Kin Structure." *American Naturalilst* 135: 398–413.

Selten, R., and P. Hammerstein. 1984. "Gaps in Harley's Argument on Evolutionary Stable Learning Rules and in the Logic of TfT." *The Behavioral and Brain Sciences* 7: 115–16.

Sigmund, K., C. Hauert, and M. A. Nowak. 2001. "Reward and Punishment." *Proceedings of the National Academy of Sciences of the USA* 98: 10757–62.

Taylor, P. D. 1992. "Altruism in Viscous Populations—An Inclusive Fitness Model." *Evolutionary Ecology* 6: 352–56.

Taylor, P. D., and D. S. Wilson. 1988. "A Mathematical Model for Altruism in Haystacks." *Evolution* 42: 193–97.

Traulsen, A., and M. A. Nowak. 2006. "Evolution of Cooperation by Multilevel Selection." *Proceedings of the National Academy of Sciences of the USA* 103: 10952–55.

Traulsen, A., and H. G. Schuster. 2003. "Minimal Model for Tag-Based Cooperation." *Physical Review E* 68, 046129.

Trivers, R. L. 1971. "The Evolution of Reciprocal Altruism." *The Quarterly Review of Biology* 46: 35–57.

Wedekind, C., and M. Milinski. 2000. "Cooperation through Image Scoring in Humans." *Science* 288: 850–52.

West, S. A., I. Pen, and A. S. Griffin. 2002. "Cooperation and Competition between Relatives." *Science* 296: 72–75.

Williams, G. C., and D. C. Williams. 1957. "Natural Selection of Individually Harmful Social Adaptations among Sibs with Special Reference to Social Insects." *Evolution* 11: 32–39.

Wilson, D. S. 1975. "A Theory of Group Selection." *Proceedings of the National Academy of Sciences of the USA* 72: 143–46.

Wilson, E. O. 1975. *Sociobiology*. Cambridge, MA: Harvard University Press.

Wilson, E. O., and B. Hölldobler. 2005. "Eusociality: Origin and Consequences." *Proceedings of the National Academy of Sciences of the USA* 102: 13367–71.

5

Mathematical Models of Cooperation

Christoph Hauert

Cooperation is a conundrum that has challenged researchers across disciplines and over many generations. In mathematical models, cooperation simply refers to behavioral actions that benefit others at some cost to an actor. Prima facie, defectors always outperform cooperators, but groups of cooperating individuals fare better than groups of noncooperating defectors. However, each individual faces the temptation to defect in order to avoid the costs of cooperation while free-riding on the benefits produced by others. This generates a conflict of interest between the individual and the group known as a social dilemma (Dawes 1980; Hauert et al. 2006).

Social dilemmas are abundant in nature. For example, musk oxen create defense formations to protect their young from wolves (Hamilton 1971). However, each ox would be better off and avoid potential injury by standing in the second line—but if every individual behaves in this way, their defense formation breaks down and the group becomes prone to attacks by wolves. A similar conflict of interest occurs in sentinel behavior in meerkats (Clutton-Brock et al. 1999): a few individuals are on the lookout for predators and warn foraging group members of impending danger. Spotting the predators first returns a direct benefit to the sentinels because they get a head start on finding shelter. At the same time, the costs for keeping watch depend on the individuals' hunger. From time to time sentinels abandon their position and get replaced by other individuals. Other prominent examples of social dilemmas occur in predator inspection behavior in fish (Milinski 1987; Magurran and Higham 1988; Pitcher 1992), in phages competing for reproduction (Turner and Chao 1999, 2003), or in microorganisms producing extra cellular products such as enzymes in yeast (Greig and Travisano 2004), biofilms (Rainey and Rainey 2003), or antibiotic resistance (Neu 1992).

Social dilemmas also occurred on an evolutionary scale and life could not have unfolded without the repeated incorporation of entities of lower complexity or degrees of self-organization into higher-level entities. Major transitions such as the formation of chromosomes out of replicating DNA molecules, the transition from single cells to multicellular organisms, or the change from individuals to societies all require cooperation (Maynard-Smith and Szathmáry 1995). Finally, humans have taken the problem of cooperation to yet another level (Hardin 1968) when it comes to social welfare such as health care or pension plans and, even more importantly, to global issues concerning natural resources such as drinking water, clean air, fisheries, or climate change (Milinski et al. 2006).

In order to analyze individual behavior in social dilemmas or other types of interactions, the economist Oscar Morgenstern and the mathematician John von Neumann developed a mathematical framework termed game theory (von Neumann and Morgenstern 1944). The most prominent game to study cooperation in social dilemmas is the prisoner's dilemma (Flood 1958; Axelrod and Hamilton 1981). In the prisoner's dilemma, two individuals decide whether to cooperate or to defect. Cooperation incurs costs, c, to the actor while the benefits of cooperation, b, accrue exclusively to the opponent with $b > c$. Defection does not incur costs to the actor and produces no benefits for the opponent. Thus, if both players cooperate, each receives $b - c$, whereas if both defect, neither receives anything. If only one cooperates and the other defects, the cooperator is left with the costs, $-c$, while the defector receives the full benefit, b. In this situation, defection should dominate because individuals are better off defecting, irrespective of the opponent's decision. Consequently, two rational players will opt for defection and end up with nothing, as opposed to the more favorable reward, $b - c$, for mutual cooperation.

Evolutionary Dynamics

The predicted dominance of defection in social dilemmas, however, contrasts with abundant evidence of cooperation in nature. Ever since Darwin (Darwin 1859), the evolution, and maintenance of cooperation has posed a major challenge to evolutionary biologists and social scientists. The theoretical foundation for addressing the problem of cooperation rests on Hamilton's kin selection theory (Hamilton 1964) and Maynard Smith's adaptation of game theory to evolutionary scenarios (Maynard Smith and Price 1973), by linking game theoretical payoffs with biological fitness. In evolutionary biology, fitness denotes the single determinant of evolutionary success and essentially reflects the reproductive output of individuals over their life span. Thus, under Darwinian selection, behavioral traits (or strategies) of individuals with high fitness are more likely to be passed on to future generations.

Replicator Dynamics

Consider a population with a fraction x cooperators (and $1-x$ defectors). If individuals randomly engage in prisoner's dilemma interactions, the average payoffs for defectors is $f_D = xb$ (with probability x the defector interacts with a cooperator and obtains the benefit b); for cooperators it is $f_C = xb - c$ (with probability x the cooperator interacts with another cooperator and receives b but always pays the costs of cooperation c); and for the entire population $\bar{f} = xf_C + (1-x)f_D = x(b-c)$. Selection prescribes that strategies that perform better than the population on average increase in abundance. In the simplest case this leads to the replicator equation (Hofbauer and Sigmund 1998):

$$\dot{x} = x(f_C - \bar{f}) \tag{1}$$

which states that the change in frequency of cooperators ($\dot{x}$ denotes the time derivative of the fraction of cooperators) is proportional to the payoff difference between cooperators and the population average. Since $f_C - \bar{f} = -c(1-x) < 0$, cooperators decrease over time and eventually disappear ($x^\star = 0$ is the only stable fixed point).

In order to model the promotion and maintenance of cooperation among unrelated individuals, a variety of approaches have been proposed over the last few decades. In particular, cooperation can be established through conditional behavioral rules in repeated encounters (Trivers 1971). If behavioral actions are reflected in an individual's reputation, repeated interactions are not required to establish cooperation (Nowak and Sigmund 1998). Moreover, punishment of defectors (Yamagishi 1986; Sigmund, Hauert, and Nowak 2001), or voluntary interactions (Hauert et al. 2002, 2007), also promote and sustain cooperation. Finally, spatial extension and limited local interactions can enhance cooperation, which is the topic of the second part of this chapter.

Another complementary approach to address the problem of cooperation is to reconsider the mathematical implementation of the social dilemma. In contrast to the theoretical effort expended on studying the prisoner's dilemma, it receives surprisingly little support from empirical evidence in biological systems. In fact, in all of the prominent examples of social dilemmas listed above, it remains largely unresolved whether individuals indeed engage in prisoner's dilemma type interactions. Instead, another game, called the snowdrift game (Sugden 1986; Hauert and Doebeli 2004), seems to be a biologically appealing alternative to the prisoner's dilemma (Doebeli and Hauert 2005). The anecdotal story behind the snowdrift game states that two drivers are caught in a blizzard and trapped on either side of a snowdrift. Each driver has the option to remove the snowdrift and start shoveling or to remain in the cozy warmth of the car. If both cooperate and shovel, they both get home while sharing the labor, but if only one shovels, again both get home, but the cooperator has to do all the work. If no one shovels, neither gets anywhere, and they have to wait for spring to melt the snowdrift. In contrast to the prisoner's dilemma, the best strategy now depends on the co-player's decision: if the other driver shovels, it is best to shirk, but when facing the potential for a lazy counterpart, it is better to start shoveling instead of remaining stuck in the snow.

The snowdrift game potentially seems to apply whenever individuals generate a valuable public resource. For example, in antibiotic resistance (Neu 1992) bacteria *(Staphylococci)* secrete an enzyme, β-lactamase, that destroys penicillin. The production of this enzyme is costly to the bacterium, while the resulting protection represents a public resource that benefits not only the enzyme-producing bacterium but also its fellow bacteria. Thus, in the vicinity of enzyme producers it pays for a bacterium to throttle enzyme production and increase replication, but in the absence of the enzyme, protection against penicillin becomes vital.

The evolutionary dynamics of the snowdrift game can again be analyzed using the replicator equation. For mutual cooperation the costs are shared and each individual receives $b-c/2$, whereas a cooperator facing a defector obtains $b-c$. The payoffs for defectors remain the same as in the prisoner's dilemma, with b against a cooperator and zero for mutual defection. Thus, the average payoff of cooperators becomes $f_C = b - c(1-x/2)$, and the average payoff for defectors remains $f_D = xb$. A short calculation shows that, in the snowdrift game, the replicator equation admits another fixed point that is stable (obtained by setting $f_C = \bar{f}$) at $x^\star = 1 - r$, where $r = c/(2b-c)$ denotes the cost-to-net-benefit ratio of mutual cooperation. Consequently, cooperators and defectors coexist in the snowdrift game. Nevertheless, the conflict of interest persists because a population at $x^\star$ is still worse off than if everybody had cooperated. Hence, the snowdrift game represents a relaxed social dilemma as compared to the prisoner's dilemma.

Adaptive Dynamics

In nature, the problem of cooperation may not always be adequately addressed by limiting the analysis to two distinct strategic types—the cooperators and the defectors. Instead, in many situations, it might be more appropriate to consider continuous degrees of cooperation such as time and effort expended in producing a public resource. In such continuous games, the strategy or trait u of an individual denotes its cooperative investment and can vary between zero and an upper limit u_{max}. The fitness costs and benefits are determined by the increasing functions $C(u)$ and $B(u)$, respectively, such that an increase in the cooperative trait u both raises costs and increases benefits. Moreover, no cooperation ($u=0$) neither provides benefits nor incurs costs, $C(0) = B(0) = 0$.

In the continuous prisoner's dilemma, the payoff of an individual with trait u interacting with a trait v individual is written as $Q(u,v) = B(v) - C(u)$. This means that the benefits are determined by the opponent's trait, whereas the costs are determined by the individual's own trait. Thus, the only way to improve the payoff is to lower the costs, and hence the degree of cooperation, because $C(u) < C(u + \Delta u)$ for $\Delta u > 0$. Consequently, evolution, when modeled on the dynamics of continuous prisoner's dilemma interactions, selects lower investors such that cooperation gradually declines and eventually disappears.

The situation is rather different in the continuous snowdrift game (Doebeli, Hauert, and Killingback 2004). In the previous section the snowdrift game was introduced by assuming constant benefits and costs that are shared among cooperators. Mathematically, this is equivalent to assuming that costs are constant and benefits depend on the number of cooperators. In the continuous snowdrift game, the payoff to an individual with trait u interacting with a trait v individual then becomes $P(u,v) = B(u+v) - C(u)$. If $B(u) > C(u)$ holds at least for small u, it seems that evolution should select intermediate degrees of cooperation. However, it turns out that the evolutionary dynamics of the continuous snowdrift game are much richer.

The evolution of the trait u can be analyzed using the adaptive dynamics framework (Dieckmann and Law 1996; Metz et al. 1996; Geritz et al. 1998). This assumes a homogeneous resident population with trait u. Occasionally, an initially rare mutant trait v appears and attempts to invade. The fitness of the mutant type is given by $P(v, u)$—interactions with its own type, $P(v,v)$, can be neglected because the mutant is rare. According to the replicator equation, the mutant increases in abundance if $P(v, u)$ exceeds the average population payoff—namely, the fitness of the resident, $P(u, u)$—interactions of the resident with the mutant are again neglected because the mutant is rare. Thus, the growth rate of the mutant v is given by $f_u(v) = P(v, u) - P(u, u)$ and is called the invasion fitness because for $f_u(v) > 0$ invasion succeeds, but fails for $f_u(v) < 0$. If mutations are small, such that v is close to u, it follows that whenever invasion succeeds ($f_u(v) > 0$), the mutant v eventually takes over and becomes the new resident (Geritz et al. 1998). Mutations are assumed to be rare, such that mutants always face a homogeneous resident population. Under these assumptions, the selection gradient $D(u)$ determines whether more or less cooperative mutants can invade.[1] For $D(u) > 0$ mutants with $v > u$ can invade and cooperation increases, but it decreases for $D(u) < 0$. Thus, the evolutionary change of cooperation is given by the canonical equation of adaptive dynamics, $\dot{u} = D(u)$ (Metz et al. 1996). Of particular interest are singular traits $u^\star$, for which the selection gradient vanishes, $D(u^\star) = 0$, and hence denote fixed points of the adaptive dynamics ($\dot{u} = 0$). If no singular trait $u^\star$ exists in the interval $(0, u_{max})$, then either cooperation

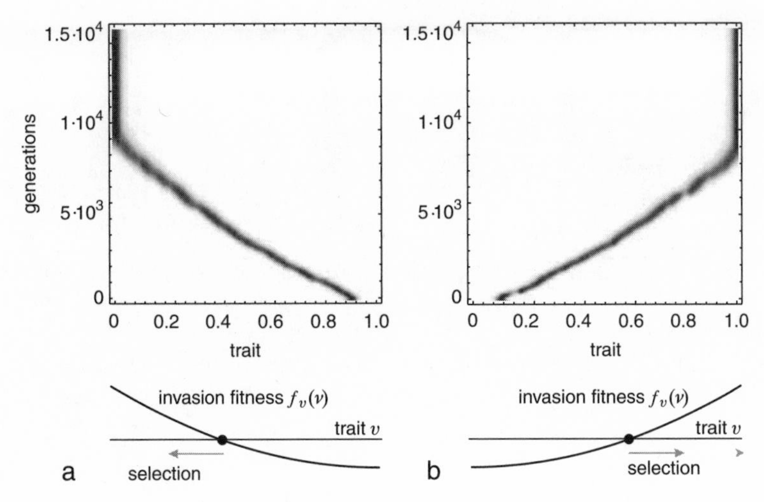

Figure 5.1 Dynamics in the continuous snowdrift game in the absence of singular traits. Simulation results for the trait distribution in the population over time (top row) and schematic illustration of the fitness profile in the population (bottom row). The benefit and cost functions are quadratic, $B(u)=b_2u^2+b_1u$, $C(u)=c_2u^2+c_1u$, such that $C(u)$, $B(u)$ are saturating and strictly increasing over the trait interval [0, 1]. **a** The selection gradient is always negative, $D(u)<0$, and evolution always favors less cooperative individuals until cooperation vanishes. The qualitative features of the invasion fitness $f_u(v)$ do not change as u changes over time. **b** Counterpart to **a**: $D(u)>0$ always holds and evolution selects more cooperative individuals until the upper bound is reached. Parameters: $b_2=-1:5$, $b_1=7, c_2=-1$ and **a** $c_1=8$; **b** $c_1=2$.

keeps decreasing, $D(u)<0$, as in the continuous prisoner's dilemma, or it keeps increasing until u_{max} is reached. The latter refers to situations where the social dilemma is fully relaxed and cooperation merely evolves as a by-product (Connor, 1996). Both cases are possible in continuous snowdrift games (see Figure 5.1).

The Origin of Cooperators and Defectors

The dynamics of cooperation become more interesting if the continuous snowdrift game admits singular traits. A singular trait $u^\star$ can be convergent stable such that traits in the vicinity of $u^\star$ converge to $u^\star$ (see Figure 5.2a and b), or unstable and traits near $u^\star$ evolve away (see Figure 5.2c). Interestingly, convergent stability of $u^\star$ does not necessarily imply that $u^\star$ represents

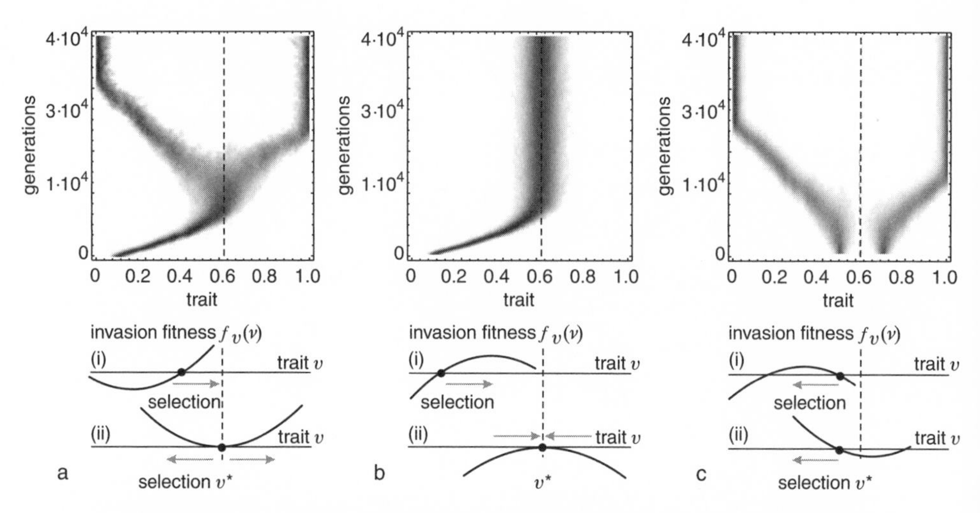

Figure 5.2 Dynamics in the continuous snowdrift game in the presence of a unique singular trait $u^\star$. Simulation results for the trait distribution in the population over time (top row) and schematic illustration of the fitness profile in the population (bottom row). $u^\star$ is marked by a vertical dashed line. As in Figure 5.1, the benefit and cost functions are quadratic, saturating and strictly increasing in [0, 1].
a *Evolutionary branching*—the singular strategy is convergent stable and the trait distribution approaches $u^\star$ but it is not evolutionarily stable and the population branches into two distinct phenotypic clusters. Evolution selects individuals with higher fitness [bottom panel (i)], which in turn changes the profile of the invasion fitness $f_u(v)$ such that the fitness minimum catches up at $u^\star$ [bottom panel (ii)] and mutants with both higher and lower v can invade. **b** *Evolutionary stability*—the singular strategy is not only convergent stable but also evolutionarily stable. As the population converges to $u^\star$ [bottom panel (i)] the profile of $f_u(v)$ changes and at $u^\star$ the trait catches up with the maximum of $f_u(v)$ [bottom panel (ii)] and no mutants are able to invade. **c** *Evolutionary repellor*—the singular strategy is an evolutionary repellor such that the traits evolve away from $u^\star$. Two separate simulation runs are shown: when starting below $u^\star$ cooperation disappears but if initial cooperative contributions are sufficiently high they keep increasing until the maximum is reached. In this case it is irrelevant whether $u^\star$ is evolutionarily stable [bottom panel (i)] or an evolutionary branching point [bottom panel (ii)] because evolution never reaches $u^\star$. Parameters: **a** $b_2=-1.4$, $b_1=6$, $c_2=-1.6$, $c_1=4.56$; **b** $b_2=-1.5$, $b_1=7$, $c_2=-1$, $c_1=4.6$; **c** $b_2=-0.5$, $b_1=3.4$, $c_2=-1.5$, $c_1=4$.

an evolutionary end state because u^* either represents a maximum or a minimum of the invasion fitness.[2] If it is a maximum, then u^* is not only convergent stable but also evolutionarily stable and u^* indeed denotes stable intermediate degrees of cooperation (see Figure 5.2b). If, however, u^* denotes a fitness minimum, then both, more *and* less cooperative mutants can invade. In this case u^* is called an evolutionary branching point and the population undergoes a spontaneous division into two distinct trait groups of cooperators and defectors (see Figure 5.2a). Thus, the evolutionary dynamics recovers the original snowdrift game discussed above. The continuous snowdrift game therefore suggests an evolutionary pathway for social diversification and for the origin of cooperators and defectors (Doebeli, Hauert, and Killingback 2010). All scenarios can be further explored using the *EvoLudo* (Hauert 2012), a growing collection of interactive tutorials on evolutionary dynamics.

Two Tragedies

Social dilemmas potentially occur whenever individuals supply a public resource, such as in the case of antibiotic resistance (Neu 1992), or whenever individuals consume a public resource, such as in Hardin's *Tragedy of the Commons* (Hardin 1968). Individual interests cause public resources to become overexploited. In the context of humans, this is hardly surprising—Aristotle (384–322 BC) already drew the same conclusion: "That which is common to the greatest number has the least care bestowed upon it." The continuous snowdrift game serves as a model of social dilemmas and may equally apply to communal enterprises in humans. However, the spontaneous separation into coexisting cooperators and defectors could additionally raise a *Tragedy of the Commune* (Doebeli, Hauert, and Killingback 2004), which states that evolution may not favor egalitarian contributions to the common good but instead promote highly asymmetric commitments. However, differences in cooperative contributions bear a formidable risk for escalating conflicts based on the accepted notion of fairness.

Spatial Games

According to the replicator equation, Eq. (1), cooperators are doomed and disappear in the prisoner's dilemma in the absence of supporting mechanisms, whereas in the snowdrift game coexistence of cooperators and defectors is expected. These predictions are based on the assumption that individuals randomly interact with other members of the population. This is a convenient assumption because it admits a full analysis. However, more realistic scenarios should take spatial extension and local interactions into account. In order to model this, individuals are arranged on a rectangular lattice and each individual interacts only with neighbors on its four adjacent sites. The population is then updated according to a spatial analogue of the replicator equation: first, a focal individual is randomly selected and its payoff, f_f, is determined through interactions with all its neighbors. Second, a neighbor of the focal individual is randomly chosen and its payoff, f_n, is determined in the same way. Third, the focal individual adopts the strategy of the neighbor with a probability proportional to the payoff difference, provided that the neighbor performs better, and sticks to its own strategy otherwise. This procedure is repeated many times in order to determine the equilibrium frequency and configuration of cooperators and defectors. Note that for increasing population and neighborhood sizes, this microscopic update rule recovers the replicator equation (Traulsen, Claussen, and Hauert 2005). Unfortunately, the dynamics of spatial systems is no longer analytically accessible, and results are either based on a technique called pair approximation (Matsuda et al. 1992; van Baalen and Rand 1998; Szabó and Hauert 2002a), or, as in the following, on simulation data.

In the spatial prisoner's dilemma, cooperators are able to survive by forming clusters (Nowak and May 1992; Ohtsuki et al. 2006; Taylor, Day, and Wild 2007). Compact clusters increase interactions with other cooperators, while reducing exploitation by defectors. However, the clustering advantages are limited, and the equilibrium fraction of cooperators decreases when costs are increased or benefits are lowered (see Figure 5.3a). Eventually, cooperators are unable to survive and go extinct. Effects of spatial structure, and hence the characteristic features of spatial configurations, are most pronounced near the extinction threshold of cooperators (see Figure 5.3b).

Upon approaching the extinction threshold, the population exhibits interesting dynamical features as it undergoes a critical phase transition (Szabó & Hauert 2002b)—a well-studied phenomenon in statistical mechanics. Critical phase transitions are exciting for physicists but may not seem exceedingly important in biologically relevant scenarios. However, they do

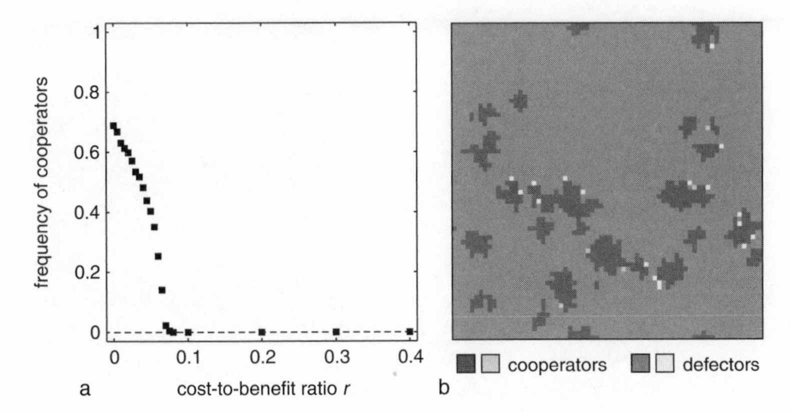

Figure 5.3 Spatial prisoner's dilemma on a square 100×100 lattice with four neighbors. **a** Equilibrium fraction of cooperators (solid squares) as a function of the cost-to-benefit ratio of mutual cooperation $r = c/(b-c)$. For small r cooperators persist but disappear for $r > r_c \approx 0.076$. In unstructured populations, cooperators cannot survive (dotted line). **b** Snapshot of a typical lattice configuration near the extinction threshold r_c. Spatial clustering enables cooperators to persist through more frequent interactions with other cooperators while reducing exploitation by defectors.

have substantial implications with far reaching consequences. For example, small changes in the costs or benefits can have disastrous effects on the equilibrium state of the population. Figure 5.3a illustrates that variations in the cost-benefit ratio r near the extinction threshold of cooperators result in big changes in their equilibrium frequency. Moreover, for populations that are prone to extinction, this indicates intrinsic difficulties for the empirical assessment of the current state because spatial and temporal variation increases when approaching the critical threshold.

Based on the results of the prisoner's dilemma in spatially extended settings, it has become widely accepted that spatial extension with limited local interactions is beneficial for cooperation. However, this does not necessarily apply to social dilemmas in general (Hauert and Doebeli 2004; Hauert 2006). In particular, in the spatial snowdrift game, the equilibrium proportion of cooperators tends to be lower than in unstructured settings with random interactions (see Figure 5.4a). Only for very low costs or high benefits does spatial structure support cooperation, but for most values cooperation is reduced and even gets eliminated altogether if costs are high or benefits are low.

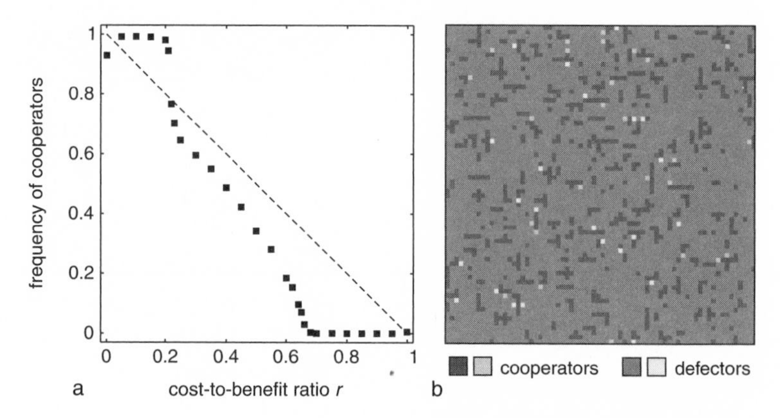

Figure 5.4 Spatial snowdrift game on a square 100 × 100 lattice with four neighbors. **a** Equilibrium fraction of cooperators (solid squares) as a function of the cost-to-benefit ratio of mutual cooperation $r = c/(2b - c)$. In unstructured populations, cooperators and defectors coexist (dotted line). With the exception of small r, spatial structure inhibits cooperation and for $r > r_c \approx 0.68$ cooperators even disappear. **b** Snapshot of a typical lattice configuration near the extinction threshold r_c. The equilibrium fraction of cooperators lies around 20% as compared to 40% cooperation in the absence of spatial structure.

The fact that in the snowdrift game the best option of one individual depends on the opponent's behavior, such that it is best to choose a strategy that is different from the opponent, often prevents cooperators from forming compact clusters as in the spatial prisoner's dilemma. Instead of minimizing the boundary between cooperators and defectors to avoid exploitation, the characteristics of the snowdrift game tend to maximize this boundary, which results in dendritic or filament-like cluster shapes. These cluster shapes become most apparent near the extinction threshold of cooperators (see Figure 5.4b). An intuitive approximation of the extinction threshold is obtained by considering the payoff of a single isolated cooperator in the spatial snowdrift game. As soon as the cooperator's payoff drops below the payoff of its defecting neighbors it perishes. Note, however, that this scenario underestimates the actual extinction threshold because of larger patches of cooperators. On average, the filament-like structures generate an advantage for defectors because of increased exploitation along the fractal-like boundary between cooperators and defectors. This results in an overall reduction of cooperators when compared to unstructured populations.

The *EvoLudo* site (Hauert 2012) encourages further interactive explorations and comparisons of the fascinating spatiotemporal dynamics in different types of spatial games.

Conclusions

The most prominent mathematical model to study the evolution of cooperation is the prisoner's dilemma. Only more recently has the snowdrift game attracted increasing attention as a viable and biologically interesting alternative for modeling cooperation dynamics. For example, *RNA* phages engage in prisoner's dilemma interactions in cells (Turner and Chao 1999), but selection alters the payoff structure, leading to the stable coexistence of cooperating and defecting types in a snowdrift game (Turner and Chao 2003). Unfortunately, differentiating between the two games can be challenging in real populations because the determination of payoffs is notoriously difficult. For example, predator inspection in sticklebacks is an often-cited application of the prisoner's dilemma (Milinski 1987). Sophisticated experimental setups confirmed the payoff ranking for three out of the four possible behavioral outcomes (Milinski et al. 1997). However, only the missing ranking of the payoff for mutual defection would enable one to discriminate between the prisoner's dilemma and snowdrift games. Other well-known examples of potential snowdrift games include alarm calls in meerkats (Clutton-Brock et al. 1999), and fighting in large ungulates (Wilkinson and Shank 1977). Cooperation is ubiquitous in meerkats, whereas serious escalations of fights seem to be common in musk ox. Because the costs of alarm calls are small, whereas the costs of forgoing reproduction are high, both observations are in agreement with the spatial snowdrift game, which promotes cooperation for low cost-to-benefit ratios but suggests more frequent escalations for high ratios of cost-to-benefit (Hauert and Doebeli 2004).

The significance and scientific value of game theoretical models for understanding the evolution of cooperation does not primarily lie in their predictive power for particular applications to specific scenarios. Instead, these games represent a conceptual framework to highlight and to emphasize the rich and often-unexpected dynamics generated by simple models that capture the essence of biologically and socially relevant interaction patterns. For example, the snowdrift game demonstrates that spatial structure may not be as universally beneficial for cooperation as previously thought, based on results of the spatial prisoner's dilemma. Similarly, evolutionary branching in the continuous snowdrift game illustrates that distinct behavioral traits may easily originate in a continuum of behavioral options. Game theoretical results delineate evolutionary constraints that are critical in resolving the problem of cooperation.

In theoretical and evolutionary biology, the ambiguity and limited consensus on the usage of the terms "cooperation," "altruism" and "helping" can cause some disputes but generally pose less severe problems than in the interdisciplinary discourse of this book. In all theoretical investigations, the starting point as well as the concluding results are unambiguous mathematical statements. Misunderstandings are readily averted by resorting to the mathematics. Instead, the core challenge lies in the proper translation of biological questions into tractable mathematical models as well as in interpreting the mathematical results in meaningful biological terms.

Notes

1. The selection gradient $D(u)$ denotes the slope of the invasion fitness $f_u(v)$ at $v=u$ and is defined as $D(u)=\partial f_u(v)/\partial v|_{v=u}=B'(2u)-C'(u)$ where the primes indicate the derivatives with respect to u (for details see Doebeli, Hauert, and Killingback 2004).
2. $u^\star$ is convergent stable if $dD(u)/du|_{u=u^\star}=B''(2u^\star)-C''(u^\star)<0$ and evolutionarily stable if the invasion fitness $f_{u^\star}(v)$ has a fitness maximum at $v=u^\star$, i.e. if the second derivative $\partial^2 f_{u^\star}(v)/\partial v^2\ |_{v=u^\star}=2B''(2u^\star)-C''(u^\star)<0$ is negative. Hence the two stability criteria are not identical.

References

Axelrod, R., and W. D. Hamilton. 1981. "The Evolution of Cooperation." *Science* 211: 1390–96.

Clutton-Brock, T. H., M. J. O'Riain, P. N. M. Brotherton, D. Gaynor, R. Kansky, A. S. Griffin, and M. Manser. 1999. "Selfish Sentinels in Cooperative Mammals." *Science* 284: 1640–44.

Connor, R. C. 1996. "Partner Preferences in By-Product Mutualisms and the Case of Predator Inspection in Fish." *Animal Behaviour* 51: 451–54.

Darwin, C. 1859. *The Origin of Species*. Cambridge, MA: Harvard University Press. Reprinted in 1964.

Dawes, R. M. 1980. "Social Dilemmas." *Annual Review of Psychology* 31: 169–93.

Dieckmann, U., and R. Law. 1996. "The Dynamical Theory of Co-Evolution: A Derivation from Stochastic Ecological Processes." *Journal of Mathematical Biology* 34: 579–612.

Doebeli, M., and C. Hauert. 2005. "Models of Cooperation Based on the Prisoner's Dilemma and the Snowdrift Game." *Ecology Letters* 8: 748–66.

Doebeli, M., C. Hauert, and T. Killingback. 2004. "The Evolutionary Origin of Cooperators and Defectors." *Science* 306: 859–62.

Flood, M. 1958. "Some Experimental Games." *Management Science* 5: 5–26.

Geritz, S. A. H., E. Kisdi, G. Meszéna, and J. A. J. Metz. 1998. "Evolutionarily Singular Strategies and the Adaptive Growth and Branching of the Evolutionary Tree." *Evolutionary Ecology* 12: 35–57.

Greig, D., and M. Travisano. 2004. "The Prisoner's Dilemma and Polymorphism in Yeast SUC Genes." *Biology Letters* 271: S25–26.

Hamilton, W. D. 1964. "The Genetical Evolution of Social Behaviour." *Journal of Theoretical Biology* 7: 1–16.

———. 1971. "The Geometry of the Selfish Herd." *Journal of Theoretical Biology* 31: 295–311.

Hardin, G. 1968. "The Tragedy of the Commons." *Science* 162: 1243–48.

Hauert, C. 2006. "Spatial Effects in Social Dilemmas." *Journal of Theoretical Biology* 240: 627–36.

———. 2012. *EvoLudo: Interactive Tutorials on Evolutionary Game Theory,* http://www.evoludo.org.

Hauert, C., S. De Monte, J. Hofbauer, and K. Sigmund. 2002. "Volunteering as Red Queen Mechanism for Cooperation in Public Goods Games." *Science* 296: 1129–32.

Hauert, C., and M. Doebeli. 2004. "Spatial Structure often Inhibits the Evolution of Cooperation in the Snowdrift Game." *Nature* 428: 643–46.

Hauert, C., F. Michor, M. Nowak, and M. Doebeli. 2006. "Synergy and Discounting of Cooperation in Social Dilemmas." *Journal of Theoretical Biology* 239: 195–202.

Hauert, C., A. Traulsen, H. Brandt, M. Nowak, and K. Sigmund. 2007. "Via Freedom to Coercion: The Emergence of Costly Punishment." *Science* 316: 1905–7.

Hofbauer, J., and K. Sigmund. 1998. *Evolutionary Games and Population Dynamics.* Cambridge, MA: Cambridge University Press.

Killinback, T., M. Doebeli, and C. Hauert. 2010. "Diversity of Cooperation in the Tragedy of the Commons." *Biological Theory* 5/1: 3–6.

Magurran, A. E., and A. Higham. 1988. "Information Transfer Across Fish Shoals under Predator Threat." *Ethology* 78: 153–58.

Matsuda, H., N. Ogita, A. Sasaki, and K. Sato. 1992. "Statistical Mechanics of Populations." *Progress of Theoretical Physics* 88(6): 1035–49.

Maynard-Smith, J., and G. Price. 1973. "The Logic of Animal Conflict." *Nature* 246: 15–18.

Maynard-Smith, J., and E. Szathmáry. 1995. *The Major Transitions in Evolution*. Oxford: W. H. Freeman and Co.

Metz, J. A. J., S. A. H. Geritz, G. Meszena, F. J. A. Jacobs, and J. S. van Heerwaarden. 1996. "Adaptive Dynamics: A Geometrical Study of the Consequences of Nearly Faithful Reproduction." In *Stochastic and Spatial Structures of Dynamical Systems*, ed. S. J. van Strien and S. M. Verduyn Lunel. Amsterdam: North Holland, 183–231.

Milinski, M. 1987. "Tit for Tat in Sticklebacks and the Evolution of Cooperation." *Nature* 325: 433–35.

Milinski, M., J. H. Lüthi, R. Eggler, and G. A. Parker. 1997. "Cooperation under Predation Risk: Experiments on Costs and Benefits." *Proceedings of the Royal Society B* 264: 831–37.

Milinski, M., D. Semmann, H. Krambeck, and M. Marotzke. 2006. "Stabilizing the Earth's Climate Is Not a Losing Game: Supporting Evidence from Public Goods Experiments." *Proceedings of the National Academy of Sciences of the USA* 103: 3994–98.

Neu, H. C. 1992. "The Crisis in Antibiotic Resistance." *Science* 257: 1064–73.

Nowak, M. A., and R. M. May. 1992. "Evolutionary Games and Spatial Chaos." *Nature* 359: 826–29.

Nowak, M. A., and K. Sigmund. 1998. "Evolution of Indirect Reciprocity by Image Scoring." *Nature* 393: 573–77.

Ohtsuki, H., C. Hauert, E. Lieberman, and M. Nowak. 2006. "A Simple Rule for the Evolution of Cooperation on Graphs and Social Networks." *Nature* 441: 502–5.

Pitcher, T. 1992. "Who Dares, Wins—The Function and Evolution of Predator Inspection Behavior in Shoaling Fish." *Netherlands Journal of Zoology* 42: 371–91.

Rainey, P. B., and K. Rainey. 2003. "Evolution of Cooperation and Conflict in Experimental Bacterial Populations." *Nature* 425: 72–74.

Sigmund, K., C. Hauert, and M. Nowak. 2001. "Reward and Punishment." *Proceedings of the National Academy of Sciences of the USA* 98: 10757–62.

Sugden, R. 1986. *The Economics of Rights, Cooperation and Welfare*. Oxford and New York: Blackwell.

Szabó, G., and C. Hauert. 2002a. "Evolutionary Prisoner's Dilemma with Optional Participation." *Physical Review E* 66: 062903.

———. 2002b. "Phase Transitions and Volunteering in Spatial Public Goods Games." *Physical Review Letters* 89: 118101.

Taylor, P. D., T. Day, and G. Wild. 2007. "Evolution of Cooperation in a Finite Homogeneous Graph." *Nature* 447: 469–72.

Traulsen, A., J. C. Claussen, and C. Hauert. 2005. "Coevolutionary Dynamics: From Finite to Infinite Populations." *Physical Review Letters* 95: 238701.

Trivers, R. L. 1971. "The Evolution of Reciprocal Altruism." *Quarterly Review of Biology* 46: 35–57.

Turner, P. E., and L. Chao. 1999. "Prisoner's Dilemma in an RNA Virus." *Nature* 398: 441–43.

———. 2003. "Escape from Prisoner's Dilemma in RNA phage Φ6." *American Naturalist* 161: 497–505.

van Baalen, M., and D. A. Rand. 1998. "The Unit of Selection in Viscous Populations and the Evolution of Altruism." *Journal of Theoretical Biology* 193: 631–48.

von Neumann, J., and O. Morgenstern. 1944. *Theory of Games and Economic Behaviour.* Princeton, NJ: Princeton University Press.

Wilkinson, G. S., and C. C. Shank. 1977. "Rutting-Fight Mortality among Musk Oxen on Banks Island, Northwest Territories, Canada." *Animal Behaviour* 24: 756–58.

Yamagishi, T. 1986. "The Provision of a Sanctioning System as a Public Good." *Journal of Personality and Social Psychology* 51: 110–16.

6

• • • •

Economics and Evolution

Complementary Perspectives on Cooperation

Johan Almenberg and Anna Dreber

"No man is an island." These famous words by the poet John Donne are an apt description of economics. It is hard to conceive of markets or prices without referring to social interactions, and in economic interactions, individuals weigh the costs and benefits of choosing a certain action. There is also much room for cooperative behavior, in which one person bears a cost in order for another person to reap a benefit. Behaving generously toward others in this manner, and reciprocating the generous behavior of others, through mutual trust, for example, is an integral part of economic activity.

In this chapter we describe how economists use game theory in order to understand human behavior in social interactions. This is a large area of study. We deal with one very important part: cooperation in economic games. We describe some of the basic games and show how they are being used in economics as well as other fields in order to study both human and nonhuman behavior.

Economists study human behavior with regard to scarce resources. If a resource is unlimited in supply, there is no conflict between your material well-being and mine. But resources are usually limited in supply, and the wants of many people must be reconciled, be it through informal institutions such as norms, or formal institutions such as markets.

When economists study human behavior, they usually look at de facto displays of preferences, such as choosing one commodity over another with the same price. Observed outcomes are interpreted as the result of an individual optimization process, as a *rational choice* reflecting the individuals' true preferences (which may, for example, be other-regarding in the sense that they place value on the well-being of others). Economists then design simple models that are consistent with such "revealed" preferences.

Much economic modeling makes use of three simplifying assumptions. Individuals are assumed to be fully rational—and hence able to calculate an optimal strategy even in highly complex settings. They are assumed to have perfect self-control. In addition, it is assumed that they are only concerned with their own material payoff from a game.

In the last couple of decades, experimental economics in conjunction with cognitive psychology has produced a large body of evidence showing that these standard assumptions frequently fail to generate accurate predictions. We begin by focusing on the last of the aforementioned assumptions: that individuals simply maximize their own material payoffs according to selfish goals. Later, we discuss how violations of the first assumption may also explain some findings from experimental economics.

Economic Games: A General Background

To begin with, let us define what we mean by economic games. The games described in this chapter are sometimes referred to as economic games. They are based on standard game theory used in economics, and the participants play for real money. But the underlying structure of the game applies to a number of different contexts, ranging from evolutionary biology to international relations.

A game is a simple, stylized way of representing a strategic interaction. It abstracts from specific details and attempts only to distill the essential, general structure of an interaction. A game consists of three elements: (1) the players; (2) the actions that the players choose from; and (3) the payoff functions that map each action to a payoff for an individual player, given the choices made by the other players.

In the standard game theory mostly used by economists, the payoffs are measured in terms of *utility,* defined as a subjective level of well-being.[1] More precisely, each player is assumed to have a so-called von Neumann-Morgenstern utility function (von Neumann and Morgenstern 1944). This is a function that maps each outcome in the game to a utility level for that player.

This differs from the evolutionary game theory used by, for example, biologists—and also by a number of economists. In evolutionary games, the payoffs are usually measured in terms of *fitness,* in other words, reproductive potential. We will see below that the two approaches are more closely linked than it might seem at first sight.

The theoretical predictions of game theory are based on the assumption that individuals choose actions in order to maximize their own utility. But game theory has little to say about exactly what goes into the utility function. By contrast, in an experimental setting the payoffs are typically in the form of money. While money is clearly a source of utility for most people, few, if any, economists believe it to be the only source.

The difference between the utility payoffs in game theory and the material payoffs used in economic games is crucial. The experimental findings outlined in this chapter are sometimes taken as a wholesale rejection of (a) game theory, or (b) the economists' paradigm. As we will see, both conclusions are incorrect.

We never observe an individual's utility function, only his or her actions, and hence cannot know what goes into it. For this reason, we must be careful in what conclusions we draw from observing people's behavior in an experimental setting. For example, a person's "utility," as here understood, may contain elements of both altruism and spite. That some players fail to maximize their monetary payoff does not imply that they fail to maximize their "utility."

What we can reject, however, is the notion that individuals only seek to maximize their own material payoffs in games. This will be explored extensively below.

Four Common Games

We will now describe four of the most commonly used games. These games belong to *noncooperative* game theory. This does not mean that the players never cooperate. It simply means that, both experimentally and theoretically, each individual makes up his own mind—that is, maximizes his own self-interest. This self-interest can be broadly defined. For example, it may include a concern for the well-being of others.

In this section, we will be referring exclusively to *one-shot* games—namely, games that are played in a nonrepeated setting. In addition, most of the results mentioned are from games where the players are anonymous from one another. Anonymity enhances the one-shot nature of an experiment. If the players know each other, they are more likely to be concerned about how their reputations outside the lab will be affected, or about other possible repercussions. In this case, the game played in the lab becomes part of a larger game involving the world outside the experiment. This can make it hard to draw conclusions based solely on the behavior observed in the lab.

Later in the chapter, we argue that humans may not be well equipped fully to comprehend anonymous one-shot interactions. Our actions are influenced by instincts and emotions that evolved in an environment where we mostly interacted with the same, moderately sized group of people day in and day out. It is tenuous to think of anonymous interactions in such an environment. This point becomes important in the following section when we introduce more complexity—and realism—by looking at games that are played in a repeated setting.

The Dictator Game

Consider the following two-player game. Two participants enter the lab. They do not know each other, and they will not meet during the experiment. Their roles are assigned at random, for example, by flipping a coin. Player 1 receives an amount of money from the experimenter. Some fraction between zero and one of this money may be given by player 1 to player 2. The remainder is kept by player 1. Player 2 has no choice but to accept the "gift." In other words, only player 1 makes a decision, but the decision affects both players. For this reason, the game is known as the dictator game, and player 1 as the dictator (Kahneman, Knetsch, and Thaler 1986). Note that if individuals were only concerned with maximizing their own material payoffs in the game, dictators would keep all the money.

Since the first experiment in 1986, a vast number of dictator game experiments have taken place. A couple of key findings have proven to be highly robust. A simple economic model based solely on individuals maximizing their own material payoff is rejected by the fact that we frequently observe dictators giving away some fraction of the money. The mean amount transferred is about 20%, and giving as much as 50% is not uncommon (Camerer 2003).[2] When dictators forego part of their potential material payoff in this manner, it means that they incur a cost for themselves in order for somebody else to receive a benefit. That we commonly observe such cooperative behavior in this simple game has been interpreted as an indication of other-regarding preferences.

The Ultimatum Game

Next, let us add another stage to the game we have just described, making it a strategic interaction. Suppose that after I (player 1) have chosen what fraction of the money to give away, you (player 2) get to make the following choice: either accept my proposed gift, or reject it. If you choose to reject the gift, neither of us will get any money at all. Thus, in a world where we simply want to maximize our monetary payoff, you, the responder, will accept any offer that I, the proposer, give you, since something, even if it is just one cent, is better than nothing. I, knowing this, will offer you the smallest possible amount, such as one cent.

This game is known as the ultimatum game due to its take-it-or-leave-it character (Güth, Schmittberger, and Schwarze 1982).[3] Once more, a standard economic model based on the assumption that we simply maximize our own material payoff does a poor job of predicting what actually happens. It is frequently observed that player 1 offers more than the minimum amount. Mean offers of 40–50% are typical (Camerer 2003), which is higher than in the dictator game. This comes as no surprise, since player 1 now needs to worry about the risk of player 2 rejecting the offer, whence both players end up with payoffs of zero. However, we know from dictator game experiments that even when player 2 cannot retaliate, player 1 frequently gives something to the other player. Hence, it would be incorrect to attribute generosity shown by player 1 entirely to strategic concerns.

The most striking feature of the ultimatum game is that it is common for player 2 to reject low offers. Given that player 2 is likely to reject a low offer, the strategy in which player 1 gives the minimum amount is no longer optimal. Rather, player 1 must try to give the smallest amount required to get Player 2 to accept the offer. The likelihood of rejection is about 50–50 at offers of less than 20%, and even larger for smaller offers (Camerer 2003). Thus, a sizeable share of the population makes quite generous offers as proposers, and a sizeable share rejects offers that are perceived to be selfish or unfair, even though it is costly for the responder to do so.

The rejection of low offers in the ultimatum game gives us an important insight: some individuals prefer an outcome in which neither player gets anything over an outcome in which they have been "wronged." In terms of material payoffs, player 2 has nothing to gain by rejecting a nonzero offer, no matter how small it is, in a situation like this. Yet many of us are willing to incur a cost by foregoing the small offer in order to reduce the material payoff of selfish players. Thus, the observed behavior in ultimatum games appears to be yet another example of social preferences manifesting themselves in the desire for perceived fairness.

The Prisoner's Dilemma

In the dictator game, player 1 can behave cooperatively toward player 2 by giving a nonzero amount. In the ultimatum game, we added the threat of rejection. This introduced a new element of reciprocity: be nice to me, and I will be nice to you in return. This reciprocity is sequential: the first player can be rewarded by the second player once the second player has observed the first player's choice.

We can also think of a similar game, but one in which the players simultaneously face the decision of whether or not to play cooperatively. The Prisoner's Dilemma (PD) is such a game (see the previous chapters by Nowak and Hauert). The essence of the one-shot PD is that the payoff to each player is higher when both cooperate than when both defect, but the most profitable choice is to defect when the other player cooperates. Theoretically, the temptation to defect generates a unique equilibrium (in pure strategies): both players defect. This is an example of a so-called Nash equilibrium, a key concept in game theory. A Nash equilibrium is when every player chooses her optimal strategy, given that every other player also chooses her optimal strategy (Nash 1950). In a two-player game, this means that the action I choose is my best response to the action you choose, which in turn is your best response to my action. Neither of us can unilaterally improve our payoffs from the game by choosing some other action.

However, in the lab, players do not always defect in a one-shot PD. Many individuals act in a way that strongly indicates a concern for the well-being of the other player. They are willing to forego part of their material payoff in order to benefit the other player. Once again, this shows that a simple theoretical economic model in which individuals are only concerned with maximizing their own material payoffs does a bad job of predicting actual behavior.[4]

The Public Goods Game

Another popular type of game in economic experiments is the public goods game. In economic jargon, a public good is a good that is (1) nonrivalrous, and (2) nonexcludable. The first requirement means that my benefit is unaffected by how many others also benefit. We can think of listening to beautiful music as an example. The second requirement means that it is difficult to prevent others from consuming the good. If the beautiful music is played in a concert hall, and it charges an entry fee, the music is less of a public good than if it were played on the street outside, where I cannot easily be prevented from enjoying it. Few goods are pure public goods—meaning that both criteria are fully applicable—but there are many examples of goods that have these characteristics to some extent.

In a typical public goods game, each player receives an amount of money and then chooses how much to keep for himself and how much to contribute to a public good. The sum of the contributions is multiplied by a factor larger than one and split up equally among all players. Thus, the players receive a benefit from the public good that depends only on the total amount of contributions and not on how much they contributed individually. The game is usually set up so that everyone is better off when all players contribute, compared to when they keep the money for themselves. But each individual has an incentive to play selfishly, hoping to benefit from the contributions of others without contributing himself. Such behavior is sometimes referred to as free-riding.

Suppose all of the other players contribute in a public goods game. Then the best strategy, in terms of material payoffs, is for you to contribute nothing. You cannot improve your outcome by choosing any other level of contribution. The same reasoning applies for each of the other players. It turns out that when individuals only maximize their own material payoffs, the unique Nash equilibrium of the standard public goods game (which is an n-player PD) is that nobody contributes anything to the public good.

Making a nonzero contribution in a public goods game is also a form of cooperation. An additional dollar contributed gives the individual one dollar less than in increased total payoff. The difference between the two is a benefit to other players, at a cost to the individual. On average, about 40–60% of subjects cooperate in the one-period public goods game. The distribution tends to be bimodal, with a majority of subjects contributing either all or nothing (Camerer and Fehr 2004).

"The Shadow of the Future"

So far, we have talked about games that are played just once. In a one-shot PD there are no repercussions if a player chooses to defect. But suppose that the same two people were to play the game again the following day. Would this give the players incentives to play differently? In theory, no. For, on the second day, both players face the usual one-shot game. If individuals only maximize their material payoffs, a player who defects in the nonrepeated game would defect here too. Knowing this, the players have no incentive to cooperate on the first day either.

However, suppose that the game was played not once or twice but several times. If there is a known last round, the optimal strategy in the last round will be the same as when the game is played just once. If, for example, the PD is played with a certain number of finite rounds, no matter how many, the optimal strategy in the final round is still to defect. Knowing this, players will also defect in the penultimate round, and knowing this, they will defect in the round before that, and so on. Iterating backward in this manner will cause the game to "unravel," and the unique equilibrium is once again to defect in every single round of the game.

Instead, however, think of the game as continuing onto another round with a certain probability p. This turns out to make a crucial difference. A strategy choice for each player that constitutes a Nash equilibrium in the one-shot PD is also a Nash equilibrium when the game is repeated, but the repeated game also has equilibria that involve cooperation. Suppose you and I play a PD. Let R be the payoff I get when we both cooperate. If I defect but you cooperate, I get T, and you get S. If both of us defect, we each get the payoff P. Assume that $T > R > P > S$. This allows us to describe a condition on p for sustaining cooperation: $p > (T - P)/(T - R)$. Imagine that I interact with a tit-for-tat player—namely, somebody who cooperates with me if I cooperated with him in the previous period but defects if I defected. If the probability of continued play is sufficiently high, the possibility of a long sequence of rewarding mutual cooperation in the future makes it more attractive for me to cooperate in the current period more. In other words, the threat of receiving the lower P payoff in the future enables us to cooperate today.

It turns out that for a game with this form, cooperation can be sustained indefinitely if p, the probability of the game continuing, is sufficiently large. This so-called folk theorem casts a different light on cooperative play: I might incur a cost today in order to reap a benefit tomorrow.

This simple mechanism points to an important principle: in a repeated setting, cooperation in a single round of the game can also be the strategy that maximizes utility for individuals who only seek to maximize their own material payoffs. In the context of evolution, this distinction is crucial for two reasons. First, humans evolved in environments where the notion of a one-shot game is not applicable because interactions were likely to be repeated (Hagen and Hammerstein 2006). Second, in evolutionary time, access to material resources is crucial for survival and reproductive success. In the economist's experiment, we can make assumptions about exogenous preferences—namely, we can ignore the question of why we would have certain preferences. In evolutionary time, we cannot. In addition, even in a modern economy, many transactions are of a repeated-game nature.

The "shadow of the future"—the value of future interactions—can have dramatic effects. Dal Bó (2005) found that behavior in the lab is remarkably consistent with the theoretical prediction. In Dal Bó's experiment only a small fraction of the participants cooperate in a one-shot PD. In a finitely repeated game of four rounds, the cooperation rate in the last round is about the same as in the one-shot game. By contrast, when there is a three-quarter probability of the game continuing into another round (making the average length of this game and the finitely repeated game the same), the rate of cooperation is about four times higher in the fourth round than in the one-shot game. The conclusion from the experiment is clear: the shadow of the future can be a very powerful mechanism for achieving cooperation.

This has important consequences for how we interpret behavior in one-shot games. Suppose the way people play in one-shot games is influenced by instincts that have evolved in a repeated-game setting where reciprocity can be direct and indirect, for example, through reputation. Then we would observe cooperation in one-shot games too. We will get back to this point later in the chapter.

Economics and Evolution

In the preceding analysis we have briefly pointed to an important difference between economic and evolutionary approaches. Evolutionary theorists actively seek to explain how our preferences might have evolved. Economists, by contrast, are primarily concerned with what preferences are consistent with observed behavior.

These, we submit, are two complementary perspectives. Evolutionary modeling is gaining ground in economics. Traditionally, economists have relied on standard game theory, in which individuals choose strategies in order to maximize expected payoffs as measured in terms of utility. In evolutionary game theory, by contrast, strategies are represented by subpopulations that always play that strategy. Payoffs are measured in terms of the reproductive success—the *fitness*—of that subpopulation. Since more successful strategies reproduce more quickly, the whole population may converge toward a single strategy.[5]

Biological evolution is a process of mutation and selection. Simple organisms (we speak anthropomorphically here) cannot be expected to calculate the optimal strategy. Rather, mutation makes some members of a species play a different strategy, and if this strategy is more advantageous than that played by other members of the species, we expect it to grow in frequency as the mutants reproduce faster.

In human economics, we can replace mutation with "trial-and-error." Individuals with limited cognitive capacities—and let us face it, even the brightest of us have such limits—may opt for trying new strategies in the hope of improving their lot without much preceding calculation.

In addition, we can replace selection with imitation. Suppose a firm tries a new strategy. It turns out to be wildly profitable. Then it is quite reasonable to expect other firms to seek to adopt the same strategy. The assembly-line production process introduced by Henry Ford in 1913 generated extraordinary profits for Ford, and it was soon copied all over the world. The same reasoning applies to individuals: we mimetically adopt behavior that is advantageous to us.[6]

Evolutionary game theory can also shed light on which equilibrium is likely to be selected when there are multiple equilibria, and under what conditions a mutation is *not* likely to succeed. A strategy that would be far more profitable if everybody adopted it might still not be adopted if it is individually disadvantageous when played by only one or a few individuals. Social norms are examples of such equilibria: a wasteful or destructive social norm can keep a community in poverty because it is too costly for one or a few individuals to deviate from it. But if the whole community could be made to abandon the norm, it might benefit everybody.

An Evolutionary Perspective on Economic Games

One common explanation for behavioral findings in anonymous one-shot games is that the observed behaviors reflect the preferences of some non-negligible share of the population who fully understand, both cognitively and emotionally, the anonymous one-shot setup.

Another view on the experimental results, which comes mainly from evolutionary theorists, is that cognitive constraints are important determinants of the observed behavior. According to this view, we do not fully comprehend anonymous one-shot games because humans have not evolved in settings where this is a meaningful notion (Hagen and Hammerstein 2006). Unlike the rational, utility-maximizing brain of the economist, the biologist's brain consists of different specialized mechanisms that do not always work optimally together. Cognitive constraints are precisely that: *constraints* on optimization. Economists too are increasingly thinking along these lines and have sought to develop game theoretical models of decision making featuring individuals that are *boundedly rational* (Simon 1957). Such individuals seek to maximize their utility but face limits on their ability to identify the optimal strategy.

From this perspective repeated games are the most natural setting for experiments. Analyzing one-shot experiments from this perspective may shed more light on observed behavior. In particular, the repeated-game paradigm allows for an easier (evolutionary) rationalization of observed cooperative play. If our instincts have evolved in small-scale societies with little chance of anonymous interactions, where interactions were repeated with high probability and where cooperation was hence a rational "choice," we might not be well equipped to understand the notion of a one-shot game.

In line with this reasoning, experiments suggest that we may not have evolved in a setting with anonymous encounters, and that we have difficulties "switching off" this evolved reasoning. We continue to react to subtle cues that either purposely or accidentally inform us about the possibility of reputational effects. For example, dictators (in dictator games) were found to be more generous when the experimental instructions were adorned with drawings that vaguely resembled eyes (Haley and Fessler 2005). Showing the dictators photos of the recipients or vice versa had similar effects (Burnham 2003). The presence of a robot with humanoid eyeballs leads to increased contributions in a public goods game (Burnham and Hare 2007). These experiments show that seemingly irrelevant, subtle, personalized cues make people play more generously or cooperatively. It is argued that this is evidence of a reputation mechanism. This mechanism may be hard to ignore: even when the subtle cues are not introduced on purpose, subjects may seek them through, for example, eye contact with the experimenter.

We need to emphasize that we do not question the finding that many people really choose to give generously in dictator games, or reject low offers in ultimatum games, and that they do so because they feel this is the right thing to do. The question is why they would have evolved to feel that this is the right thing to do?

Biological and Cultural Evolution

The economic experiments described in this chapter suggest that humans have a strong propensity to cooperate in economic games with repeated and potentially indefinite interactions. Has this always been the case? One way of answering that question is to conduct similar experiments in traditional, small-scale societies. There are a number of such studies, using dictator games, ultimatum games, public goods games, and other games.

Existing studies show that individuals in traditional societies rarely act as predicted by a standard economic model in which players simply maximize their own material payoff. There are, however, large differences between these cultures with regard to how they deviate from the predicted behavior.

Moreover, these studies point to an apparent connection between the economic organization of traditional societies and the observed behavior in experiments. Cultures with more cooperative economic activities, such as hunting in groups, show higher rates of cooperation in the experiment, as do cultures with a higher degree of market integration (a common language and cash markets for both labor and crops) (Henrich et al. 2001).

We should not conclude from this that individuals cooperate more in some societies *because* there is cooperative economic activity. The reverse causality is equally plausible: collective endeavors are more likely to emerge in societies where individuals have a high propensity to cooperate.

We can, however, conclude that deviations from the standard Western economic model seem to apply not only to student populations in Western countries (where most lab experiments are performed) but also to traditional societies across the globe. Despite a number of cultural differences, cooperation appears to be a nearly universal characteristic of humans. The experiments also show that experimental economics provides useful tools for anthropologists and others interested in cross-cultural differences.[7]

The Use of Experiments to Compare Ourselves with Nonhuman Primates

As we have shown, humans clearly show a strong propensity to cooperate, even in anonymous one-shot games. The research to date in experimental economics suggests that cooperation is manifest across cultures, including traditional hunter-gatherer societies.

Is this high degree of cooperation—such as kindness toward strangers—a defining *human* characteristic? In order to shed more light on this question, the toolkit from experimental economics has been applied to other (nonhuman) primates.

Cooperative behavior in nonhuman primates is largely limited to kin and familiar individuals (Silk 2005), and even then, generosity may be very limited: experimental evidence suggests that when chimpanzees *(Pan troglodytes)* can provide a benefit to another—familiar—chimpanzee, at no cost to themselves, they typically fail to make use of this opportunity.

In one set-up, chimpanzees could choose between two outcomes: either (1) receive a reward, or (2) receive exactly the same reward *and* costlessly generate a reward for another chimpanzee. Chimpanzees, it turned out, are indifferent between these two options. In fact, not a single one of the chimpanzees in the experiment was more likely to choose the second, cooperative option (Silk et al. 2005). Another study explores to what extent chimpanzees are altruistic, spiteful, or willing to help another individual receive a benefit at the same time as they will get it (Jensen et al. 2006; Jensen, Call, and Tomasello 2007a). Across treatments, the salient finding is that chimpanzees care only about their own personal gain, though they are willing to seek revenge by punishing direct theft. Chimpanzees also appear to be "rational maximizers" in the ultimatum game: they make very low offers as proposers, and accept very low offers as responders (Jensen, Call, and Tomasello 2007b). Our distant relatives, it seems, are somewhat similar to the selfish and rational beings characterized by the standard economic model of behavior.

Stronger indications of cooperation have been reported among common marmoset monkeys *(Callithrix jacchus)*. These monkeys—which are not primates—provide food spontaneously to nonreciprocating and genetically unrelated individuals (Burkart et al. 2007). Marmosets are more distant relatives of humans than are chimpanzees, but they are similar to humans in the sense that both species are cooperative breeders.[8] Thus, it could be that social preferences evolved after the split with our closest relatives—the chimpanzees—and that cooperative breeding, implying both high levels of mutual interdependence and social tolerance, may explain why we observe prosocial behavior among humans and marmosets but not among chimpanzees (Burkart et al. 2007).

A large component of our cooperativeness is likely to be biological. This view is supported by recent twin studies (Wallace et al. 2007; Cesarini et al. 2008). But what makes us more cooperative than our nonhuman primate relatives may also be due to the cultural evolution of social norms that promote cooperation. Such norms might have evolved in part thanks to the welfare gains (the potentially large social benefits) from mitigating social dilemmas. By exploring how the social preferences of nonhuman primates and monkeys differ from those of humans, we can further our understanding of how these preferences originated and shed more light on the role of cultural evolution in shaping human societies.

Concluding Remarks

In this chapter we have attempted to present a snapshot of cooperation in economic games. The experimental research in this area shows that cooperation is frequent even when this is not an equilibrium strategy in terms of the material payoffs of the experimental game.

We wish to emphasize the following three points in closing. First, these findings do not contradict game theory, *tout court.* Second, these findings constitute an improvement of our understanding of economics, and not a rejection of economic theory. Third, research on evolution—theoretical and empirical—is an important means to help us understand what motivates the behavior that we observe today. The findings from experimental economics have offered a wealth of possible refinements to economic models that can help economists design better models with the hope of making better predictions about human behavior.

These simple games offer deep insights into human decision making. Using these games and incorporating this improved understanding of behavior is now an ongoing project in both the social and the natural sciences.[9]

Notes

1. Utility, in the sense used here, is simply the psychological measure that individuals seek to maximize. The standard economics framework does not require or offer a closer characterization of this utility but merely treats it as a way of ranking the relative desirability of different outcomes. It is not observable, but under certain conditions it can be inferred from observable actions. If an individual can afford both X and Y but chooses X, economists typically infer that the individual, at that point in time, gets a larger incremental increase in utility from X than from Y.
2. While the average amount transferred is high, there are large differences between individuals. Typically, about half of the participants give nothing.
3. This game was originally referred to as the ultimatum bargaining game.

4. In economics, a repeated version of this framework is used to study the problem of *preventing* cooperation. In many countries, it is illegal for firms to collude when setting their prices, since this reduces competition and thereby harms consumers. When colluding firms can be fined for colluding but also rewarded if they inform authorities about the firms they have been colluding with, a PD arises.
5. The two game theoretical approaches are more closely related that it might seem at first sight. The population interpretation was first suggested by Nash himself in his PhD thesis. The concept of an evolutionary stable strategy was formalized by Maynard Smith and Price (1973), who also showed the equivalence between the Nash equilibrium of standard game theory and the ESS of evolutionary game theory. Evolutionary game theory was originally concerned with biological evolution. Beginning with Axelrod (1984), this framework has also been applied to the process of cultural evolution. For a broader discussion of cultural evolution, see, for example, Cavalli-Sforza and Feldman (1973), Dawkins (1976), or Boyd and Richerson (1988).
6. For a technical discussion on how to apply evolutionary game theory to social imitation, see Weibull, 1995, Chapter 4.
7. This has also been pointed out in, for example, Camerer and Fehr (2004).
8. Cooperative breeding is a social system where the young are cared for also by individuals who are not their parents.
9. We are grateful for excellent comments from Magnus Johannesson, David G. Rand, and Zachary Simpson.

References

Axelrod, R. 1984. *The Evolution of Cooperation.* New York: Basic Books.

Boyd, R., and P. J. Richerson. 1988. "An Evolutionary Model of Social Learning: The Effects of Spatial and Temporal Variation." In *Social Learning: A Psychological and Biological Approach,* ed. Thomas Zentall and Bennett G. Galef, Jr. New Jersey: Lawrence Erlbaum Associates, 29–48.

Burkart, J. M., E. Fehr, C. Efferson, and C. P. van Schaik. 2007. "Other-Regarding Preferences in a Non-Human Primate: Common Marmosets Provision Food Altruistically." *Proceedings of the National Academy of Sciences of the USA* 104(50): 19762–66.

Burnham, T. C. 2003. "Engineering altruism: A theoretical and Experimental Investigation of Anonymity and Gift Giving." *Journal of Economic Behavior & Organization* 50(1): 133–44.

Burnham, T. C., and B. Hare. 2007. "Engineering Cooperation: Does Involuntary Neural Activation Increase Public Goods Contributions?" *Human Nature* 18(2): 88–108.

Camerer, C. 2003. *Behavioral Game Theory,* Princeton, NJ: Princeton University Press.

Camerer, C., and E. Fehr. 2004. "Measuring Social Norms and Preferences Using Experimental Games: A Guide for Social Scientists." In *Foundations of Human Sociality: Economic Experiments and Ethnographic Evidence from Fifteen Small-Scale Societies,* ed. Joseph Henrich, Robert Boyd, Samuel Bowles, Colin Camerer, Ernst Fehr, and Herbert Gintis. Oxford: Oxford University Press, 55–95.

Cavalli-Sforza, L. L., and M. W. Feldman. 1973. "Cultural versus Biological Inheritance: Phenotypic Transmission from Parent to Children (A Theory of the Effect of Parental Phenotypes on Children's Phenotype)." *American Journal of Human Genetics* 25: 618–37.

Cesarini, D., C. Dawes, J. Fowler, M. Johannesson, P. Lichtenstein, and B. Wallace. 2008. "Heritability of Cooperative Behavior in the Trust Game." *Proceedings of the National Academy of Sciences of the USA* 105: 3271–76.

Dal Bó, P. 2005. "Cooperation under the Shadow of the Future: Experimental Evidence from Infinitely Repeated Games." *American Economic Review* 95(5): 1591–1604.

Dawkins, R. 1976. *The Selfish Gene.* Oxford: Oxford University Press.

Güth, W., R. Schmittberger, and B. Schwarze. 1982. "An Experimental Analysis of Ultimatum Bargaining." *Journal of Economic Behavior and Organization* 3: 367–88.

Hagen, E. H., and P. Hammerstein. 2006. "Game Theory and Human Evolution: A Critique of Some Recent Interpretations of Experimental Games." *Theoretical Population Biology* 69: 339–48.

Haley, K. J., and D. M. T. Fessler. 2005. "Nobody's Watching? Subtle Cues Affect Generosity in an Anonymous Economic Game." *Evolution and Human Behavior* 26: 245–56.

Henrich, J., R. T. Boyd, S. Bowles, C. Camerer, H. Gintis, R. McElreath, and E. Fehr. 2001. "In Search of Homo Economicus: Experiments in 15 Small-Scale Societies." *American Economic Review* 91(2): 73–79.

Jensen, K., J. Call, and M. Tomasello. 2007a. "Chimpanzees are Vengeful but not Spiteful." *Proceedings of the National Academy of Sciences of the USA* 104: 13046–51.

———. 2007b. "Chimpanzees Are Rational Maximizers in an Ultimatum Game." *Science* 318: 107–9.

Jensen, K., B. Hare, J. Call, and M. Tomasello. 2006. "Are Chimpanzees Spiteful or Altruistic When Sharing Food?" *Proceedings of the Royal Society B.* 273: 1013–21.

Kahneman, D., J. Knetsch, and R. H. Thaler. 1986. "Fairness and the Assumptions of Economics." *Journal of Business* 59(4): 285–300.

Maynard Smith, J., and G. R. Price. 1973. "The Logic of Animal Conflict." *Nature* 246, 15–18.

Nash, John F. 1950 "Equilibrium Points in *n*-Person Games." *Proceedings of the National Academy of Sciences of the USA* 36:48–49.

Silk, J. B. 2005. "The Evolution of Cooperation in Primate Groups." In *Moral Sentiments and Material Interests: The Foundations of Cooperation in Economic Life,* ed. Herbert Gintis, Samuel Bowles, Robert Boyd, and Ernst Fehr. Cambridge, MA: The MIT Press, 43–74.

Silk, J. B., S. F. Brosnan, J. Vonk, J. Henrich, D. J. Povinelli, A. S. Richardson, S. P. Lambeth, J. Mascaro, and S. J. Schapiro. 2005. "Chimpanzees are Indifferent to the Welfare of Unrelated Group Members." *Nature* 437: 1357–59.

Simon, H. 1957. *Models of Man, Social and Rational: Mathematical Essays on Rational Human Behavior in a Social Setting.* New York: John Wiley and Sons.

von Neumann, J., and O. Morgenstern. 1944. *Theory of Games and Economic Behavior.* Princeton, NJ: Princeton University Press.

Wallace, B., D. Cesarini, P. Lichtenstein, and M. Johannesson. 2007. "Heritability of Ultimatum Game Responder Behavior." *Proceedings of the National Academy of Sciences of the USA* 104: 15631–34.

Weibull, J. W. 1995. *Evolutionary Game Theory.* Cambridge, MA: The MIT Press.

III

Psychology, Neuroscience, and Intentionality in the Cultural Evolution of Cooperation

7

Social Prosthetic Systems and Human Motivation

One Reason Why Cooperation Is Fundamentally Human

Stephen M. Kosslyn

One of the central questions in psychology is why some things motivate us whereas others do not. In this brief chapter I consider a uniquely human motivation and point out the direct implications of this idea for theories of cooperation. In keeping with the goals of this volume, I will characterize "cooperation" as "a form of working together in which one individual pays a cost (in terms of fitness, whether genetic or cultural) and another gains a benefit." My argument leads me to suggest that at least some forms of cooperation are inextricably linked to a major source of human motivation.

The study of motivation has a long and illustrious history in psychology, but I want to argue that previous theories have overlooked a fundamental type of human motivation. This type of motivation grows out of the fact that the human brain has limitations, and that none of us is equipped to solve all tasks entirely on our own. In the first section below I explore the idea that we humans rely on the environment to help us compensate for our limitations. Following this, I suggest that other people are a key part of the environment that we rely on for help and review the idea of social prosthetic systems (Kosslyn 2006). I then contrast this type of relationship with other sorts of relationships. Finally, I point out the implications of social prosthetic systems both for the nature of motivation and of cooperation.

Environmental Crutches

A recent study showed that a young chimp can store information in short-term memory much more effectively than humans (Inoue and Matsuzawa 2007). Aside from the blow to our pride-of-place in the animal kingdom, this finding underscores an important fact about the human mind: it has severe limitations. In fact, we are strikingly limited not only in our memory abilities (e.g., Chua, Rand-Giovannetti, and Schacter 2004), but also in our perceptual (Coulter and Coulter 2007; Gibb 2007) and reasoning abilities (Tversky and Kahneman 1983). For example, we cannot memorize a newspaper after a single reading, cannot hear the rustle of a butterfly's wings, cannot always reason logically without (or even with!) effort, and so forth. Given our obvious limitations in other domains (we cannot leap over buildings in a single bound), such cognitive limitations— which may ultimately grow out of fundamental limitations in how biological mechanisms operate—should not be surprising. To be human is to be limited, and our limitations are pervasive.

The fact of our limitations leads us to engineer the environment in order to extend the capacities of our brains. For example, let us consider the nature of crosswalks. Do you remember the cover of the Beatles' *Abbey Road* album? The four of them were walking across a street on a zebra-striped crosswalk. The first time I ever saw a British crosswalk like that was on this album cover. I remember noticing it, because at that time in Los Angeles, California, crosswalks looked like a pair of lines that stretched across the street, not a set of dashes that were parallel to the curb. However, the zebra-stripe kind of crosswalk is objectively better for human beings, given the way that our perceptual systems work. From the point of view of drivers in cars there is a greater reflective surface, and thus it is easier for them to see the corridor for pedestrians, and from the point of view of pedestrians, there is more material on the road to tell us where it's safe to cross.

An effective crosswalk is not a trivial matter, as you may know if you have lived in Boston; many of us cross streets at random locations, and this is not always entirely safe given the limitations of human judgment. The crosswalk is a kind of crutch, something in the world to help us overcome some of our limitations in judgment.

But a well-constructed crosswalk is more than just a well-marked corridor across the street. Go back and look at the distance between the legs of each Beatle in that photo. The length of their strides almost perfectly mirrors the distance between the stripes. This is not an accident. For example, consider an illuminating finding with patients who have Parkinson's disease. This disease depletes the levels of the neurotransmitter dopamine in the frontal lobes, which play a key role in organizing and directing voluntary movements. The dopamine deficit in the frontal lobes leads these patients to have difficulty moving smoothly. Indeed, these patients typically shuffle when they walk. However, such problems disappear if the patients are asked to step over a set of two-by-fours, placed like railroad ties in front of them. They may begin by walking with their usual shuffle, but as soon as they begin to step over the boards, they walk absolutely smoothly. Without the boards, they cannot walk well, but when they have to step over the boards, all of a sudden they look normal. What is going on? A different part of the brain, the cerebellum (which is concerned, in large part, with motor control and timing), operates to help them go over the obstacles automatically. The cerebellum, unlike the frontal lobes, does not rely on the depleted neurotransmitter dopamine, and hence is not affected by the disease.

Returning to the Beatles, they apparently were unconsciously calibrating their strides to correspond to the distances between the stripes. Notice what happens next time you walk across such a crosswalk. I seriously doubt that the designers of the zebra-stripe crosswalk knew much about the cerebellum and how it leads us to walk automatically, but by making the stripes that wide and positioned in that particular way, the crosswalk automatically engages the brain (specifically, the cerebellum) to encourage us to hustle across the street. The world is actually helping us in situations in which we might exhibit poor judgment! Such crosswalks not only tell us where to cross but also help us to accomplish the task most effectively.

In many cases, we have devised devices to fill in for our human limitations. For example, let us say that I ask you to multiply two six-digit numbers together in your head, such as 123456 times 654321. Not so easy, is it? You would probably prefer to have a pad of paper and a pencil, or better yet, a calculator. These are "cognitive crutches," which you can lean on when your mental legs just are not strong enough. We have limitations on how much we can hold in our short-term memories, and how long we can work on that information. A calculator is a kind of prosthetic device that fills in for our cognitive limitations. Some theorists, notably Clark and Chalmers (1998) and Clark (2003), have taken such ideas to the point where they suggest that each of us is, in fact, a cyborg, a biological entity that is intimately joined with inanimate devices.

I have extended this line of reasoning one step further, arguing that a major cognitive prosthetic we rely on is other people, as I summarize in the following section.

Social Prosthetic Systems

According to the theory of social prosthetic systems, we humans rely on other humans as extensions of ourselves, which allows us to overcome some cognitive and emotional limitations. First, let me be clear on what I mean by the term "prosthesis." I'll begin by noting what I do not mean, and then will explain what I do mean. Often people think that a prosthesis is a replacement for something a person once had. So, for example, if you had to have part of a leg removed, you would receive a replacement, the contemporary metal-and-plastic version of a wooden leg. That is a prosthesis. Well, what about thalidomide babies, who are born with truncated limbs? We can still give them prostheses. In this case, the prosthesis clearly is not a replacement for something they once had, and yet the term fits well. A prosthesis is not simply a replacement for a missing part but, rather, fills in for something that the person needs, given a particular set of challenges. Even though those babies never had complete limbs in the first place, to function well in the world like other members of the species, they need limbs. In the present context, I will treat a prosthesis as an extension of ourselves that can help us overcome a lack in order to accomplish a particular task.

We rely on other people as prosthetics in many ways. We rely on other people to help us extend our intelligence. For example, we rely on other people as sounding boards to help us try things out. They provide feedback, noting what we did not realize. But more than this, other people help us to recognize and regulate our emotions. For instance, when you are feeling distraught, you may want to talk to somebody; if you choose the right person and she sees that you are getting agitated, she can help you recognize your state of mind and calm you down.

The idea is that when people serve as a prosthesis for you, they are helping you to accomplish a certain task, and in doing so they are literally lending you cognitive resources. They are lending you part of their brains, in two senses: first, they allow you to rely on their brains at a particular point in time, and second, in order to help you effectively, the person acting as your prosthesis must learn about you—she must learn how best to help you. By participating as a social prosthesis, she is proactively engaged in grappling with your problem, not simply waiting for questions or requests. In interacting with you, she must learn how best to apply her knowledge, abilities, and skills so that she can help you as effectively as possible. When a person augments your initial knowledge, abilities, and skills in this way, you become something more than you were at the outset. In short, when someone acts as your social prosthesis, she is cooperating with you, paying a cost in order to confer a benefit for you. In so doing, the self "emerges" from the interactions between two people.

In this view, the self is not just in the head but is extended into the world, particularly the social world. When someone lends you part of his or her brain to help you accomplish a task, the two of you form what I call "a social prosthetic system," or SPS. These systems are extensions of one person: it is you extending into the world, into other people. You engage other people so that they will help you. Your "self" depends on which SPS you are in; your "self" arises from interactions with other people, and any given particular version of your "self" is primarily suitable in a particular context. We develop a whole raft of SPSs, with different ones being relied upon for different purposes. We have somebody we may go to when we are sad, somebody we may go to when we need help dealing with a family crisis, somebody else if we have financial issues, and somebody else if we need help buying a new car. By analogy, think of these people as tools in a toolbox, where we have a saw for cutting, a hammer for pounding nails, a tape measure, a screwdriver, and so on.

A key notion here is that we are actually different people when we are functioning within our different SPSs. When you pick up that screwdriver,

Table 7.1 Characteristics of social prosthetic systems (SPS) and similar relationships and concepts

	SPS	Reciprocal Altruism	Friend	Team Member	Social Network	Social Capital	Colleague	Employee	Distributed Cognition	Social Support
Extended Self	X									
Anticipatorily Assists	X		X	X			X			X
Endogenous Motivation	X	X	X		X	X			X	X
Exogenous Motivation	X			X	X	X	X	X	X	X
Context-Dependent	X			X			X	X		X
Task Specific	X			X			X			
Agenda Setter	X							X		
Extend Intelligence	X								X	X
Regulate Emotions	X	X								X
Typically Dyad	X	X	X					X		X
Provide Company		X	X							X
Shared Goal				X			X	X	X	
Roles Assigned				X				X	X	
Simultaneous Payoff				X			X		X	
Transcends Individuals				X	X	X			X	
No. links Crucial					X	X				
Shared Environment				X			X	X	X	X
Shared Mission				X			X	X	X	X
Divided Responsibilities				X				X	X	X
Affected by Genetic Relatedness		X								
Balanced Interests		X	X	X			X		X	

you have a different function in the world than when you heft that saw–and when you are interacting with somebody who knows you well enough to help you calm down, that is different than when you are interacting with somebody who knows when you need help in organizing your work week. Let me stress that this is not just a one-way relationship, in which you exploit other people; an SPS is not like a wooden leg that hangs off your stump: it is not a passive device that you use. It is a system. The other member of the SPS is an active participant who anticipates and behaves proactively. Such a person is "part of you."

It is worth being more explicit about what an SPS is so that I can turn to the next two sections of this chapter—which use the theory to discuss motivation—and then turn to cooperation per se. Thus, consider Table 7.1. Notice first two defining characteristics of SPSs not shared by any of the other relationships. First, an SPS is an extended self: when you interact with one person who functions as your social prosthesis, you are a different person than when you interact with another person who functions as your social prosthesis in a different domain. Second, SPSs are context-dependent: depending on which social prosthesis you are with, the system functions differently. For example, if you are with someone who helps you recognize and regulate your emotions, you function differently than if you are with someone who helps you figure out how to organize your weekend plans. Now let us note how SPSs differ from other relationships.

Reciprocal altruism specifies that you will give back to others in exchange for their helping you. Notice that simply keeping someone company can trigger reciprocal altruism, whereas SPSs are always defined relative to accomplishing a specific task. Reciprocal altruism includes many situations in which goods or services are provided but in which one is not "extending the self" or "lending" cognitive or emotional resources. For example, you might lend a friend a small amount of money (e.g., to pay for lunch), with the expectation that not only would she pay you back but also that you could ask her to make a loan to you at some point in the future. Also notice that both genetic relatedness and balancing interests are important for reciprocal altruism, but not SPSs.

Friendships are not task-specific, in sharp contrast to SPSs. You might comfort a friend who is upset because she broke up with her boyfriend (by hugging her and saying, "it is going to be okay"), but this is not the same as functioning in a SPS with that person—on the other hand, helping her analyze her relationship patterns would imply that a SPS has been established. Moreover, interests are balanced with friends; if either one is receiving too much from the other, the relationship starts to feel exploitative.

A *team member* is distinct from a SPS in numerous ways, as is evident in the table. Of particular note is that all members of the team share the same goal, roles are assigned, and all members of the team at the same time receive a payoff from achieving the goal. None of these are true of SPSs; in fact, a social prosthesis may not even be aware of your goal, and may not receive a benefit until much later, well after a service was provided.

A *social network* is a structure that specifies particular types of links (representing relations such as friendship, trade, or transmission of disease) between individuals or organizations. Social networks involve more than two people, whereas SPSs typically involve only two people at a time. Social networks transcend individuals: a network can remain the same even when some particular people depart and others join. In addition, the number of links among members is crucial in social networks (in fact, the study of social networks often focuses on the effects of the density or distribution of links), whereas this is not a factor in SPSs. Furthermore, little research has addressed the individual differences that would lead a particular person to evolve a certain number or strength of ties in a network. However, that said, a SPS can, in principle, form an elementary component of a broader social network. A SPS is not the same thing as a social network, but they can be related.

Social capital rests on the idea that "social networks have value. Just as a screwdriver (physical capital) or a college education (human capital) can increase productivity (both individual and collective), so too, social contacts affect the productivity of individuals and groups" (Putnam 2000, 18). We would add that SPSs are another part of social capital. Social capital is the value of social networks, and as such shares the same characteristics as social networks in the table—and is distinguished from SPSs in the same ways.

Colleagues, as we use the term here, are coworkers with whom one regularly interacts. These people share goals, receive a payoff (or promise of payoff) at the same time when they work on the same project, have a shared environment (which is not necessarily a shared physical venue), shared mission, and balanced interests. None of these characteristics necessarily applies to SPSs.

Being an *employee* also requires one to share the same goals as the employer (almost by definition, given that the employer sets the goals), entails a specific assigned role, and typically involves a shared environment and mission with the employer. Moreover, an employee is given a piece of the project, which is allocated on the basis of a prior analysis of the task and what must be accomplished. This analysis leads to divided responsibilities being assigned, where the employee knows that he or she is in charge of accomplishing one aspect of the task. In contrast, a member of a SPS may have no idea of the agenda, and may not have any responsibilities per se.

Distributed cognition occurs when different aspects of a task are assigned to different people who work together to accomplish a shared goal. Unlike an SPS, distributed cognition does not require an agenda setter (the group can self-organize), nor is it task-specific.

Finally, *social support* does not require you to be the agenda setter; others can offer social support, even if you do not want it. Roles are clearly understood in a context in which one person is offering social support to another (and the support is likely to be comfort or "moral support" rather than the "loaning" of cognitive or emotional resources), and, notably, the understanding that one person cares about another is critical for social support, but not for SPSs.

So much for what an SPS is not. Now let me reiterate what an SPS is: At its core, it is an extended self that emerges from the interactions of the agenda setter and the person serving as the social prosthesis. Depending on which social prosthesis you interact with, you are—at least in part—a different person.

Any given person can participate in more than one kind of relationship, and hence the various relationships are not mutually exclusive. However, because the relationships are different, there may be tension if the same person often shifts from one role to another. For example, the requirements of a team member (e.g., balanced interests) are not the same as those for being in a SPS, and thus intermittently relying on a fellow team member as an SPS may require a delicate balancing act—with the roles being clearly demarcated.

A Uniquely Human Motivation

This theory of social prosthetic systems leads me to characterize a uniquely human motivation. This motivation is unlike hunger, thirst, sex, and the other kinds of motivations that are shared by other species. Before characterizing the motivation, I must set the stage by asking a simple question: "Why would other people lend you part of their brains?" Consider three possible reasons.

First, they may expect reciprocity in the future: if you have abilities or skills that they perceive as useful, they may agree to help you with the promise (often implicit) that when the time comes, you will help them. This is a version of the old adage, you scratch my back now, I shall agree to scratch yours later—but we are talking about more nuanced and challenging cognitive or emotional contributions, not literally back scratching.

Second, once they have learned how to help you, people serving as social prostheses have new skills that they can use toward their own ends. For example, consider executive assistants, who typically work extraordinarily hard for their bosses. Such assistants start off simply as employees but often evolve into social prostheses as well—they become part of a system, they are actively engaged. To be an effective social prosthesis, the assistants need to learn about their employer (such as his or her needs, goals, and strategies). One reason they may be motivated to do so is that they get to use the acquired knowledge, skills, and abilities for their own ends. An anecdote might help to illustrate this concept. I know an executive who has a team of assistants. When he first heard this hypothesis, he paused for a long moment and then mused that this observation might explain why he has to let go of his assistants after ten years or so, the time typically required to become an expert in a domain. From the point of view of the theory, as they incorporated more of him into their brains, the boundaries would tend to get blurred over time, and the assistants would start getting confused about what was in their purview and what was not. This is an inevitable result of this sort of SPS relationship, where the prosthesis is strongly motivated to learn as much as possible from the agenda setter.

Third, if you are perceived as competent in some domain, there may be a measure of "reflected glory" in being associated with you, and that reflected glory in turn makes the person who functions as your prosthesis more attractive to other people, who in turn are more likely to be willing to serve in his or her SPSs. Indeed, the fact that you rely on certain people indicates that you find them competent in some regard, which contributes to their reputation (which in turn makes them more valuable to others, which again increases the likelihood that others will be willing to serve as their SPSs).

I am sure that there are other possible reasons, but focus on these three because they imply a uniquely human motivation: people are driven to become competent so that they have something to offer other people. In fact, I would go so far as to suggest that people find it pleasurable to acquire and exercise competence, even if it is in a domain that has no immediate relevance for their lives. This is not to say that we are conscious of this motivation; in fact, all we may be conscious of is that we enjoy acquiring and exercising competences. According to this theory, our brains are wired so that we find it fun to acquire and exercise competences because this allows us to become valuable or useful for other people—which then in turn motivates others to serve as our SPSs. The more accomplished you become, the more potentially valuable you are to other people. To return to my toolbox metaphor, by becoming competent, it is as if you become a really good saw that cuts with ease and precision, and if you have something to offer others, that motivates them to want to do things for you because they expect that you will help them in the future. For example, I find that I acquire seemingly random skills, such as learning to play bass guitar. Why would I learn to play bass guitar? It turns out there are a lot of people who play music, and there are not very many people who play bass guitar. So, this is a way to contribute something that is valuable to a certain kind of person, and it turns out that there are a lot of very interesting people my age who have been playing music since high school.

Another Sort of Cooperation

This theory implies a specific form of cooperation where people devote themselves to doing something for another person. That is, although cooperation implies that *"one individual pays a cost (in terms of fitness, whether genetic or cultural) and another gains a benefit,"* cooperation at one point in time may set up the cooperator to gain a benefit at a later point in time.[1] Although the focus is on grappling with your problem (you set the agenda), people who function as your social prostheses are active and willing partners—based in part on their notions about the potential gains they may garner subsequently.

In my view, the human brain evolved to be part of a social matrix, and we are wired to cooperate with each other to form SPSs. One part of the underlying mechanism may have been revealed by a recent discovery, although others have not interpreted it this way. Specifically, Rizzolatti and his colleagues (e.g., Rizzolatti 2005) discovered what they dubbed "mirror neurons." They were recording activity from single neurons in a monkey's frontal cortex. These neurons are involved in controlling the animal's movements. Strikingly, they noticed that the neurons were activated not simply when the animal moved, but also when it watched another person (or monkey) move in the same way. For example, if the neuron fired when the animal reached, it also fired when the animal saw another monkey reach. These neurons apparently have something to do with imitation, with allowing us to produce actions that we have seen others do. Moreover, some researchers have proposed that they play a key role in empathy, allowing us to feel as others do (Iacoboni 2007).

What is the connection between mirror neurons and SPSs? The link between the two may involve what psychologists call our "theory of mind" (e.g., Schülte-Ruther et al. 2007). A theory of mind is what allows us to infer what someone else is thinking and why they behave in specific ways. Researchers have proposed two classes of theories of how such a theory of mind is used. According to the "theory theory" (e.g., Matsumura 1994), people have derived a set of abstract principles that they store and use to generate predictions. This approach posits that a theory of mind is like a mathematical theory in which axioms and postulates are used to prove theorems. In contrast, the "simulation theory" (e.g., Gallese and Goldman 1998) posits that people imagine themselves "in the other person's shoes," and—perhaps via mental imagery—observe "what would happen" if they were in that person's situation. In fact, in tasks that require using a theory of mind, areas of the brain are activated that have previously been shown to be involved in visual mental imagery (Kosslyn and Thompson 2003).

If so, then mirror neurons may play a crucial role in helping us to construct mental simulations, which in turn allow us to anticipate whether someone else would function well in an SPS with us. That is, such simulations not only would allow us to identify people who might be able to help us, but also would allow us to anticipate how they would respond to overtures to them for help.

This point of view puts the mirror neurons, and the brain networks in which they are embedded, squarely in the business of helping to set up, rely on, and maintain SPSs. If so, then mirror neurons are—at least in part—about cooperation.

Conclusions

There are four key ideas in the preceding analysis of SPSs. One is that we rely on other people to complete us as necessary to perform a task. To take an extreme case, if you have Asperger's syndrome, you are sometimes going to need help from someone who has high emotional intelligence (in fact, this is probably true even if you do not have Asperger's!). An SPS is a system in which the people who function as your prostheses are willing, proactive participants, in spite of the fact their behavior leads them to pay a cost at the time they produce it. Depending on what the task is, you may need help from different people who have different skill sets; we have different SPSs that we call into play in different contexts. In fact, I have argued that we are actually different people when working with our different SPSs; we have essentially externalized ourselves into other people. So when we are with a person who acts as a social prosthesis, there is an emergent self that is created by interacting with him or her and that is different from the self that emerges when we're with a member of a different SPS.

The second idea answers the question: What is in it for them? Why does someone actively sign on to being in your SPS? By cooperating, he or she pays a cost at the time of his or her cooperation. The theory posits several possible reasons. Namely, if these people help you now, they expect you to help them in the future; they get to use the part of their brain that is "you" for themselves, even when not interacting with you; and they become more valued by the mere fact that you value them.

The third idea follows directly from the second. We are motivated to develop and exercise skills, abilities, and capacities that will attract others to serve in our SPSs.

The final idea is that our brains are specifically wired for cooperation in order to establish and use SPSs. We not only are predisposed to identify others who can help us but also to anticipate the best ways of interacting with others—which is crucial for setting up and maintaining SPSs.[2]

Notes

1. See Dreber's and Almenberg's account of repeated effects in reputational games in Chapter 6.
2. This chapter was improved by comments from, and conversations with, Debo Dutta, Jeffrey Epstein, Richard Hackman, Dominic D. P. Johnson, Robin Rosenberg, Alexandra Russell, Jennifer Shephard, and Anita Woolley. None of these people necessarily endorses what I have written.

References

Chua, E. F., E. Rand-Giovannetti, and D. L. Schacter. 2004. "Dissociating Confidence and Accuracy: Functional Magnetic Resonance Imaging Shows Origins of the Subjective Memory Experience." *Journal of Cognitive Neuroscience* 16: 1131–42.

Clark, A. 2003. *Natural-Born Cyborgs: Minds, Technologies, and the Future of Human Intelligence.* New York: Oxford University Press.

Clark, A., and David J. Chalmers. 1998. "The Extended Mind." *Analysis* 58: 10–23.

Coulter, K. S., and R. A. Coulter. 2007. "Distortion of Price Discount Perceptions: The Right Digit Effect." *Journal of Consumer Research* 34: 162–73.

Gallese, V., and A. Goldman. 1998. "Mirror Neurons and the Simulation Theory of Mind-reading." *Trends in Cognitive Sciences* 2: 493–501.

Gibb, R. 2007. "Visual Spatial Disorientation: Revisiting the Black Hole Illusion." *Aviation, Space, and Environmental Medicine* 78: 801–8.

Iacoboni, M. 2007. "Face to Face: The Neural Basis of Social Mirroring and Empathy." *Psychiatric Annals* 37: 236–41.

Inoue, S., and T. Matsuzawa. 2007. "Working Memory of Numerals in Chimpanzees." *Current Biology* 17: R1004–5.

Kosslyn, S. M. 2006. "On the Evolution of Human Motivation: The Role of Social Prosthetic Systems." In *Evolutionary Cognitive Neuroscience,* ed. Steven M. Platek, Todd K. Shackelford, and Julian P. Keenan. Cambridge, MA: The MIT Press, 541–54.

Kosslyn, S. M., and William L. Thompson. 2003. "When Is Early Visual Cortex Activated during Visual Mental Imagery?" *Psychological Bulletin,* 129: 723–46.

Matsumura, N. 1994. "Child's Theory of Mind and the Theory Theory: Arguments with the Simulation Theory." *Japanese Psychological Review* 37: 92–107.

Putnam, R. 2000. *Bowling Alone: The Collapse and Revival of American Community.* New York: Simon and Schuster.

Rizzolatti, G. 2005. "The Mirror Neuron System and Imitation." In *Perspectives on Imitation: From Neuroscience to Social Science,* ed. Susan Hurley and Nick Chater. Vol. 1 of *Mechanisms of Imitation and Imitation in Animals.* Cambridge, MA: The MIT Press, 55–76.

Schulte-Rüther, M., H. J. Markowitsch, G. R. Fink. 2007. "Mirror Neuron and Theory of Mind Mechanisms Involved in Face-to-Face Interactions: A Functional Magnetic Resonance Imaging Approach to Empathy." *Journal of Cognitive Neuroscience* 19: 1354–72.

Tversky, A., and D. Kahneman. 1983. "Extensional Versus Intuitive Reasoning: The Conjunction Fallacy in Probability Judgment." *Psychological Review* 90: 293–315.

8

The Uniqueness of Human Cooperation

Cognition, Cooperation, and Religion

Dominic D. P. Johnson

Game theory examines the evolution of cooperation using imaginary "agents" with simple decision rules. Consequently, the outcomes and predictions may apply to any kind of agent, whether human, animal, or toy robot. This approach is powerful because it can reveal the emergence of complex behavior from simple starting conditions. But it is also potentially misleading because we know that human brains are vastly more complex than toy robots. Therefore, we should not necessarily expect the predictions of game theory to map neatly onto human behavior.

Here is an example. A behavior that looks like "forgiveness" emerges in computer agents who play iterated prisoner's dilemmas (Axelrod 1984; Nowak 2006). Agents that forgive (or, from a decision-making standpoint, simply forget!) that a partner exploited them in the past outcompete unforgiving agents, because the latter shirk cooperation forever while forgivers are able to reap great benefits once mutual cooperation is reestablished. At one level, this is inspiring: forgiveness is a basic building block of social interaction that may stretch back to the origins of life. At another level, however, it is somewhat bizarre, because it means there may be nothing special about human forgiveness (see McCullough 2008 for an excellent exploration of human forgiveness).

Human behavior may, empirically, appear to match the behavior of computerized agents. But this does not mean they arise from the same proximate or ultimate mechanism (indeed, one can argue this is rather unlikely given the complexity of human brains and the simplicity of automata). Therefore, we must consider the possibility that the predictions for human cooperation emerging from game theory may be: (1) wrong; (2) correct, but for the wrong reasons; or (3) at the least missing some key elements of cognition and behavior unique to humans (and impossible in animals or toy robots).

In this chapter, I will focus on factors that make human cooperation different from all other species. I argue that there are two crucial characteristics of the brain that make human cooperation "unique": (1) a sophisticated theory of mind, and (2) the capacity for complex language. The evolution of these cognitive traits made selfish behavior more costly than at any previous point in our evolutionary history, because theory of mind and language increase both the *probability* and the *severity* of negative consequences for selfish actions. Natural selection would therefore have favored mechanisms that reduced overt selfishness. This moment in evolutionary history may have turned humans from relatively selfish beings into relatively cooperative ones, pushing cooperation to a significantly higher level than it had been at any previous stage of evolution. Not coincidentally, this argument suggests a key role for indirect reciprocity in the evolution of human cooperation (Trivers 1971; Alexander 1987; Nowak and Sigmund 2005), and one that operates on humans very much more powerfully than it could on other animals.

While I argue that game theoretical models fall short in omitting key aspects of human cognition, I also argue that unique features of human brains—such as theory of mind and complex language—could, in principle, be built into game theoretical models. Doing so may lead to novel insights and hypotheses. Even apparently nonmaterial preferences and outcomes, such as religious beliefs, can be modeled mathematically and, as long as the *consequences* in the payoff matrix are based on real-world costs and benefits, such a model should meet the standards even of skeptical scientific atheists. Rather than the more typical bottom-up approach to game theory, which starts with very simple decision rules, my suggested approach starts by including quite complex cognitive capacities and asking what it predicts about the success of alternative strategies in competition with each other. Although they are complex, the advantage is that they represent what we know *is* empirically true about human cognitive capacities, and we can explore what behaviors emerge as a result. The paper concludes that human cooperation is qualitatively unique in the animal kingdom and reaches higher levels as a result.

The Supernatural Punishment Hypothesis

The field of evolutionary religious studies has proposed a diverse range of adaptive benefits of religion, spanning health, group commitment, solidarity, cooperation, collective action, reproduction, group survival, and many others (Wilson 2002; Sosis 2004; Bering 2006; Bulbulia et al. 2008). Here I focus on just one such adaptive theory of religion: the *supernatural punishment hypothesis* (Johnson and Kruger 2004; Bering and Johnson 2005; Johnson 2005, 2009, 2011; Johnson and Bering 2006; Schloss and Murray 2011). Simply put, the argument is that a belief in *supernatural* punishment is adaptive because it helps to avoid the *real-world* costs of selfish actions (e.g. retaliation, rejection, or punishment from victims or other members of one's social group). Drawing on interdisciplinary research from evolutionary biology, psychology, neuroscience, and anthropology, the supernatural punishment hypothesis offers a number of novel implications for the evolution of cooperation and religion. The logic is set out below.

Selfishness as Maladaptive

Ever since the origins of life on Earth, selfish genes tended to translate into selfish behavior. Of course, there are complications and caveats to this—Darwin himself noted widespread instances of self-sacrifice in nature—but a century of research has shown that such apparent self-sacrifice can simply be selfish genes' way of propagating themselves more effectively, through mechanisms such as kin selection, reciprocity, or signaling (Wilson 1975; Dawkins 1976; Dugatkin 1997). After 3.5 billion years of evolutionary history, however, humans ran into a significant danger created by our own selfish behavior.

Everything changed for us the moment that we evolved two of our unique cognitive features: (1) a sophisticated "theory of mind," and (2) complex language (Bering 2002; Bering and Shackelford 2004). "Theory of mind"—as used in psychology rather than philosophy—is our ability to imagine what other people are thinking, to know that they know things, and to recognize their intentions. No other species on the planet have these two cognitive abilities. Chimpanzees and some other high-intelligence mammals may have a rudimentary theory of mind, but its level of sophistication is extremely basic compared to that of humans (Povinelli and Bering 2002). Moreover, although nearly all animals have some form of "language," broadly defined, they have nothing like our *complex* language that is able to transmit abstract thoughts and ideas (Pinker 2002). As will become clear, these qualifications are interesting but irrelevant to the supernatural punishment hypothesis.

The reason theory of mind and language are so important here is because of their colossal implications for *selfishness*. Suppose Tom has sex with Harry's wife while Harry is away on business. If they were chimpanzees, Harry will never know—Tom's selfishness can pay rich evolutionary dividends. By contrast, if they are both humans, Harry is inordinately more likely to find out, one way or another. He or someone else is likely to figure out or learn what happened. Harry can also exploit language to gain information, recruit allies, and seek revenge. Adultery is far more risky for modern humans than it was for our ancestors. The same logic applies to numerous other self-interested behaviors.

The implications are vast. Evolutionarily ancient self-interested desires such as sex, hunger, and dominance were fashioned by natural selection to lead an organism to maximize reproductive fitness, and these traits will spread at the expense of the chaste, restrained, or meek (that is exactly why selfish desires evolved in the first place—it is evolution's way of making organisms do whatever behaviors maximize their genetic replication in future generations). However, these selfish drives present a special problem for us humans because they can lead us into enormous trouble in our socially transparent society—a society brimming with language and theory of mind. And they do so every day. Pick up any newspaper and you will find plentiful examples of scandal, greed, and competition for power and people paying the consequences.

There was, therefore, a defining moment at some point in our evolutionary history where the entire cost-benefit calculus of selfishness was more or less thrown out of the window (Bering and Shackelford 2004). Evolution no longer operated on us in the same way it did on other animals. The best way to illustrate this is with a snapshot comparison of life before and after the advent of our "big-brother" brains.

Before the evolution of theory of mind and complex language, selfish behavior would be consistently favored by natural selection as long as it brought net fitness benefits (even when these behaviors occurred in full view of other individuals, since they could not tell anyone else). For example, chimpanzees can rape and steal in front of other chimpanzees without their behavior being discovered by or reported to absent others—such as the alpha male of the group, or relatives of the victim. There can therefore be *no negative repercussions from absent third parties.* Absent individuals could not entertain others' knowledge states, nor could they learn such complex information by communication. Sherlock Holmes would have little to do in chimp society: he would have no interviewees, and he would have no concept of witnesses or motive.

After the evolution of theory of mind and complex language, by contrast, it was now in our genes' interests to *avoid* selfish behavior in many contexts that might bring negative repercussions for fitness. Now—for the first time in evolutionary history—one had to worry about the consequences of other actors, wholly removed from the scene of the "crime," learning of the act and reacting later. People could hear about, discover, infer, remember, report, hypothesize, plan, and act on others' behavior—even long after the event. The risk of punishment for selfish behavior therefore became both *more likely* (because selfish acts were more likely to be discovered), and *more severe* (because selfish acts were more likely to be punished by groups of people rather than just individuals). Group punishment thus became especially cheap for the punishers and especially costly for the perpetrator.

There is much we do not know about our distant evolutionary past, which often leads to speculation. But there is something that we do know happened at some point or other for certain: as soon as we evolved theory of mind and complex language, *selfishness became significantly more costly than at any time in evolutionary history.*

Evolutionary Adaptations to the New Costs of Selfishness

Evolution responds rapidly to stamp out costly behavior, so we would expect natural selection to have favored corrective mechanisms that guarded against the discovery and punishment of selfish behavior. In our new socially transparent world, if I carried on behaving like a chimpanzee without regard for the fact that others might find out what I have been doing, while you modified your behavior to avoid the wrath of absent others, people like you would leave more descendents than people like me.

However, it is tricky to figure out what that corrective mechanism was. The problem is that simply "tuning down" our overall level of selfishness may not be effective, because evolutionarily ancient selfish behavior cannot be easily controlled. Basic desires such as sex, hunger, and dominance have been wired into our neural system over many millions of years, long before we developed into *Homo sapiens* (Barkow et al. 1992; Buss 2005). Indeed, physiological and psychological causes of these desires stem from the limbic system, an ancient part of the brain that we share with pigs, rats, and lizards. Cool-headed rational calculations that allow us to restrain selfish desires, by contrast, come from the neocortex—the modern part of the brain that has developed much more recently in evolutionary history. Although these brain areas are complex and interconnected, they tend to offer broadly conflicting recommendations for action, and it is hard to know which one will prevail, especially in heated or emotional situations (Damasio, 1994; Davidson, Putnam, and Larson, 2000; Sanfey et al., 2003; also see Lee, Chapter 9 in this book). Moreover, the limbic responses influencing hormones, brain chemistry, and behavior remain powerful and fitness enhancing (we rarely need reminding, for example, to have sex, eat, or defend ourselves). They are not only physiologically entrenched but are highly adaptive and sometimes life saving (or offspring promoting). We cannot simply switch these proximate mechanisms off whenever it is convenient. It is hard to do, and potentially counterproductive (Damasio 1994; Pinker 2002).

What we need is not something to tune down our (essential and adaptive) selfish desires. Instead, we need a mechanism that makes us more prudent about *when to follow* the recommendations of our selfish desires, and *when to suppress them*. It needs to be able to control passions and urges, and it needs to be sensitive to context. Whatever that mechanism is, it has obviously been partially successful—for example, we do not usually have sex in public places, stuff ourselves during job interviews, or attack our bosses when they insult us. The important point, however, is that there are often failures. There are many people languishing in prison for crimes of passion (Goldstein 2002), for example, and most of us can recall social gaffes where our limbic system rode roughshod over someone else's neocortical sensibilities. When humans acquired theory of mind and complex language, it thrust them into a social minefield. What adaptive solutions emerged as a result?

Religious Beliefs as an Adaptive Solution

One solution is religious beliefs. Religious beliefs offer a way to reform the cheating ape inside us so that we can avoid the retribution of the hairless and brainy apes around us. To anticipate my argument, those that developed a *fear of supernatural punishment* for moral transgressions would have experienced lower real-world costs than more indiscriminately selfish individuals who did not care what God (or some other supernatural entity) thought of their actions (Johnson and Bering 2006). Fearing supernatural retribution makes us consider every action especially carefully—*even* when we think we are alone. After all, God is always watching (this idea varies across religions, but some form of omniscience and omnipotence in relevant domains is a common theme; see Bering and Johnson 2005). While the average human, therefore, may have an evolutionarily ancient devil on one shoulder recommending selfishness and its various pleasures, we have a modern angel on our other shoulder: an adaptive inclination to fear the consequences of our actions. Evolution may have favored a belief in supernatural punishment as a mind guard that ensured prudence in order to navigate safely and successfully through our transparent social environment.

A growing body of research in the anthropology and cognitive psychology of religion strongly corroborates this view. For example, humans across cultures have an innate disposition to perceive intentional agency in the environment whether it is real or not. This is the so-called "hyperactive agency detector device" (Guthrie 1993; Boyer 2001; Atran 2004; Atran and Norenzayan 2004; Barrett 2004). Gods or other supernatural agents are commonly seen as responsible for natural events (especially where they have significant fitness consequences, such as floods, crop failures, or illness). We also have an innate tendency to attribute positive and negative life events as happening for a reason—namely, God did this *because* I was bad, or good (Bering 2002; Bering and Johnson 2005). And we have an innate mind-body dualism: a tendency to believe in life after death (Bering and Bjorklund 2004; Bering, Mcleod, and Shackelford 2005). Diverse religions from across the world's cultures, past and present (irrespective of widely different types of supernatural entities), share a common attention to these features. Human brains, it seems, have a deeply ingrained tendency to believe that their actions are scrutinized by some kind of supernatural observer, and that they will—in this life or the next—pay a price for selfishness. Discounting supernatural causes and consequences can take real mental effort, especially when the stakes are high.

A Game Theoretical Framework

The supernatural punishment hypothesis suggests that there are adaptive advantages to holding certain beliefs rather than others. Although the beliefs themselves seem complex, they could be modeled with a game theoretical framework that pits alternative strategies against each other. No formal modeling has been done, but below I lay out what the basic alternative strategies would look like, consider their key differences in terms of individual costs and benefits, and thereby identify the conditions under which a belief in supernatural punishment may evolve (Johnson and Bering 2006).

God-Fearing Strategies

While selfish behaviors might have paid off in the simpler social life of our evolutionary forbears, many of them (or too many of them) would bring a net fitness loss in a cognitively sophisticated, gossiping society. The advent of theory of mind and complex language increased the likelihood of public exposure for selfish behavior, and could bring high costs of retaliation by other group members (including shunning, social sanctions, seizure of property, physical harm, ostracism, imprisonment, punishment of kin, or death).

According to Bering and Shackelford (2004), theory of mind led to the selection of traits that militate against public exposure, or rescue inclusive fitness *after* the individual committed a social offence in this new "big-brother" society (e.g., cognitive processes underlying confession, blackmail, killing witnesses, suicide, and so forth [Bering and Shackelford 2004]). However, these post-hoc strategies tax reproductive success, so natural selection should favor more efficient traits that constrain selfishness to some extent in the first place (indeed we see such traits in human interaction every day—restraint, self-control, sacrifice, sharing, patience, etc.). Those that carried on being indiscriminately selfish would be outcompeted by prudent others who were able successfully to inhibit their more ancient selfish desires and refrain from breaching social norms to begin with.

Because the temptation to cheat remained, however, I propose that something extra—a belief in *supernatural* punishment—was an effective way to caution oneself against transgressions and thereby avoid real, worldly retribution by other group members (Johnson 2009, 2011; Johnson and Bering 2006). God-fearing people may, therefore, have had a selective advantage over nonbelievers because the latter's more indiscriminately selfish behavior carried a higher risk of real-world punishment by the community.

Machiavellian Strategies

So far we have focused on the *disadvantages* of theory of mind and complex language—selfish actions bring an increased risk of detection and retaliation. However, these cognitive innovations also brought opportunities: selective pressures for traits that *exploit* them (perhaps analogous to Kosslyn's notion of "social prosthetic systems," Chapter 7). One can manipulate others' knowledge as well as suffer from it (as a result of these two mechanisms, the overall selective effect might be expected to be quite strong, effectively "pushed" and "pulled" simultaneously in the same direction by evolution). As an example of manipulation, one can help to conceal the transgressions of kin, or preferentially cooperate with those who have established a good reputation with others—examples that hint at significant implications for the evolution of kin selection and direct or indirect reciprocal altruism among humans. In short, these new cognitive abilities gave humans, for better or worse, a new currency to trade in—social information. Our ancestors became highly invested in this exchange because it exerted a significant influence on reproductive success. Profits came from effectively gathering, retaining, and regulating (through whatever means possible, including deception, threats, and violence) the flow of social information that had the potential to impact inclusive fitness. One may therefore postulate Machiavellian strategies that exploited the human theory of mind and complex language for personal gain, but that were not God-fearing.

Which Strategy Wins?

Table 8.1 compares the performance of the above two strategies (God-fearing and Machiavellian) and the ancestral state, following the advent of theory of mind and complex language. Machiavellians would clearly out-compete ancestral individuals because, while everything else is identical between them, ancestrals cannot exploit these new cognitive features for personal gain. More important, however, the table indicates that God-fearing strategists can outcompete Machiavellians. They differ in just two respects: God-fearing strategists have a lower probability of detection and punishment but miss out on some opportunities for selfish gains. Therefore, God-fearing strategists will outcompete Machiavellians *as long as* the total expected costs of punishment—namely, the probability of detection (p) multiplied by the cost of punishment (c)—is greater than the cost of missed opportunities for selfish gains (m). In other words, when the inequality $pc > m$ is true. This would occur wherever the gains from selfishness were relatively small compared with the costs of public exposure (which may include shunning, social sanctions, seizure of property, physical harm, ostracism, imprisonment, punishment of kin, or death). Even a small probability of detection can mean selfishness does not pay on the average.

Summary of the Model

Humans often act on selfish motives—and sometimes inadvertently, due to emotionally charged situations. Such acts, thanks to theory of mind and complex language, carry a far greater chance of social exposure than in previous stages of evolution. If the costs of exposure were high enough during human evolution, individuals who were more likely to refrain from selfish behavior for fear of supernatural agents—who are believed to reward altruism and punish selfishness—could have outreproduced otherwise equal but more indiscriminately selfish individuals. Of course, Machiavellian, nonbelieving cheats who do not get caught would do best of all, but only if $m > pc$. We suggest that the heightened probability and costs of exposure by virtue of human cognitive sophistication favored the evolution of traits that

Table 8.1 Three strategies come into competition with the advent of theory of mind (TOM) and complex language (CL). Grey shading indicates consequences that act *against* genetic fitness. Machiavellians outcompete ancestral individuals, and God-fearing strategists outcompete Machiavellians as long as $pc > m$.

Strategy	TOM and CL present?	Can exploit TOM and CL for personal gain?	Probability of detection (p)	Cost of punishment (c)	Cost of missed opportunities (m)	Payoff
Ancestral	No	No	High	Same	None	Lowest
Machiavellian	Yes	Yes	High	Same	None	Highest (if $pc < m$)
God-fearing	Yes	Yes	Low	Same	Some	Highest (if $pc > m$)

suppress selfish behavior and favored instead the kind of moralistic behavior that is, after all, empirically common among human societies (Trivers 1971; Alexander 1987; Hauser 2006; Haidt 2007).

Criticisms and Extensions

A common reaction at this point is this: Why invoke the evolution of supernatural beliefs instead of the evolution of, say, *conscience* (defined here as the sense of right and wrong that governs our thoughts and actions)? Is that not a simpler way for evolution to calibrate selfish behavior to a socially sophisticated world? It may be simpler, but several lines of evidence point to a theoretical, empirical, and historical superiority of God or gods over mere conscience in overcoming selfish temptation (Johnson 2011).

First, since the original problem of social transparency arose through theory of mind, it is not surprising that theory of mind might also feature in the solution—as soon as human brains evolved to perceive agency in other humans, they would also begin to perceive, and seek explanations for, agency in the natural world.

Second, billions of people actually alter their everyday behavior because of the perceived consequences of supernatural observation and punishment, so a good theory should ideally account for this empirical phenomenon (a theory based solely on conscience leaves a great deal of human beliefs and behavior unexplained). Recent experimental work supports the anthropology: people are significantly less selfish in laboratory experiments when surreptitiously primed to believe that supernatural agents are watching them (Bering, McLeod, and Shackelford 2005; Shariff and Norenzayan 2007, 2011).

Third, conscience can be shifted to fit any norm. Nazi concentration-camp guards could experience a guilty conscience if they did not kill as many Jews as ordered. A system of moral right and wrong needs some kind of intrinsic reference point. Gods and other supernatural entities of many religions offer exactly such a reference point, typically close to what we might consider universal moral principles (Hauser 2006; Whitehouse 2008). Of course, religious morality has sometimes been horrendously perverted in history, but these cases are the prominent exceptions rather than the norm.

Fourth, a belief in supernatural punishment may be a *better* method of avoiding real-world retribution for selfish actions than other possible methods because of an asymmetry in the possible decision errors that may be made (Johnson 2009). The consequences of our selfish behavior are often underestimated or ignored (after all, it is often the limbic system prompting self-interested behaviors which has little regard for the long-term future; see Lee, Chapter 9). Perhaps, therefore, we need an especially sensitive morality monitor to compensate in the other direction. Support for this idea comes from signal detection theory. Signal detection theory shows that, wherever the costs of false positive and false negative errors have been asymmetric over evolutionary history, we should expect natural selection to favor a bias that avoids whichever is the more costly error (Nesse 2001, 2005; Nettle 2004; Haselton and Nettle 2006). For example, we often assume a stick is a snake, but we do not assume snakes are sticks—a highly adaptive bias. Similarly, when we set the sensitivity of a smoke alarm, it is better to err on the side of caution, because the costs of being burned to death in a real fire are infinite, whereas the costs of a false alarm are negligible (however annoying it may seem at the time). One should thus bias the smoke alarm to go off a bit "too often." Whenever the true probability of some event—such as the detection and punishment of selfish behavior—is uncertain (and thus cannot be precisely predicted), a biased decision rule can be better than an unbiased one. An *unbiased* decision rule targeting the true probability of the event will make both false positive and false negative errors—it will have some false alarms, and it will fail to go off in some real fires. Only a *biased* decision rule will err on the side of caution (allowing some false positives, but more importantly avoiding all false negatives). In the language of the model presented above, error management theory predicts that, where $pc > m$, we should expect *exaggerated* estimates of p to outperform *accurate* estimates of p, because the latter will engender more mistakes in a situation of judgment under uncertainty (Haselton and Buss 2000; Nettle 2004; Haselton and Nettle 2006). Applying this logic to the dangers of our social minefield, an exaggerated belief that one is constantly being watched and judged by supernatural agents—the fabled hyperactive agency detector device—may be an especially effective mind guard against careless selfishness.

The supernatural punishment hypothesis focuses attention on the human mind. But it also offers some interesting insight into the mind of *God*. The logic of the supernatural punishment hypothesis is reflected in both the similarities and differences between godly and human minds. First, God is similar in being envisaged to have a largely human-like theory of mind—essential to God's being is the fact that God can entertain other people's knowledge states and intentional actions (just like humans, Bering 2002). Second, God is different in being more powerful in theory, and yet more limited in practice. For example, the Christian God—according to doctrine—knows anything and everything, but controlled laboratory experiments reveal that people tend to believe that God knows only a certain subset of things. Contrary to common teachings, people believe that God is primarily interested in *strategic social information*—the information that applies to *me* and *my* moral actions—rather than all knowledge. For example, people tend not to think God knows the contents of their fridge, but God will *if* the fridge contains, say, something stolen from someone else (or anything else revealing behavior that may be right or wrong; Bering and Johnson 2005). Given these cognitive propensities, it may be no coincidence that God, gods, and other supernatural entities, while they may be very different from each other, are assumed to share a similar type of brain to us. It may also be no coincidence that we think of them as interested in particular types of information—information that surrounds and reveals our moral actions. God's brain, at least, appears to be made in the image of the human brain.

What about atheists? Do they not prove that we do not need a supernatural mind guard to steer us away from selfish behavior? This is far from obvious. A number of *secular* concepts also lead people—with or without religion—to expect some kind of supernatural punishment for selfish behavior: namely superstition, folklore, karma, or informal notions such as getting one's "comeuppance" or "just desserts." A series of psychological experiments by Melvin Lerner showed that even atheists tend to have a gnawing belief and expectation that people who do wrong will somehow be punished by subsequent life events—a phenomenon known as just world theory (Lerner 1980). These various forms of supernatural belief work in exactly the same way as a fear of punishment from God, and thus fit within the general supernatural punishment theory. Irrespective of the source of punishment, if we fear the consequences of our actions, we will think twice about carrying them out. All such beliefs may thus have been favored by natural selection.

Conclusions

The evolution of theory of mind and complex language made selfishness uniquely costly for humans, and the question arises how natural selection responded to this newly transparent social world. The supernatural punishment hypothesis offers a number of advantages over alternative theories. First, it shows that, contrary to many recent authors (Harris 2004; Dawkins 2006; Dennett 2006), religious beliefs can be adaptive for individual fitness and may thus have evolved (or have been co-opted) by natural selection. Second, the hypothesis ties up human cognition with theology—offering a theory for God's mind as well as our own. Third, the idea that religious beliefs and practices have adaptive advantages for humans is a much better fit with the broadly positive influence (and enormous following) of the world's religions.

As the science-versus-religion debate rages on, it is odd that the science *of* religion is such a small and understudied area. Evolutionary scientists themselves are divided over the origins, functions, and utility of religion, and there remains much to explore (Bulbulia et al. 2008). It is ironic that Dawkins, the great prophet of evolution, does not see the many potential adaptive advantages of religion. Evolutionary theories of religion suggest that we cannot get rid of religion as easily as Dawkins might like. God will not go away, because our brains have been expressly fashioned by natural selection to perceive supernatural agency, judgment, and punishment—and such beliefs can increase Darwinian fitness. God can never be dead in a world in which every child is born with a brain predisposed to see supernatural agency at work all around them. It is actually very difficult to let go of intrinsic religious or superstitious beliefs—atheism takes years of training. And even then, we often return to religious beliefs and reasoning when our lives hit tough times (or very good times). Religion offers an account of our moral place in the world, one that is likely to have been highly adaptive in human evolution, as it often still is today. While religion may sometimes fan the flames of conflict or discourage freedom of thought, these costs may be insignificant compared to the less visible but far more pervasive benefits of supernatural beliefs in the everyday lives of billions of believers.

References

Alexander, R. D. 1987. *The Biology of Moral Systems.* Aldine, NY: Hawthorne.

Atran, S. 2004. *In Gods We Trust: The Evolutionary Landscape of Religion.* Oxford: Oxford University Press.

Atran, S., and A. Norenzayan. 2004. "Religion's Evolutionary Landscape: Counter-intuition, Commitment, Compassion, Communion." *Behavioural and Brain Sciences* 27: 713–30.

Axelrod, R. 1984. *The Evolution of Cooperation.* London: Penguin.

Barkow, J. H., L. Cosmides, and J. Tooby, eds. 1992. *The Adapted Mind: Evolutionary Psychology and the Generation of Culture.* Oxford: Oxford University Press.

Barrett, J. L. 2004. *Why Would Anyone Believe in God?* Lanham, MD: Altamira Press.

Bering, J. M. 2002. "The Existential Theory of Mind." *Review of General Psychology* 6: 3–24.

———. 2006. "The Folk Psychology of Souls." *Behavioural and Brain Sciences* 29: 453–62.

Bering, J. M., and D. F. Bjorklund. 2004. "The Natural Emergence of Reasoning about the Afterlife as a Developmental Regularity." *Developmental Psychology* 40: 217–33.

Bering, J. M., and D. D. P. Johnson. 2005. " 'Oh Lord, You Hear my Thoughts from Afar': Recursiveness in the Cognitive Evolution of Supernatural Agency." *Journal of Cognition and Culture* 5: 118–42.

Bering, J. M., K. A. McLeod, and T. K. Shackelford. 2005. "Reasoning about Dead Agents Reveals Possible Adaptive Trends." *Human Nature* 16: 360–81.

Bering, J. M., and T. K. Shackelford. 2004. "The Causal role of Consciousness: A Conceptual Addendum to Human Evolutionary Psychology." *Review of General Psychology* 8: 227–48.

Boyer, P. 2001. *Religion Explained: The Evolutionary Origins of Religious Thought.* New York: Basic Books.

Bulbulia, J., R. Sosis, C. Genet, R. Genet, E. Harris, and K. Wyman, eds. 2008. *The Evolution of Religion: Studies, Theories, and Critiques.* Santa Margarita, CA: Collins Foundation Press.

Buss, D. M, ed. 2005. *The Handbook of Evolutionary Psychology.* New York: Wiley.

Damasio, A. R. 1994. *Descartes' Error: Emotion, Reason and the Human Brain.* New York: Avon.

Davidson, R. J., K. M. Putnam, and C. L. Larson. 2000. "Dysfunction in the Neural Circuitry of Emotion Regulation: A Possible Prelude to Violence." *Science* 289: 591–94.

Dawkins, R. 1976. *The Selfish Gene.* Oxford: Oxford University Press.

———. 2006. *The God Delusion.* New York: Houghton Mifflin.

Dennett, D. C. 2006. *Breaking the Spell: Religion as a Natural Phenomenon*. New York: Viking.

Dugatkin, L. A. 1997. *Cooperation in Animals*. Oxford: Oxford University Press.

Goldstein, M. A. 2002. "The Biological Roots of Heat-of-Passion Crimes and Honor Killings." *Politics and the Life Sciences* 21: 28–37.

Guthrie, S. E. 1993. *Faces in the Clouds: A New Theory of Religion*. New York: Oxford University Press.

Haidt, J. 2007. "The New Synthesis in Moral Psychology." *Science* 316: 998–1002.

Harris, S. 2004. *The End of Faith: Religion, Terror, and the Future of Reason*. New York and London: W. W. Norton and Co.

Haselton, M. G., and D. M. Buss. 2000. "Error Management Theory: A New Perspective on Biases in Cross-Sex Mind Reading." *Journal of Personality and Social Psychology* 78: 81–91.

Haselton, M. G., and D. Nettle. 2006. "The Paranoid Optimist: An Integrative Evolutionary Model of Cognitive Biases." *Personality and Social Psychology Review* 10: 47–66.

Hauser, M. D. 2006. *Moral Minds: How Nature Designed our Universal Sense of Right and Wrong*. New York: Ecco.

Johnson, D. D. P. 2005. "God's Punishment and Public Goods: A Test of the Supernatural Punishment Hypothesis in 186 World Cultures." *Human Nature* 16: 410–46.

———. 2009. "The Error of God: Error Management Theory, Religion, and the Evolution of Cooperation." In *Games, Groups, and the Global Good*, ed. S. A. Levin. Oxford: Oxford University Press, 169–80.

———. 2011. "Why God is the Best Punisher." *Religion, Brain and Behavior* 1/1: 77–84.

Johnson, D. D. P., and J. M. Bering. 2006. "Hand of God, Mind of Man: Punishment and Cognition in the Evolution of Cooperation." *Evolutionary Psychology* 4: 219–33.

Johnson, D. D. P., and O. Kruger. 2004. "The Good of Wrath: Supernatural Punishment and the Evolution of Cooperation." *Political Theology* 5(2): 159–76.

Lerner, M. J. 1980. *The Belief in a Just World: A Fundamental Delusion*. New York: Plenum Press.

McCullough, M. 2008. *Beyond Revenge: The Evolution of the Forgiveness Instinct*. San Francisco: Jossey-Bass.

Nesse, R. M. 2005. "Natural Selection and the Regulation of Defenses: A Signal Detection Analysis of the Smoke Detector Problem." *Evolution and Human Behavior* 26: 88–105.

———. 2001. "Natural Selection and the Regulation of Defensive Responses." *Annals of the New York Academy of Sciences* 935: 75–85.

Nettle, D. 2004. "Adaptive Illusions: Optimism, Control and Human Rationality." In *Emotion, Evolution and Rationality,* ed. D. Evans and P. Cruse. Oxford: Oxford University Press, 193–208.

Nowak, M. 2006. "Five Rules for the Evolution of Cooperation." *Science* 314: 1560–63.

Nowak, M., and K. Sigmund, K. 2005. "Evolution of Indirect Reciprocity." *Nature* 437: 1291–98.

Pinker, S. 2002. *The Blank Slate: The Modern Denial of Human Nature.* New York: Penguin Putnam.

Povinelli, D. J., and J. M. Bering. 2002. "The Mentality of Apes Revisited." *Current Directions in Psychological Science* 11: 115–19.

Sanfey, A. G., J. K. Rilling, J. A. Aronson, L. E. Nystrom, and J. D. Cohen. 2003. "The Neural Basis of Economic Decision-Making in the Ultimatum Game." *Science* 300: 1755–58.

Schloss, J. P., and M. Murray. 2011. "Evolutionary Accounts of Belief in Supernatural Punishment: A Critical Review." *Religion, Brain and Behavior* 1/1: 46–99.

Shariff, A. F., and A. Norenzayan. 2007. "God Is Watching You: Supernatural Agent Concepts Increase Prosocial Behavior in an Anonymous Economic Game." *Psychological Science* 18: 803–9.

———. 2011. "Mean Gods Make Good People: Different Views of God Predict Cheating Behavior." *International Journal for the Psychology of Religion* 21/2: 85–96.

Sosis, R. 2004. "The Adaptive Value of Religious Ritual." *American Scientist* 92: 166–72.

Trivers, R. L. 1971. "The Evolution of Reciprocal Altruism." *Quarterly Review of Biology* 46: 35–57.

Whitehouse, H. 2008. "Cognitive Evolution and Religion: Cognition and Religious Evolution." In *The Evolution of Religion: Studies, Theories, and Critiques,* ed. J. Bulbilia, R. Sosis, C. Genet, R. Genet, E. Harris, and K. Wyman. Santa Margarita, CA: Collins Foundation Press, 19–29.

Wilson, D. S. 2002. *Darwin's Cathedral: Evolution, Religion, and the Nature of Society.* Chicago: University of Chicago Press.

Wilson, E. O. 1975. *Sociobiology: The New Synthesis.* Harvard: Belknap Press.

9

Self-Denial and Its Discontents

Toward Clarification of the Intrapersonal Conflict between "Selfishness" and "Altruism"

Maurice Lee

According to the definitions that are providing points of common reference for this volume, *cooperation* is "a form of working together in which one individual pays a cost and another gains a benefit," and *altruism* is "a form of cooperation in which an individual is motivated by goodwill or love for another." These definitions imply that—as the chapter in which they appear explicitly affirms—we may regard altruism as being "prepared for" by, or as "building" or "layering upon," cooperation.

A key issue, then, is this: does the definition of *altruism* as a form of *cooperation* in behavioral and psychological terms reflect a deeper structural and functional relationship between the two? Does altruistic behavior build on cooperative behavior in the sense that the same—no doubt exceedingly complex—biological mechanism, developed over a common evolutionary history, structures both? And to what does talk of "motivation" correspond in that mechanism and history? Is it the addition or emergence of something novel, or the unmasking or activation of something latent? Simply put: how do we get from cooperation to altruism? At least two scientific tasks are posed by this pair of definitions. The first is to describe the relationship between cooperation and altruism at the biological level so that it becomes thematic how species—most significantly, of course, the human—capable of altruism are both linked to and distinguished from others. The second task is to specify what, precisely, it is to be "motivated by goodwill or love for another," in order to investigate how this capacity, like other aspects of life on earth, emerged within and as part of evolutionary history.

To clarify some of the issues involved in these tasks, I will consider a feature of human social life that is, in a sense, the exact opposite of the phenomenon studied by Stephen M. Kosslyn in his contribution to this volume. Kosslyn identifies "self-extension"—the creation of "social prosthetic systems" in which one appropriates others' skills and acts to help accomplish one's own intentions and purposes—as a basis of human cooperation: "we find it fun to acquire and exercise competences because this allows us to become valuable or useful for other people—which then in turn motivates others to serve as our [social prosthetic systems]" (Kosslyn, Chapter 7). The "object" of a social prosthetic system—the one whose competences are being "used" by the "extended" self—is not necessarily aware of subordinating her own goals to those of the other; indeed, she may be pursuing her goals in parallel or even quite independently. In contrast, I shall talk explicitly about *self-denial*—the voluntary withdrawal or disavowal of one's own interests and projects, in favor and support of others, such that certain desirable benefits to oneself are deferred or even lost.

Formulated in this way, self-denial is not integral to the definitions of altruism and cooperation mentioned above; the definitions are not incompatible with an actor's both paying a cost *and* gaining a benefit in the course of delivering benefit to another. Yet in the tradition of moral discourse to which our concept of altruism is clearly intended to stand in relation—with talk of "love" and "goodwill"—self-denial plays an important role and indeed has engendered passionate debate (see Post 1987, and Jackson, Chapter 16). To speak of self-denial for the sake of another is a fortiori to speak of altruism. Self-denial, then, may be considered a species or form of altruism, in the same way that altruism is a species or form of cooperation. Whatever its moral status might be, there is no doubt that self-denial *happens,* and so it will be my focus in the following.

The human act of helping another person or persons—particularly when it puts one's own gain, comfort, safety, or life at risk—often follows a conscious decision. Faced with another's need or desire, the potential helper must *choose:* either to decline to assist the other, preferring instead to advance or maintain her own welfare, or else to set aside (at least temporarily) her self-directed intentions, and to accept inconvenience, pain, or even death for the sake of the other's good.

It is interesting that such decisions are often subjectively *difficult.* When it comes to helping someone else at probable cost to oneself, many of us do not always simply either plunge in or turn away without a second thought. We waver, worry, argue with ourselves; we weigh our obligations, recall our convictions, project the consequences. It is not always immediately clear to us either what we *want* (more) to do, or what we *should* do, and it can take time and effort to come to a conclusion. As John Calvin asserts with typical vigor in the *Institutes of the Christian Religion:*

> Now, in seeking to benefit one's neighbor, how difficult it is to do one's duty! Unless you give up all thought of self and, so to speak, get out of yourself, you will accomplish nothing here. For how can you perform those works which Paul teaches to be the works of love, unless you renounce yourself, and give yourself wholly to others? . . . If this is the one thing required—that we seek not what is our own—still we shall do no little violence to nature, which so inclines us to love of ourselves alone that it does not easily allow us to neglect ourselves and our possessions in order to look after another's good, nay, to yield willingly what is ours by right and resign it to another. (Calvin [1559]1960, 3.7.5)

How is this phenomenon—the difficulty with deciding about self-sacrifice to benefit another, the struggle over altruistic self-denial—to be accounted for? How and why does it happen? This seems to be a uniquely human experience: in no other species, as far as we know, do individuals debate *with themselves* about whether or not to help another. Understanding more about it would deepen our account of the development and the distinctiveness of human altruism.

But what would constitute "understanding" the phenomenon in question? Phenomenological, theological, philosophical, sociological, and biological approaches all have their place here. Arguably, the biological study of human behavior, as the youngest discipline among those named, has the most work to do. But there are at least two complications. First, how does one examine from the "outside," in "objective" mode, something originally known—at least in the way I have described it above—from the "inside," in "subjective" mode? We have to assume that what we are speaking of as the feeling of difficulty or struggle over self-denial is reflected in, subserved by, structures and processes amenable to scientific analysis. This by no means implies that the phenomenon in question can be *exhaustively* explained by such analysis, that there is nothing worth knowing about the phenomenon other than what can be known scientifically. Biological investigation is only a component in a full-orbed appreciation and understanding of the phenomenon of altruism: a crucial component, but only a component. Second, even acknowledging that a biological approach is possible and important, are the tools and knowledge needed to pursue this work up to the task? In other words, is our technology precise enough, and our grasp of how biological mechanisms underlie psychological dynamics clear enough, to make a scientific inquiry tractable? We cannot make blanket judgments here; whether we can realistically achieve what we are after will depend on what, exactly, we are after. But we should not forget that our most up-to-date understanding of the vastly complex machinery constituting—in this case—the human brain is still very much in the beginning stages, still very provisional, and that our most advanced analytical techniques are still remarkably coarse, still quite "low-resolution" compared to the intricacies to which they are being applied. I shall draw attention to these limitations repeatedly in what follows. Yet with these caveats in mind, I propose that there is a wide field for experimental investigation into the difficulty of the decision for altruistic self-denial. I shall consider two areas for such work, more precisely, two groups of questions in search of answers.

First, what is the *neural structure* of the altruistic decision-making process? Which brain areas are activated when people make decisions about self-denial for the sake of others? How are these areas connected? Are the areas involved when the decision is made in one's own favor (i.e., against self-denial) different from those involved in the decision for the other? Are different areas activated by self-denying decisions for private purposes (e.g., dieting), as opposed to self-denying decisions for purposes of helping others?

Investigation along these lines could, in the long run, lead to insights into how self-denial and altruism are "represented" and "computed" at the neural level, including new understanding of what corresponds, in the way the mind works, to our language of "love" and "goodwill." The distance in comprehension between watching regions of the brain light up on a scanning monitor and describing how the brain instantiates altruism should not be underestimated, of course. We do not even know how long "representation" and "computation" will continue to be useful metaphors. Yet understanding how the brain works is, inescapably, the goal of neuroscience research.

The sorts of issues involved may be seen in recent experimental work using moral dilemmas to study human moral cognition. In these studies, subjects must decide whether certain actions are morally appropriate in response to (hypothetical) life-or-death situations. One form of this decision matrix involves (a) killing one person in order to save many others, as opposed to (b) refusing to cause that one person's death and so permitting the others to die. (The situations are very contrived, to be sure, but they have the advantage of making the choices starkly clear [see, for example, Greene et al. 2004, Koenigs et al. 2007.]) While subjects are deciding whether option (a) is appropriate under the specified circumstances, their brains are scanned using functional magnetic resonance imagery (fMRI) to detect localized neural activation. (Functional MRI technology uses differences in the response to a magnetic field between oxygenated blood and deoxygenated blood to map where in the brain oxygen is being supplied by the blood at greater rates; this is thought to be correlated with higher neuronal activity in those brain areas.)

Many subjects take longer to make such decisions than if no dilemma is involved, or than if the dilemma is less extreme. The difference in reaction time presumably reflects the mental negotiation of conflicting or confused moral intuitions—although it would be interesting and useful to have actual subjects report on how they felt while making their decisions. In other words, such decisions—even in hypothetical situations—are *difficult* for many people, since either option involves a morally repugnant outcome. Nevertheless, subjects are required to decide one way or the other, and it often takes them some time, several seconds, to do so.

What is happening in the brain while these decisions are being made? In one report of such experiments, decisions in which option (a) was judged appropriate—we can call these "utilitarian," since they accept one person's hypothetical death in order to secure the survival of a greater number of others—were found to be correlated with greater activity in the dorsolateral prefrontal cortex (Greene et al. 2004). In another set of experiments, subjects with lesions in the ventromedial prefrontal cortex were found to be more likely than nonlesioned subjects to make the same "utilitarian" choice (Koenigs et al. 2007). In other words, *suppression* of ventromedial prefrontal cortex (VMPFC) and *activation* of dorsolateral prefrontal cortex (DLPFC) are both correlated with more "utilitarian" moral decision making. Greene et al. interpret these results to suggest that moral decisions normally involve both "social-emotional" processes (representing, e.g., revulsion at directly causing another person's death) and "cognitive-utilitarian" processes (representing, e.g., a preference for the survival of as many people as possible, whatever the means). The VMPFC, in this picture, is implicated in "social-emotional" processes, and the DLPFC is implicated in "cognitive-utilitarian" processes. Some decisions, then, might be experienced as difficult partly because these different processes, underlying conflicting moral intuitions, are both activated.

This leads naturally to the question of how the conflict between the two kinds of intuitions is resolved, and the decision finally made. Does a particular portion of the brain specialize in general conflict processing and adjudication, receiving and disposing of requests for decisions like a judge? Or is a conflict-negotiation algorithm run "locally" wherever incompatible results are produced, like a police officer empowered to make an on-the-spot determination? In the experiments with moral dilemmas, as well as in experiments involving other sorts of mental conflicts, conflict detection—and possibly also resolution—is associated with activity in the anterior cingulate cortex (ACC). The fact that a particular part of the brain reliably "lights up" to signal internal moral conflict does not, of course, answer the question of how that conflict is resolved. But other studies have suggested that some forms of activity in the ACC might lead to the suppression of activity in other areas, such as the amygdala, which in turn might be correlated with a reduction in emotional responsivity (Etkin et al. 2006). One possibility, then, is that activation in the ACC "turns off" one of the conflicting inputs.

An analogous hypothesis for the case of the decision about self-denial immediately suggests itself: that one brain area "recommends" the self-serving action, and another area "recommends" the self-giving action, with perhaps a third area (or group of areas) crucial for bringing the decision-making process to a conclusion.

What areas would these be? Only experiments can tell, of course. But self-preservation is an ancient biological imperative, and one might predict that human "instincts" for self-benefit would be partially realized by neural activity in evolutionarily old brain areas—areas that have been part of the brain for much longer than the neocortex. For example, the limbic system is associated with basic individual survival drives (Mega et al. 1997). It would not be particularly surprising to find signals being "passed up" from the limbic system to the cortex so as, at least partially, to represent the interests of the self.

On the other hand, cooperation is also evolutionarily ancient (Nowak 2006), and it would not be any more surprising to find signals associated with kin (or other group member) detection or assessment of the other's need—or their precursors—emerging from more "primitive" regions and feeding into the decision-making process. Moreover, representations of self and self-benefit, which become the particular targets of conscious consideration—the only level at which evaluative notions such as "selfish" can properly be applied—are surely a principally cortical and recent modification. To divide the altruistic decision-making process into "ancient" self-interested and "modern" other-oriented components is highly problematic.

There are other factors to be considered. For example, empathy or compassion has been proposed to be deeply embedded in human altruistic behavior (Batson 1998). Compassionate emotional responses are complex and not well understood at the brain level but no doubt involve an interaction of perceptual, conceptual, and affective factors. Jorge Moll and his colleagues have proposed a specific group of areas engaged in this interaction: prefrontal cortex, superior temporal sulcus, anterior temporal cortex, and the limbic system (Moll et al. 2005).

It would be useful to know more about the *time dimension* of decision making about altruism. Is there a clear correlation between the experienced difficulty of decisions and the time it takes to make them? Do neural activations show significant changes over time during a difficult decision? Are there areas that fail to be activated, or areas that are only activated when decisions are made relatively quickly, with little struggle? Time considerations offer a window onto the variegated nature of the decision about altruism. Numerous factors potentially affect the course and result of the decision, and the level and kind of difficulty experienced in conjunction with it. We need data on what is happening mentally as the time-to-decision increases. Are subjects "revisiting" the same options over and over? Are they considering the possible ramifications of each action? Are they casting about for fresh information or insight? Work on the time dimension in decision making exists (e.g., McClure et al. 2004) but has not focused on the finer-grained, time-varying details of the decision *process*.

As our understanding of the neural basis of altruistic decision making grows, we will need to revisit the common notion of altruism as pure, unadulterated, other-directed concern, without any admixture of self-directed motives. This has had its staunch defenders from the perspective of religious ethics (Jackson 1999), while being seeming philosophically problematic to others (Adams 2006). In light of the vast neural goings-on underlying the decision-making process, the debate will deepen. The "purist" account cannot merely be dismissed as simplistic from a biological standpoint—since we are not even sure what a "pure" motivation amounts to, neurally—but the complexity of computations, motivations, and reinforcements going into real altruistic decisions must be considered as we ask what it means for a human person, using her or his brain, to deny her or his *self* for the sake of another.

Second, what is the *evolutionary history* of the altruistic decision-making process? How do we characterize the emergence through natural selection, first, of a decision-making mechanism that has a hard time making certain decisions, and, second, of an uncomfortable, sometimes paralyzing, subjective tracking of such difficulties? What is the biological significance of this feeling of struggle—a common-enough feature of the experience of decision making? There are a number of possible scenarios for the evolution of difficult decision making. At the current stage of our knowledge, of course, they can be described only in outline. First, it could be—if some sorts of decision making involve not just different areas of the cortex but structures as different as the neocortex and the limbic system—that processes and programs already "in play" (so to speak) before the invention of the cortex, and processes and programs developed specifically for use within the cortex, do not always "mesh" together perfectly. The limbic system might make "recommendations" more or less the way it has for 150 million years, while the much younger, more complex neocortex might "see" things differently. If this were the case, we would expect to find some broad similarities in limbic system responses to decision-making situations across phylogenetic lines.

Alternatively and more generally, it could be that the conflict that sometimes arises in decision making is not a selected feature, but rather a necessary by-product of the system. An increased chance of incompatibilities in decision making under certain circumstances might be an inevitable consequence of the increased complexity of the machinery underwriting flexibility in decision making; occasional low-level conflict might be the unavoidable price of consistent high-powered function. As in politics, there might be no way to get perfectly unitary decision making—say, via a "monarchical" or "totalitarian" scheme—without losing key advantages—say, the ability to deal rapidly with changing conditions. The implication would be that this capacity for bugs or glitches in the process has not imposed a large-enough fitness cost for the whole system to be selected out.

Finally, it could be that under the constraints of finite computational and environmental resources, an arrangement of relatively independent modules that "do their own thing," even if at times those things conflict with one another—a neural "free market," as it were—was actually more economical and useful than a system constructed to find the optimal, "top-down" solution to every problem (Livnat and Pippenger 2006). In this picture, determining the best trade-off among competing demands is usually algorithmically complicated and biologically expensive, and it is cheaper to incorporate (somehow) the conflict itself into the decision-making process. On such an "adaptive" account, conflict could be interpreted not as a breakdown or malfunction of the process, but as one of its evolutionary features.

And the consciousness of difficulty, the subjective feeling attending conflicted decisions? This, too, might be a "by-product" of the neural mechanics of decision making—that is, not a trait with a distinct genetic specification but something we get "for free" (so to speak) as an emergent property of the machinery of cognition. On the other hand, the consciousness of mental conflict might also be a naturally selected way to encourage the bringing of other cognitive resources to bear on difficult decisions—a "slow-down-and-think-about-it" mechanism, a means of decreasing the probability that decisions made under conditions of conflict are made too quickly or simply randomly. Furthermore, in the specific case of decisions about altruistic self-giving, another set of selective pressures can be imagined: perhaps those who gave up their lives easily, spontaneously, for others, on the one hand, and those who consistently avoided helping others and so failed to build their social capital, on the other hand—in other words, those who experienced less subjective difficulty with their decisions—both ended up at a reproductive disadvantage.[1]

Can we lift these possibilities above the level of plausible speculation? Certainly we cannot reproduce the "ancestral" environmental conditions under which the phenomenon in question presumably evolved, so the issue of what is "experimentally" or "empirically" verifiable is a tricky one. Perhaps, as the mathematical and computational tools for the study of evolutionary dynamics become more sophisticated, we will feel able—as seems to have happened for neuroscience and cosmology, for example—to put more trust in simulations of evolutionary events. At least such simulations might enable us to rule out possibilities. Here, as with the study of the neural basis of altruistic self-denial, we are still very much at the beginning. Is speaking of the evolutionary origins of cognition and consciousness a sign of capitulation to a mechanistic, naturalistic view of human life, to the exclusion of a consideration of the moral and religious dimensions often associated with decisions about self-denial? It is a central thesis of this book that even the more sophisticated typical arguments over the compatibility of Darwinian evolution with Christian theism have been hampered until now by insufficient awareness of the depth to which the principle of cooperation is embedded in the evolutionary history of life. As the link between the neural basis of behavior and thought and the evolutionary development of those aspects of human life becomes closer and more precise in our understanding, the ways in which cooperation is woven into the fabric of human nature—and so the ways in which, in theological interpretation, the groundwork for a noncompetitive, self-giving ethic is being laid—will become clearer.

We do not yet understand human altruism scientifically. To get to that point will require concentrated resources and innovative experimentation. It will also demand considerable conceptual discipline, for it is easy to imagine, given the ways in which the terminology is currently used, that there is something essentially in common, even continuous, between the human behaviors and intentions we call "altruistic"—or, for that matter, those we call "selfish"—and any interaction between organisms that is deleterious ("costly") to one and advantageous ("beneficial") to the other. But such continuity—the path from "cooperation" to "altruism"—is not something that can be presupposed: it must be demonstrated. Moreover, it necessarily begs *philosophical* questions about the mind/body matrix that this exploratory essay on the neurophysiology of altruism has only hinted at. We shall start to address such questions more explicitly in the next section of the book.[2]

Notes

1. Note here the potential parallels between self-denying decision-making processes in public goods games (see Almenberg and Dreber's chapter) and the potential for self-conflict at the neurological level.
2. Deep thanks to Sarah Coakley, Philip Clayton, Dominic D. P. Johnson, and Zachary Simpson for extremely helpful comments on earlier versions of this chapter.

References

Adams, R. M. 2006. *A Theory of Virtue: Excellence in Being for the Good.* Oxford: Oxford University Press.

Batson, C. D. 1998. "Altruism and Prosocial Behavior." In *The Handbook of Social Psychology,* ed. D. T. Gilbert, S. E. Fiske, and G. Lindzey. 4th ed. New York: McGraw-Hill, 282–316.

Calvin, J. [1559] 1960. *Institutes of the Christian Religion,* ed. John T. McNeill and Trans. Ford Lewis Battles. Philadelphia: Westminster Press.

Etkin, A., T. Egner, D. M. Peraza, E. R. Kandel, and J. Hirsch. 2006. "Resolving Emotional Conflict: A Role for the Rostral Anterior Cingulate Cortex in Modulating Activity in the Amygdala." *Neuron* 51: 871–82.

Greene, J. D., L. E. Nystrom, A. D. Engell, J. M. Darley, and J. D. Cohen. 2004. "The Neural Bases of Cognitive Conflict and Control in Moral Judgment." *Neuron* 44: 389–400.

Jackson, T. P. 1999. *Love Disconsoled: Meditations on Christian Charity.* Cambridge: Cambridge University Press.

Koenigs, M., L. Young, R. Adolphs, D. Tranel, F. Cushman, M. Hauser, and A. Damasio. 2007. "Damage to the Prefrontal Cortex Increases Utilitarian Moral Judgements." *Nature* 446: 908–11.

Livnat, A., and N. Pippenger. 2006. "An Optimal Brain can be Composed of Conflicting Agents." *Proceedings of the National Academy of Sciences of the USA* 103: 3198–202.

McClure, S. M., D. I. Laibson, G. Loewenstein, and J. D. Cohen. 2004. "Separate Neural Systems Value Immediate and Delayed Monetary Rewards." *Science* 306: 503–7.

Mega, M. S., J. L. Cummings, S. Salloway, and P. Malloy. 1997. "The Limbic System: An Anatomic, Phylogenetic, and Clinical Perspective." *Journal of Neuropsychiatry and Clinical Neurosciences* 9: 315–30.

Moll, Jorge, R. Zahn, R. de Oliveira-Souza, F. Krueger, and J. Grafman. 2005. "The Neural Basis of Human Moral Cognition." *Nature Reviews Neuroscience* 6: 799–809.

Nowak, M. A. 2006. "Five Rules for the Evolution of Cooperation." *Science* 314: 1560–63.

Post, S. G. 1987. *Christian Love and Self-Denial: An Historical and Normative Study of Jonathan Edwards, Samuel Hopkins, and American Theological Ethics.* Lanham, MD: University Press of America.

∴ IV ∴

Philosophy of Biology and Philosophy of Mind

Adjudicating the Significance of Evolutionary Cooperation

10

• • • •

Unpredicted Outcomes in the Games of Life

JEFFREY P. SCHLOSS

> A social organism of any sort whatever, large or small, is what it is because each member proceeds to his own duty with a trust that the other members will simultaneously do theirs. Wherever a desired result is achieved by the co-operation of many independent persons, its existence as a fact is a pure consequence of the precursive faith in one another of those immediately concerned . . . A whole train of passengers (individually brave enough) will be looted by a few highwaymen, simply because the latter can count on one another, while each passenger fears that if he makes a movement of resistance, he will be shot before anyone else backs him up. If we believed that the whole car-full would rise at once with us, we should each severally rise, and train-robbing would never even be attempted. There are, then, cases where a fact cannot come at all unless a preliminary faith exists in its coming. And where faith in a fact can help create the fact, that would be an insane logic which should say that faith running ahead of scientific evidence is the 'lowest kind of immorality' into which a thinking being can fall. Yet such is the logic by which our scientific absolutists pretend to regulate our lives!
>
> —William James, "The Will to Believe"

"When I first made myself master of the central idea of the *Origin,*" T. H. Huxley mentions in his reflections on the reception of Darwin, "[I thought] [h]ow extremely stupid not to have thought of that!" (Huxley 1900, 183).

Indeed, the central idea of *On the Origin of Species*—that if some heritable variations confer to their possessors the ability to leave more offspring than others, their proportion in a population will increase over time—does seem simple, even obvious. John Maynard Smith, a key figure in evolutionary game theory, concurs: "Darwin's idea is simple . . . It is, perhaps, the one profound idea in science that we can all readily understand" (Maynard Smith and Szathmáry 1999, 1).

Actually, even this may be something of an understatement, for the idea of natural selection is not just simple or obvious—it is seemingly undeniable. The claim that a population will change in the direction of those entities that make more copies of themselves could not possibly be false; it is a logical truism. This may be the first and only time in the empirical sciences that a necessary truth has served and persisted in a central explanatory role. To be fair, no doubt this is at least partly due to the fact that for all its apparent simplicity, Darwin's theory does not *just* state an a priori truth, but a) illuminates the various causes underlying successful reproduction, and b) predicts particular effects of differential reproduction on the features of organisms. In fact, one of the features it predicts is the lack of, or difficulty in attaining, genuine cooperation.

Much the same could be said about the recent extension of Darwinian theory by kin selection, direct/indirect reciprocity, and evolutionary game theory. Though neither simple nor obvious, there is a logical necessity to claiming that the evolutionary maintenance of an investment behavior requires its reproductive costs not exceed its benefits, or that the fitness of a particular interactive strategy depends on the frequency of that and other strategies, or that preferentially replicated strategies will, under specified conditions, come to dominate or achieve stable equilibrium in a population. In fact, it is precisely this necessitarian feature that makes game theoretical models possible apart from empirical data and that has enabled them to develop considerably ahead of experimental testing (Dugatkin 1997, 2001). But as with Darwinism itself, such models have made important contributions both to explaining what we do see and predicting what we "should" see, in areas as diverse as sex ratios, parental investment strategies, territoriality, foraging behavior, habitat selection, social status and ritualized conflict, predator–prey dynamics, and allometric scaling (Maynard Smith 1982; Dugatkin and Reeve 1998). Many of these directly or indirectly involve the issue of cooperation on which this volume focuses.

I want in this chapter to address the question of how "what we do see" relates to ideas of "what we should see." Any good theory, like headlights, illuminates a field of view, and this volume has able expositors of game theory's impressive resolving power. But a theory, particularly if it seems *necessarily* true and adequate, also directs our view and therefore may exclude from attention aspects of the world that lie outside its present illuminative range. Indeed, as the above passage from William James suggests—prescient both in its description of barriers to collective action in the commons game and also in its anticipation of recent disputes over scientific nihilism[1]—a priori commitments may exclude certain possibilities not only from consideration but also from coming into existence.

This essay therefore examines two scientific questions about cooperation that are theologically significant and that currently involve limits of game theoretical accounts. The first is the issue of central tendencies or large-scale evolutionary trends in cooperation that are neither logically entailed by nor at present fully accounted for by selection or game theory (but that may well yet be). Are there cooperative themes to evolution? The second is the question of behavioral phenotypes—specifically counterreproductive sacrifice that foreshadows intentional altruism—that seem in principle not to be explicable by traditional Darwinian accounts. Are there certain characters that appear thematically inconsistent with the evolutionary drama?

Trends in Cooperation

The internal logic of natural selection and game theory does not lead us to expect that living organisms will necessarily evidence cooperation at all, much less show an evolutionary increase in it (or in any characteristic): "It predicts only that organisms will get better at surviving and reproducing . . . or at least that they will not get worse" (Maynard Smith and Szathmáry 1999, 15). Indeed, the notion of evolutionary directionality, particularly large-scale trends in cooperation, has been outside the "headlights" of much recent work for just this reason, and also for several others that not only fail to expect, but expect failure of, cooperative trends. First, evolution and game theory have been "deeply rooted in methodological individualism," which reflects the important and quite fruitful "change of paradigm regarding the level of aggregation at which selection shows its strongest effect" (Hammerstein

1998, 3–4). Understandable though this may be, we shall see how it can beg the question of what constitutes an individual. Second, even between individuals within a given level of aggregation, Darwinian logic provides "no general reason why evolution should lead" to cooperation or interdependent complexity but also has been prominently interpreted as requiring its exclusion. In contrast to the definition of cooperation given in this volume, there is a substantial tradition of asserting that "What passes for cooperation turns out to be a mixture of opportunism and exploitation" (Ghiselin 1974, 247). Third, for a variety of theoretical and social reasons, the notion of evolutionary progress (in cooperation or other features) has been characterized as a "noxious . . . intractable idea that must be replaced if we wish to understand evolutionary history" (Gould 1988, 319). Even the more modest and less value-laden idea of evolutionary directionality has been resisted, via Stephen Gould's famous image of the stupefied drunkard's stumbling trajectory, whose "motion includes no trend whatever" (1996, 150). Among the many posited entailments of rejecting inherent trends is the theologically significant assertion that what we call higher organisms, and the mammalian capacity for social attachment, and we humans are "a momentary cosmic accident that would never arise again if the tree of life could be replanted . . ." (1996, 18).

The above points involve numerous and frequently conflated sources of debate. Are functional reductionism, exclusion of cooperation, or radical contingency necessary entailments of evolutionary theory? If there are trends, are they explicable by selection, or does selection need to be supplemented with a generative theory that accounts for the contours of possibility space within which selection acts? What is the relationship of each claim to theological notions of grace and providence? (For example, if cooperation and altruism are natural inevitabilities, is that somehow more theologically hospitable than if they are happy improbabilities or even grace-mediated "impossibilities"?) Here I want to focus on the foundational empirical question: regardless of what scientific or theological precommitments would incline one to expect, what do we see? This involves the profoundly important issue not just of understanding how the living world has come into being but being clear about what kind of world it, in fact, is that we are attempting to understand.

There are two ways in which evolutionary history may involve large-scale trends or directionality: there could be a series of thematically coherent major transitions, or there could be a more continuous sequence of ongoing enhancements. With respect to cooperation, the world we see displays each.

Major Transitions

The history of life uncontroversially displays major transitions that reformulated organic function and opened up new possibilities for adaptation: the origin of oxygenic photosynthesis and aerobic metabolism, the colonization of land, the development of internal skeletons and nervous systems, etc. Though in some vague sense these may all be seen to reflect or enable increased complexity (often undefined), there is no obvious thematic vector through such transitions. More recently, however, there have been proposals for a series of major evolutionary transitions in how information is stored and transmitted that involves a sequential escalation of cooperative complexity (Maynard Smith and Szathmáry 1995, 1999; Szathmáry and Maynard Smith 1995; Sigmund and Szathmáry 1998; Michod 2000; Williams and Da Silva 2003; Lenton, Schnellnhuber, and Szathmáry 2004; Reid 2007; Batten, Salthe, and Boschetti 2008; Salthe 2008; Vermeij 2008). There are varied descriptions of epic changes; Maynard Smith and Szathmáry (1995) suggest eight major evolutionary transitions:

- Self-replicating molecules → groups of molecules working together in protocells to replicate
- Individual replicators → physically linked replicators (chromosomes)
- RNA as both gene and enzyme (information and catalyst) → A DNA genes, protein enzymes
- Bacteria (prokaryotic cells without nuclei) → Eukaryotic cells with nuclei and organelles
- Asexually (clonally) reproducing individuals → Sexually reproducing populations of individuals
- Unicellular organisms → Multicellular organisms (plants, fungi, animals)

Solitary individuals → social groups (eusocial insects with non-reproductive castes)
Primate societies → Human societies (language as extragenetic information)

The above list is not a mere compilation of unrelated episodic shifts, but each transition a) shares the feature of complexifying how information is packaged and processed, by aggregating replicating entities into higher functional levels, and therefore b) both influences and is influenced by what comes next. However, significant though this is, more can be said related to the theme of this volume: there are several characteristics that all or most transitions have in common and that relate specifically to cooperation.

The first is simply that cooperation is instantiated in and is necessary for each level of organic complexity (see Nowak, Chapter 4). Replicating molecules "cooperate, each producing effects helping the replication of others" (Maynard Smith and Szathmáry 1999, 17); chromosomes facilitate and require cooperation between genes; individual cells work together and ultimately require one another for viability, and so forth. Across these transitions, "Cooperation is essential in the emergence of new levels . . . [and] is now seen as a primary creative force behind greater levels of complexity and organization in Biology" (Michod and Roze 2001, 2). Indeed, the major transitions involve not just coordinated exchange of benefits but cooperation as defined in this volume, because there are fitness costs to individual entities. Hence, there is also the possibility for defection. Characteristic of cooperative dilemmas, each level also shares the need to solve conflicts of interest between the individual components of the aggregate. For example, some genes on chromosomes in sexually reproducing organisms may subvert the "fairness" of meiosis by influencing their chromosome to be preferentially included in gametes (a process called meiotic drive), or worker bees may sometimes lay their own (unfertilized, male) eggs, which are less closely related to other workers providing care than sister eggs laid by the queen (Foster and Ratnieks 2001). These and other cases of cooperative defection that exist at each level might on the basis of game theory's "methodological individualism" be expected to subvert cooperative aggregation of lower into higher levels. But we see that such transitions have occurred, and have occurred serially. This is a striking aspect of life's history that is not strictly expected from, though turns out not to be inconsonant with, the bare logic of selection (Michod 2007).

Second, the above transitions represent a special kind of cooperation, involving not just symmetric exchange of the same behavior (as in food sharing or the simplest versions of PD or snowdrift games) but specialization of function and division of labor (Queller 1997; Maynard Smith and Szathmáry 1999; Michod and Roze 2001). Separation of informational and catalytic roles by DNA and proteins may be more efficient than a single molecule both carrying heredity and serving as an enzyme. Cooperative breeding birds (or nesting insects) may divide foraging and incubating or guarding activities. Isogamy (similar gametes) is rare compared to differentiated sperm and ova (Bulmer and Parker 2002, for a game theoretical approach), and specialization in multicelled organisms refines homeostatic efficiency and increases metabolic and behavioral scope. Thus over the history of evolutionary transitions, "an initially identical set of objects often becomes differentiated and functionally specialized," and although neither selection nor game theory predicts on the basis of first principles that this should happen, we might ask postdictively, "Why should selection favor such a division?" (Maynard Smith and Szathmáry 1995, 12).

One image of specialized versus nonspecialized cooperation is that of rowing games (where each team member uses one oar on one side of the boat) versus sculling games (where each oarsman rows on both sides using two oars) (Maynard Smith and Szathmáry 1995, 261). The latter can be represented as a classic PD, and invites defection. On the other hand the costs of defection in the former are much greater, and synergistic outcomes—where the advantages of cooperating exceed the additive benefits of each contribution—may both establish specialization and promote cooperation (Michod et al. 2006).

Third, the above transitions entail not just cooperation or even specialization but *obligate interdependence*. Many well-described systems of cooperation, even ones with differentiated roles, are nonobligate—namely, cleaning mutualisms in reef fish (Bshary and Grutter 2006) or algae-sea anemone symbioses (Muller-Parker and Davy 2001). But a common and very striking feature of the above major transitions is that individual entities relinquish former capacities to survive and/or replicate on their own and come to *require* the aggregate. The eukaryotic organelles (mitochondria, chloroplasts) that at one time existed as free living prokaryotes, can no longer replicate by themselves but need the whole cell. Sexual reproducers can (frequently) no longer make clonal copies of themselves and by themselves but need a population. The immediate viability of individual cells in a plant or animal, not to mention the future fate of their information, depends on all the other cells.

In fact, the major transitions are now widely referred to as "evolutionary transitions in individuality" (ETIs), which incorporate previously independent individual entities into a new functional aggregate, thereby entailing an emergent level of fitness (Michod 2005, 2007). It is the interactions not just between parts but between parts and the aggregate, or between fitness levels, that determines the properties and stability of the aggregate. In a sense the "methodological individualism" of game theory is vindicated, challenged, and reformulated. On the one hand, the logic of the individual that dominated earlier sociobiological and selectionist accounts, illuminates conflicts leading to or existing within emergent levels. On the other hand, vigilant emphasis on only one level as the locus of selection obscured recognition that entities with conflicting interests may sufficiently regulate conflicts and come to be aggregated in a way that the "individual" is reformulated. In the face of ETIs, resistance to hierarchical analysis characteristic of "well-engrained attitudes about selection . . . must be abandoned" (Michod 2000, 61). The scale-free nature of game theory can accommodate but does not anticipate this.

Fourth, transitional levels not only build on each other serially but also interact with each other synergistically. Of course the transition from prokaryotic to eukaryotic cells to multicellular organisms is an additive sequence. But the manifold diversification of cellular specialization and elaboration of morphological complexity in metazoans were also facilitated by other transitions. Chromosomal organization enabled an increase in genetic material. Sexual reproduction's employment of an egg enabled coordinated modifications of morphological development to be passed on in a single cell, which is unlikely to have been achievable by asexual budding involving many cells (Wolpert and Szathmáry 2002). A commonly repeated feature is that transitions have profound future effects that "open up new possibilities for future evolution," but these developments are not strictly predictable: natural selection does not envision the future, and it is frequently the case that a change occurs for one reason but has downstream effects for different reasons (Maynard Smith and Szathmáry 1999, 25).

For all these reasons, the evolution of cooperation leading to a series of major transitions in obligate interdependence is a crucial feature of biotic history and involves a profound theme in the development of life. Ironically, in light of recent work on ETIs, the provocative statement made by Michael Polanyi in his seminal *Science* essay on life's irreducibility—which has seemed so naïve to many over the last forty years—takes on new meaning: "We can recognize a strictly defined progression, rising from the inanimate level to ever higher additional principles of life . . . Evolution may be seen, then, as a progressive intensification of the higher principles of life" (Polanyi 1968, 1311). Indeed, referring to this very comment by Polanyi, Sigmund and Szathmáry observe that "This 'progress,' nowadays described as a series of major transitions in evolution, is often due to the emergence of new units" (1998, 439).

Evolutionary Escalation

It is not only the emergence of new units but also the intensification of cooperation within units that reveals an important evolutionary trend not strictly predicted by, though by no means incompatible with, selection and game theory. By way of illustration, I will briefly point to two examples.

First, the evolution of multicellularity represents both a major transition to a new level and an escalating continuum within a level. Multicellularity itself is not just a saltational jump but a progressive emergence that a series of intermediate steps are posited to be necessary to achieve (Kirk 2005; Michod 2007). There are natural examples of this: very "simple" organisms such as Volvocine algae manifest a continuum from unicellular species to aggregates with very few cells and no specialization or three-dimensional structure to those with thousands of cells, genuine cellular specialization, and various morphological forms (Herron and Michod 2007). And subsequent to the appearance of multicellularity—at whatever point that is judged to have occurred—there has been a dramatic evolutionary elaboration of morphological complexity and cellular differentiation. For example, there are approximately 10 different cell types in sponges, 50 in fruit flies, and 120 in vertebrates along with an increase in cell number of many orders of magnitude (Carroll 2001). Michod (2007) has developed a model relating the number of total cells (i.e., organism size) to life-history trade-offs favoring specialization. This escalating division of labor is what has allowed evolution to intensify the "higher principles of life" referred to earlier. Cooperation is not just a result of evolutionary processes but itself facilitates increases in the hallmark capacities of life to sense the external environment, to move and behave, and to maintain a constant interior regime. What we see is the progressive endowment of life with organismic utilities that, while ultimately contributing to reproduction, do not themselves monitor it—that is selection's job. To what extent the proximal utilities of the organism may wander from the ultimate utility of fitness is a question I will take up shortly.

Second, evolution has elaborated the transition to sexual reproduction in a number of respects, one of which involves parental investment. With asexual reproduction there is no distinction between parent and clonal offspring; with sexual reproduction there is. Hence there is a potential conflict between reproductive interests of parents and offspring: it is always in the progeny's interest to receive investment, but a parent must balance the benefit of investing in current offspring with the cost of reduced opportunity for future reproduction. Nurturing or protective care of young is one of many ways parents can invest in offspring, and there is a substantial game-theoretical literature on the situations in which various types of parental care should and should not be expected to emerge. Although this literature does not predict what will be the case generally, empirically we see two trends. One, parental care has evolved many times, independently in many lineages. Within lineages there is significant diversity of strategies and there are transitions in different directions, but the most common transitions are from no parental care to uniparental care, and from uni- to biparental care with few reversals back to no care (Reynolds, Goodwin, and Freckleton 2002). Across lineages in vertebrates, fish are dominated by no parental care, mammals by uniparental care, and birds by biparental care (Gross 2005). Two, there is an evolutionary trend toward lower fecundity and higher investment per offspring. Part of this may be due to physical principles that constrain allometric relationships between body size, lifespan, and fecundity (Schloss 2007). But even adjusting for allometry, investment increases as fecundity decreases. This important relationship involves a trend in the maximum and is not eliminative—namely, there are plenty of "low-investing" parents around and even more cases of no parental care. But the move away from a minimum represents yet another example of evolutionary increase in cooperative interdependence. Moreover, the term parental "care" is not just an anthropomorphism. With the mammalian limbic system, the hormonally mediated systems of social and parental attachment, and the increased ability to form enduring affective bonds in low-fecundity species, it is not unreasonable to posit the emergence of organismic utilities that quite literally involve "caring."

At this point, however, an important caveat is in order. The picture of increasing cooperative interdependence, even caring, should not be oversimplified or romanticized to mean inevitable reduction of conflict. For one thing, there is still a lot of conflict. Every new level retains the need to mediate conflict between entities within the aggregate, plus induces additional conflict between higher-level aggregates. Multicellular organisms have conflicting interests within and competition without.

For another thing, the trajectory itself may not be inevitable. In some cases, there may indeed be synergistic benefits that necessarily both propel cooperation forward and prevent regression (Frank 1995; Michod 2006). Multiple transitions to multicellularity and parental care may be examples of this. Other transitions, though, appear to involve "contingent irreversibility" in which the causes of a cooperative transition are not the same ones that make cooperation obligate, or that create downstream increases in cooperation (Szathmáry and Maynard Smith 1995). Obligate sexual reproduction, which is maintained by very different factors in different kinds of organisms, appears to be just such a situation. Thus it is not yet clear to what extent, as Williams and Da Silva propose, "the major developments from prokaryotes to eukaryotes, to multicellular organisms, to animals with nervous systems and a brain, and finally to human beings . . . were an inevitable progression" (2003, 323). What is clear is that it is not the case that selection requires all ostensible cooperation to be exploitation, or that there are no directional trends in cooperation, or that the ones described here are wholly contingent and—were we to play the tapes of evolution over again (Gould 1990)—we would not see anything approximating a rerun.

For both scientific and "larger" philosophical or theological reasons then, it is important not to trim the world to fit preconceived notions of how things must be. If romanticism and natural theology erred in one direction, recent versions of evolutionary nihilism err in another. In (of all places) his treatise on love, C. S. Lewis laments both overeager theologies and antitheologies that insist nature teach "exactly the lessons we have already decided to learn" (1991, 19). This is not to say that natural history is irrelevant to theology, for it provides a rich iconography that is simultaneously powerful and ambiguous.

On the one hand—and looking only at the one hand—there is something like an evolutionary pageant of increasing cooperation, of emerging capacities to care, and even of conflict itself, leading to "greater individuality and harmony" (Michod and Roze 2001, 6). Moreover, a recurring theme of the major transitions is gain through relinquishment, or the attainment of higher levels of aggregate function involving sacrifice of individual autonomy.

But these are thematic images, not lessons, and there are, on the other hand, countervailing themes. As the scale of cooperation increases, so too do both the scale of conflict and the kinds of costs that organisms are capable of incurring. And as organismal utilities and "desires" are elaborated, so too is the capacity to feel pain—eventuating, if not culminating, in the anguish of social loss in higher mammals. In fact, it is the ultimate capacity to experience and anticipate loss, and to choose it intentionally for the sake of another's benefit, that makes altruism—as defined in this volume—even possible to conceive of. But is it actually possible to achieve?

Counterreproductive Sacrifice

Other contributions to this volume address the topic of intentional altruism in humans (or other beings), which we are defining as "a form of costly cooperation in which an individual is motivated by goodwill or love for another." Both the motivational and consequential aspects of this phenomenon entail axes of variation. I have already commented on the evolutionary elaboration of organismal desires and capacity for affective attachment or "care," although I will leave others to discuss when desires become motives, and when motive entails "goodwill" or love. But also the notion of cost involves a continuum, from exploitation or gain at cost to another, to mutual benefit or by-product mutualism without cost, to benefiting another at cost to self, compensated by gains to kin or by direct or indirect reciprocal return (resulting in no net loss of fitness), to behaviors that are ultimately fitness-diminishing and confer benefit to others at net reproductive cost to the actor.

Both other-regarding motives and genuinely sacrificial consequences are intimately associated with what we take to be the highest forms of love, yet the latter—a necessary bridge between cooperation and altruism—seems impossible by the logic of selection. The above section described cooperative trends that are not strictly inferable from selection or game theory but can be postdictively interpreted in their light. This does not appear to be the case with counterreproductive investment in others, which legitimately runs afoul of the ostensibly necessary truth that entities failing to replicate biologically will simply not persevere. In Ghiselin's view: "If natural selection is both sufficient and true, it is impossible for a genuinely disinterested or 'altruistic' behavior pattern to evolve" (Ghiselin 1973, 967).

No biologist doubts the truth of natural selection. But it turns out there is disagreement—not between those who accept and deny evolution but among evolutionary theorists—over whether or in what sense it is sufficient, and "advances on many scientific fronts . . . have led many biologists . . . to consider selection as an incomplete explanation for the evolutionary unfolding of life" (Reid 2007; Vermeij 2008, 3; Batten, Salthe 2008; Salthe, and Boschetti 2008). In one almost trivial respect this is uncontroversial. Evolution involves both generative mechanisms and selective filters (Frank 1997), and the latter do not fully account for the former. Evolutionary game theory focuses on the latter, and it is not sufficient nor does it propose to give a complete explanation of the generation of new strategies. But this kind of insufficiency seems not to make room for sacrifice, which requires not that selection be conceptually "insufficient" to explain the generation of variation but that it be operationally insufficient to differentially retain and eliminate particular variations.

More controversially, there are several approaches to asserting just this kind of insufficiency. One is to claim that evolution proceeds not when (or primarily when) selection is most effective but when it is relaxed, as in the opening up of new habitats or niches through symbiogenesis, geologic events, ecological invasions, or ETIs (Gould, 1990; Reid 2007; Badyaev 2008). As with an opportunistically open business environment, there may be room to tolerate inefficiencies if not extravagance, even fitness deficits relative to others (though not absolute deficits)—but not for long! Another approach is to claim that generative processes are inadequately filtered, either because they overwhelm selection or because they are not reproductively transmitted and hence are not filterable by differential replication. In the first category are proposals that variations—arising by mechanisms involving not only mutation and recombination but also symbiogenesis (Margulis 1991), epigenesis (Reid 2007), ETIs (Michod 2005), and self-organization (Batten, Salthe, and Boschetti 2008)—cast up novelty more promiscuously than selection fully constrains. Like the above, this is also a nonequilibrium situation. In the second category is the observation that certain types of information or strategies may be recalcitrant to the action of selection and analysis by game theory, because they do not persist by (or primarily by) replication. This is a hard sell in the biological realm, but not when it comes to cultural innovations. Even though the diffusion of ideas and their associated strategies are subjected to game theoretical analysis assuming differential replication, it is not at all clear to what extent they actually are reproductively transmitted versus arising independently by processes of reason or native intuition. In the latter cases, it would make little sense to speak of the fitness of an idea or strategy and assess it by game theory if replication and social transmission are not modes of persistence. Yet there is a growing body of empirical evidence that some moral and religious concepts and behaviors fall into just this category (Barrett 2004).

Finally, not all traits are established for their reproductive value, and some may avoid elimination by selection even when they are fitness liabilities if they are spandrels linked to, or by-products of, other adaptive traits. At the genetic level this may be thought of in terms of pleiotropy, where one gene has multiple effects, or genetic hitchhiking, where one locus is influenced by selection for other loci with which it is associated (Barton 2000). For example, it has been found in slime molds that a single gene stabilizes costly cooperation and penalizes defection via pleiotropic effects: the costly behavior is associated with a physiological benefit that the organism is worse off without (Foster et al. 2004). And models have shown that what some call "strong altruism"—what I am here calling counterreproductive sacrifice, where benefit is conferred to another at net cost to the actor's fitness—can be promoted by hitchhiking at the level of genetic (Santos and Szathmáry 2008) and cultural (Gintis 2003) information. Some proposals for costly religious behaviors link by-product accounts with notions of cognitive innateness as mentioned above (Atran 2002).

In terms of the organism, we can think of these issues as involving interacting and sometimes competing utilities. Of course there is one ultimate evolutionary utility—fitness—but while this may be what selection "values," organisms seek proximate utilities of staying warm, getting groomed, avoiding predators, attracting mates, and so forth. Evolutionary game theory typically assesses the outcomes of different strategies for playing one interactive game. But organismic actors like chess masters often play multiple games simultaneously, and sometimes a losing strategy in one game may contribute to winning a different and more important one.

In this volume the theological ethicist Timothy Jackson (Chapter 16) favors the spandrel or by-product account as a way of navigating between, on the one hand, adaptationist reductionism that dismisses altruism by understanding every feature of life in terms of individual reproductive benefit, and, on the other hand, naïve interventionism that sees the world fashioned contrary to natural processes. This is a good start propelled by a wise instinct. But viewing altruism as a spandrel will not deliver fully the benefits he and others of kindred agapeist disposition seek—which include that we not be "limited to what can be accounted for by random mutation and natural selection," and that our ability to applaud altruism not be subverted by the recognition that it is "for the sake of self-interest."

For one thing, spandrels *are* limited to what mutation and selection establish, albeit selection operating on neighboring phenotypes. Hence pleiotropy does not purchase license to add an altruism term to Hamilton's rule, though it does enable costly, unreciprocated sacrifice for nonkin to the extent—and only to the extent—that it is associated with other fitness-enhancing phenotypes. For another thing, traits that may result in reproductive benefit—parental affection, spousal fidelity, and the loves they nurture—are not necessarily done *for the sake* of self-interest. Nor do we applaud them less because they result in enhancing organismic flourishing.

Nevertheless, a legitimate question remains about whether or how organisms sculpted by natural selection, or that have evolved by other material processes, can exhibit authentic self-relinquishment for the benefit of others. While this seems to involve tension between scientific and theological views of the world, I want to close by positing that it reflects disagreements *within* both science and theology about the ambiguous place of sacrifice in nature (Schloss 2002). Evolution exhibits a fecund and beautiful capacity to generate and amplify cooperation. Yet at every step, cooperation is constrained by conflict and by the requirement that it benefit, not diminish, fitness. Thus proposals that allow for the most radical relinquishment do so by positing a measure of "transcendence", either through nonequilibrium situations or exceptions. The most extreme affirmations—and not just religious ones—depict ". . . pure, disinterested altruism [as] something that has no place in nature." Yet somehow humans attain it, since, by theologically and scientifically unspecified means, "We, alone on earth, can rebel against the tyranny of the selfish replicators" (Dawkins 2006, 201). Indignant responses by biologists (and philosophers) to this view criticize it for viewing altruism as an imposition on, not fulfillment of, humanness, "a thin crust—something we invented rather than inherited" (de Waal 2006, 23).

This is unsatisfying to say the least. But maybe this brings us back to James's admonition that we not allow our inability to give a full account of reality to prevent us from affirming it in the midst of uncertainty. As Simone Weil reminds us (1988, 100), it is equally untenable "when men do not believe that there is infinite mercy behind the curtain of the world, or when they think that this mercy is in front of the curtain."

Notes

1. The most dramatic example of expectations legislating interpretations of issues related to this volume's theme is the all-is-selfish gene-centric triumphalism of a generation ago, described as a "revolution that has taken place in the way we think about social relationships. 'Genteel' ideas of vague benevolent mutual cooperation [have been] replaced by an expectation of stark, ruthless, opportunistic mutual exploitation" (Dawkins 1982, 55). I do not intend to tilt with this windmill but to engage more current and ecumenical approaches.

References

Atran, S. 2002. *In Gods We Trust: The Evolutionary Landscape of Religion*. New York: Oxford University Press.

Badyaev, A. V. 2008. "Evolution Despite Natural Selection? Emergence Theory and the Ever Elusive Link between Adaptation and Adaptability." *Acta Biotheoretica* 56: 249–255.

Barrett, J. 2004. *Why Would Anyone Believe in God?* Lanham, MD: AltaMira Press.

Barton, N. 2000. "Genetic Hitchhiking." *Philosophical Transactions of the Royal Society B* 344: 1553–62.

Batten, D., S. Salthe, and F. Boschetti. 2008. "Visions of Evolution: Self-Organization Proposes What Natural Selection Disposes." *Biological Theory* 3(1): 17–29.

Bshary, R., and A. S. Grutter. 2006. "Image Scoring and Cooperation in a Cleaner Fish Mutualism." *Nature* 441: 975–78.

Bulmer, M., and G. Parker. 2002. "The Evolution of Anisogamy: A Game Theoretic Approach." *Proceedings of the Royal Society B* 269(1507): 2381–88.

Carroll, S. 2001. "Chance and Necessity: the Evolution of Morpholoical Complexity and Diversity." *Nature* 409: 1102–9.

Dawkins, R. 1982. *The Extended Phenotype.* New York: Oxford University Press.

———. 2006. *The Selfish Gene.* New York: Oxford University Press.

De Waal, F. 2006. *Our Inner Ape: A Leading Primatologist Explains Why We Are Who We Are.* New York: Riverhead Books.

Dugatkin, L. A. 1997. *Cooperation among Animals: An Evolutionary Perspective.* New York: Oxford University Press.

———. 2001. "Subjective Commitment in Nonhumans: What Should We Be Looking For, and Where Should We Be Looking?" In *Evolution and the Capacity for Commitment,* ed. R. M. Nesse. New York: Russell Sage Foundation, 120–137.

Dugatkin, L., and H. Reeve, eds. 1998. *Game Theory and Animal Behavior.* New York: Oxford University Press.

Foster, K., and Francis L. W. Ratnieks. 2001. "Convergent Evolution of Worker Policing by Egg Eating in the Honeybee and Common Wasp." *Proceedings of the Royal Society B* 268(1463): 169–74.

Foster, K., G. Shaulsky, J. E. Strassmann, D. C. Queller, and C. R. L. Thompson. 2004. "Pleiotropy as a Mechanism to Stabilize Cooperation." *Nature* 431: 693–96.

Frank, S. 1995. "The Origin of Synergistic Symbiosis." *Journal of Theoretical Biology* 176: 403–10.

———. 1997. "Development Selection and Self-Organization." *BioSystems* 40: 237–45.

Ghiselin, M. T. 1973. "Darwin and Evolutionary Psychology." *Science* 179(4077): 964–68.

———. 1974. *The Economy of Nature and the Evolution of Sex.* Berkeley: University of California Press.

Gintis, H. 2003. "The Hitchhiker's Guide to Altruism: Gene-Culture Coevolution and the Internalization of Social Norms." *Journal of Theoretical Biology* 220: 407–18.

Gould, S. J. 1988. "On Replacing the Idea of Progress with an Operational Notion of Directionality." In *Evolutionary Progress,* ed. M. Nitecki. Chicago: University of Chicago Press, 319–38.

———. 1990. *Wonderful Life: The Burgess Shale and the Nature of History.* New York: Norton.

———. 1996. *Full House: The Spread of Excellence from Plato to Darwin.* New York: Harmony Press.

Gross, M. R. 2005. "The Evolution of Parental Care." *The Quarterly Review of Biology* 80(1): 37–46.

Hammerstein, P. 1998. "What Is Evolutionary Game Theory?" In *Game Theory and Animal Behavior,* ed. L. A. Dugatkin and H. Kern Reeve. New York: Oxford University Press, 3–15.

Herron, M. D., and R. E. Michod. 2007. "Evolution of Complexity in the Volvocine Algae: Transitions in Individuality through Darwin's Eye." *Evolution* 62(2): 436–51.

Huxley, L., ed. 1900. *Life and Letters of Thomas Henry Huxley.* Vol. 1. New York: D. Appleton & Co.

James, W. 2006. "The Will to Believe." In W. James, *The Will to Believe and Other Essays in Popular Philosophy.* New York: Cosimo, 1–31.

Kirk, D. L. 2005. "A Twelve-step Program for Evolving Multicellularity and Cellular Differentiation." *BioEssays* 27: 299–310.

Lenton, T. M., H. J. Schellnhuber, and E. Szathmáry. 2004. "Climbing the Co-Evolution Ladder." *Nature* 431: 913.

Lewis, C. S. 1991. *The Four Loves.* New York: Harcourt.

Margulis, L. 1991. *Symbiosis as a Source of Evolutionary Innovation: Speciation and Morphogenesis.* Cambridge, MA: The MIT Press.

Maynard Smith, J. 1982. *Evolution and the Theory of Games.* Cambridge: Cambridge University Press.

Maynard Smith, J., and E. Szathmáry. 1995. *The Major Transitions in Evolution.* New York: Oxford University Press.

———. 1999. *The Origins of Life: From the Birth of Life to the Origins of Language.* New York: Oxford University Press.

Michod, R. E. 2000. *Darwinian Dynamics: Evolutionary Transitions in Fitness and Individuality.* Princeton, NJ: Princeton University Press.

———. 2005. "On the Transfer of Fitness from the Cell to the Multicellular Organism." *Biology and Philosophy* 20: 967–87.

———. 2006. "The Group Covariance Effect and Fitness Trade-Offs During Evolutionary Transitions in Individuality." *PNAS* 103(24): 9113–17.

———. 2007. "Evolution of Individuality during the Transition from Unicellular to Multicellular Life." *PNAS* 104: 8613–18.

Michod, R. E., and D. Roze. 2001. "Cooperation and Conflict in the Evolution of Multicellularity." *Heredity* 86: 1–7.

Michod, R. E., Y. Viossat, C. A. Solari, M. Hurand, and A. M. Nedelcu. 2006. "Life-History Evolution and the Origin of Multicellularity." *Journal of Theoretical Biology* 239: 257–72.

Muller-Parker, G., and S. Davy. 2001. "Temperate and Tropical Algal-Sea Anemone Symbioses." *Inverbebrate Biology*. 120: 104–123.

Polanyi, M. 1968. "Life's Irreducible Structure." *Science* 160: 1308–12.

Queller, D. C. 1997. "Cooperators Since Life Began." *Quarterly Review of Biology* 72: 184–88.

Reid, R. G. 2007. *Biological Emergences: Evolution by Natural Experiment*. Cambridge, MA: The MIT Press.

Reynolds, J. D., N. B. Goodwin, and R. P. Freckleton. 2002. "Evolutionary Transitions in Parental Care and Live Bearing in Vertebrates." *Philosophical Transactions of the Royal Society B* 357(1419): 269–81.

Salthe, S. 2008. "An Anti-Neo-Darwinian View of Evolution." *Artificial Life* 14: 231–33.

Santos, M., and E. Szathmáry. 2008. "Genetic Hitchhiking Can Promote the Initial Spread of Strong Altruism." *BMC Evolutionary Biology* 8: 281.

Schloss, J. P. 2002. " 'Love Creation's Final Law?': Emerging Evolutionary Accounts of Altruism." In *Altruism and Altruistic Love: Science, Philosophy, and Religion in Dialogue*, ed. S. Post, L. Underwood, J. Schloss, and W. Hurlbut. Oxford: Oxford University Press, 212–42.

———. 2007. "Is There Venus on Mars?: Bioenergetic Constraints, Allometric Trends, and the Evolution of Life History Invariants." In *Fitness of the Cosmos for Life: Biochemistry and Fine Tuning*, ed. J. Barrow, S. Conway Morris, S. Freeland, and C. Harper. Cambridge: Cambridge University Press, 318–46.

Sigmund, K., and E. Szathmáry. 1998. "Merging Lines and Emerging Levels." *Nature* 392: 439–41.

Szathmáry, E., and J. M. Smith. 1995. "The Major Evolutionary Transitions." *Nature* 374: 227–32.

Vermeij, G. 2008. "Review of Biological Emergences: Evolution by Natural Experiment." *Quarterly Review of Biology* 83(1): 2–3.

Weil, S. 1988. *Gravity and Grace*. New York: Routledge.

Williams, R., and J. Frausto Da Silva. 2003. "Evolution was Chemically Constrained." *Journal of Theoretical Biology* 220: 323–43.

Wolpert, L., and E. Szathmáry. 2002. "Evolution and the Egg." *Nature* 420: 745.

II

What Can Game Theory Tell Us about Humans?

Justin C. Fisher

The preceding chapters have offered various game theoretical models of cooperation and have suggested that such models can help shed light on patterns of human behavior. One natural response to these models (already mooted by more than one of this book's contributors) is to grant that they might help us to understand computers and simple animals but to be skeptical about how much these models might tell us about humans. In this chapter I examine three potential reasons for such skepticism. Each of these potential reasons involves a feature that many people suppose sets humans apart from simpler creatures:

Mind-Body Dualism. Many people suppose human minds to involve some sort of nonphysical thinking substance, distinct from the physical substances that compose our brains and bodies. Since science usually deals only with physical things, one might worry that standard scientific approaches, including game theory, will have an especially difficult time accounting for nonphysical minds.

Free Will. Many people suppose that human actions are caused just by ourselves and not by the various factors that led us to be the way we are. Since science usually deals only with things that are fully under external causal influence, one might worry that it will have a difficult time accounting for freely chosen actions.

Complexity. Many people are struck by the rich capacity of humans to behave differently across a wide variety of circumstances. Since existing game theoretical models include only a handful of parameters, one might worry that these models will be ill-suited to account for the rich complexity of human behavior.

At first blush, these three concerns seem logically independent of each other: one can imagine possible views that accept any combination of these and reject the rest.[1] Nevertheless, many people who are moved by some of these concerns are also moved by the others. Each of these concerns might be motivated by attending to the rich diversity of human behavior, and especially to our capacity to behave in ways that are creative, inspired, and/or unpredictable. This rich diversity suggests (3) that human psychology must be very complex. This complexity might seem to recommend thinking (1) that the human mind contains something nonphysical in addition to the few hundred billion neurons and the quadrillion synapses that compose the human brain, and if our complex behavior is to be unpredictable not just in practice but in principle, this might seem to recommend thinking (2) that it is produced by some special sort of "agent causation" that comes from outside the web of physical causal relations.

My own view is that we currently have no reason to expect that the complexity or practical unpredictability of human behavior implies that human minds contain anything beyond the many billions of neurons and synapses that compose our brains. So, of these concerns, I am personally gripped only by (3). However, I recognize that (1) and (2) will also be gripping to many readers, especially adherents of popular religious traditions. This chapter takes (1), (2), and (3) seriously, and addresses the question of whether someone gripped by these should therefore have significant worries about the game theoretical approach to understanding human behavior outlined by other authors in this book. My conclusion will be cautiously optimistic: game theory is compatible with all plausible positions regarding dualism and free will, and, while there is much room for game theoretical models to improve with respect to the complexities of human cognition, these are improvements that we may expect game theorists eventually to make.

Mind-Body Dualism

There are various potential motivations for dualism. These include the other two concerns mentioned above: one might doubt whether purely physical substances could produce the rich complexity of human behavior, and/or whether they could do this in a way that is appropriately free from outside causal influences. In addition, many people have doubted whether purely physical substances would be capable of the conscious experiences that humans have.[2] Several religious commitments might also motivate dualism. If the mind is to survive the destruction of the body (to go on to heaven or hell or reincarnation or wherever), mind apparently must be distinct from body.[3] And if embryos are to be full-fledged persons and/or if they are to have "original sin," this might militate in favor of supposing that they have nonphysical minds even before they have brains.

I will not attempt to evaluate these motivations here.[4] Instead, my goal in this section is to ask whether dualism, *if it were true,* would pose a threat to the game theoretical modeling undertaken in this book.

A great deal of scientific work is geared toward understanding purely physical systems. One might worry about the potential to extend scientific models to nonphysical systems, as the dualist takes human minds to be. For example, Princess Elisabeth of Bohemia, in correspondence with Descartes, famously argued that there was no way of understanding how a material brain and an immaterial mind could interact (see Blom 1978). Such worries would be especially pressing against ambitious scientific research programs that hope eventually to show how all aspects of the world are composed of the same basic elements governed by the same basic laws.

In response to these concerns, I will argue, first, that there is no principled reason why there could not be a science of nonphysical substances, and, second, that game theoretical models in particular are especially amenable to the possibility that human minds will turn out to be nonphysical.

Let us begin with a useful analogy. Consider someone who has been completely immersed in a multiplayer online game since birth—virtual reality is the only reality she knows. Through clever experimentation, she might eventually discover all the laws governing the "physics engine" of her virtual world. However, she would also discover that the characters in the game sometimes behave in ways that are consistent and reasonable but completely unpredictable from within the physics engine. A reasonable conclusion would be that some things outside the game (namely the human players) are controlling the characters within the game. This might be the end of the road for developing a "physics" of this game, but it would not be the end of *science,* for our clever player could set up various experimental situations involving characters within the game (including herself) and see how the outside controllers for those characters react. Depending upon the richness of the game interface, our clever scientist might glean a great deal about human psychology, and perhaps even about the more general laws of the world outside her game.[5]

The dualist holds that our own predicament is very much like that of the unwitting player of a virtual reality game. The dualist holds that when we completely decipher the "physics engine" of our universe, some events in our brains will be left inexplicable. Given such findings, it would be reasonable to conclude that, like the characters of the game, our own brains and bodies are controlled by something outside them, and just as our clever game player could go on to test scientific hypotheses about what was controlling the characters in her game, we might someday go on to test scientific hypotheses about mental substances controlling our brains and bodies. Depending upon the richness of the mind–brain interface, we might glean a great deal about the structure of our minds, and perhaps even about more general laws of the world our minds inhabit.

So, there is no principled opposition between dualism and a general science of human cognition. There may, however, be a tension between dualism and particular scientific approaches to human cognition. For example, dualism is incompatible with neuroscientific approaches that aim to explain all cognition via the physical characteristics of neurons and other brain structures (see Hall, Chapter 12). Someday we shall need to choose between dualism and neuroscience. But this is a book primarily on game theory, not on neuroscience, so we can leave that choice to another day. Let us now ask whether *game theory* makes assumptions that are incompatible with dualism.

Most game theoretical models do make many assumptions about the players who play the games. Indeed, most models presume that the players have certain preferences, that they are fairly well informed about the structure of the game, and that they will be quite rational in choosing strategies that would do well to satisfy their preferences in the circumstances they believe themselves to be in.

However, these assumptions are entirely neutral about what sorts of substances the various players of the games are made out of. So long as the players have the relevant beliefs and preferences, and so long as they choose their strategies in the ways specified in accordance with the game(s), game theory will apply to them regardless of whether the players are made from flesh or silicon or ghostly ectoplasm. Unlike neuroscience, game theory considers human cognition at a level of abstraction that stakes no particular claims regarding what exactly humans are made of, and hence game theory is fully amenable to the possibility that dualism might turn out to be true.

Free Will

A distinction is commonly drawn between "compatibilist" and "libertarian" understandings of free will. According to the compatibilist, the fact that we sometimes act freely is compatible with the possibility that our world is deterministic—that, given the state of the world (including any nonphysical substances it might contain) at one time, the laws of nature fully determine how all subsequent events (including all human actions) will proceed. For instance, a compatibilist might hold that, so long as my actions are produced by healthy deliberative processes, they will count as "free actions," even if those actions (and the deliberative processes that produced them) were causally determined by my genes and my upbringing. Since compatibilism views our actions as being fully embedded in the causal structure of the world, there is no special tension between compatibilism and scientific approaches such as game theory.

In contrast to compatibilism, libertarianism holds that human actions are produced by a special sort of "agent causation" in a way that makes them not be fully caused by preceding events. Since the libertarian thinks human actions are quite different from ordinary physical events, she might worry that scientific approaches—such as game theory—which are well suited for explaining other events, would not work well at explaining human actions. Our goal in this section will be to explore these concerns.

There are several potential motivations for libertarianism. Human behavior seems spontaneous and unpredictable, and it would be very disturbing to learn that, no matter how hard we try, the current state of the world fully determines what all our future actions will be. There are also strong intuitions that we do act freely, and that our actions would not be free if some state of affairs outside our control—for example, the complete state of the world before we were born—was entirely sufficient to cause our actions to occur (see van Inwagen 1983). If we embrace both these intuitions, we must suppose that our actions are somehow free from antecedent causes outside our control.

Various religious commitments might also motivate an acceptance of libertarianism. If a world with libertarian free will would somehow be better than a world lacking it, an omniperfect creator would need to have included it (Murray 1993). Libertarianism may also be needed for the "free will defense" to the "problem of evil," the defense that blames all suffering on human free choice and thereby absolves the Creator of responsibility—this would be hard to do if the Creator's act of creation was itself sufficient to cause all these choices and the ensuing suffering (see Rota, Chapter 19 in this book).

As above, I will not attempt (at least not directly) to evaluate these motivations,[6] and will instead ask whether libertarianism, *if it were true,* would pose problems for the game theoretical models in this book.

Many game theoretical models are theoretically deterministic—they presume that whenever you put an agent with certain preferences into a certain sort of circumstance, the agent will definitely behave in a certain way. (In some models agents will definitely do the rational thing; in others they will definitely imitate their most successful neighbor.) This determinism is in apparent tension with libertarian free will, which allows that agents may freely choose to do any number of things in a given circumstance.

Game theory does have one tool for accommodating uncertainty regarding how people will behave—namely, allowing that agents might behave stochastically, choosing different strategies with different probabilities. However, determinately employing a fixed probability distribution would likely strike libertarians as being no more "free" than determinately employing a single fixed strategy (see Mele 2006). If libertarians insist that there is significantly more to free action than just randomly choosing actions in accordance with predetermined probability distributions, there will be tension between libertarianism and game theory.

In response to these concerns, I will argue (a) that game theory actually has fairly minimal commitments regarding the causal and probabilistic relations between human circumstances and human choices, and (b) that any plausible version of libertarianism must be compatible with these minimal commitments.

One of game theory's commitments is that, if a model is to explain actual instances of human behavior, the distribution of behaviors predicted by the model must match fairly closely the actual distribution of behavior in the world. For example, a model that says that 60% of people in a certain circumstance will make choice A can be a good explanation of people's actual choices only if approximately 60% of people in that circumstance make choice A. It is quite challenging to come up with models that actually do fit the complex diversity of human behavior, but that is a topic for the next section. For now we are concerned only with the question of whether doing so would be incompatible with libertarianism.

It is clear that there are facts about the probabilistic distributions of human choices. An exit poll of 10% of voters, for example, provides a very reliable estimate of how the other 90% will vote. One might take facts like this to suggest that there must be some sort of causal story, or at least some sort of probabilistic story, to be told about how all these voters make these choices, and one might worry whether such a story would be compatible with libertarianism. However, our present goal is not to make difficulties for libertarianism, so let us suppose that libertarianism can, somehow or other, accommodate these clear facts. But if libertarianism can accommodate such facts, as it apparently must, then we shall need to look elsewhere to find a tension between it and game theory.

Good scientific explanations do not just pick out interesting patterns in the distribution of various features in the world; they must also provide a guide to the causal structure of the world, and, in particular, they must provide a guide to intervening in that structure to bring about different results (Woodward 2003; Fisher 2006). So, a second commitment of game theory is that whatever factors its best models take to determine agents' behavior must be factors that actually are causally relevant to human behavior. For example, many game theoretical models hold that a game's payoff structure helps to determine what players will do. If these models are to be explanatory, it must be the case that changing the payoff structures of actual scenarios is a good way of bringing it about that agents will make different choices.

There are very strong reasons for thinking that changing payoff structures *can* change how people will act. The entire point of posting a reward for a lost pet is that doing so might cause someone to call in—and sometimes it does. One primary justification offered in favor of having a system of lawful punishments is the fact that (at least for many sorts of crimes) such systems deter people from behaving poorly. Once again, one might worry about whether libertarianism is compatible with these commonly accepted facts (see Ayer 1954). But since our present aim is not to make trouble for libertarianism, let us suppose that, somehow or other, libertarianism can be made compatible with them. So we find that this second commitment of game theory is also compatible with any plausible version of libertarianism.

Does game theory have other commitments that *would* be in tension with libertarianism? I think the answer is no. Much as game theory was neutral with respect to dualism, game theory is quite neutral with respect to exactly what sorts of processes produce human action. I have noted that game theory does presume that these processes produce certain distributions of behaviors in groups of people, and that they are sensitive to interventions on at least some external factors, such as payoff structures. But, I have argued that if there are any plausible versions of libertarianism, they would be compatible with these presumptions. Insofar as libertarians can devise a theory compatible with these presumptions, game theory will be neutral regarding the question of whether we should accept that theory or not.

Complexity

We have seen that evolutionary game theory need not conflict with either dualism or (any plausible version of) libertarianism. Let's turn, then, to the third concern listed above—namely, that game theoretical models won't be able to accommodate the complexity of human cognition.

Human choices seem to be predicated upon an incredible number of factors. Our choices apparently depend upon fine-grained beliefs about our current circumstances, upon the relative strengths of our various desires, upon habits built through past learning experiences, and upon idiosyncratic personality traits and temporary moods. In contrast with all this complexity, existing game theoretical models usually involve only a handful of parameters, and indeed, game theorists often strive to "keep it simple" and not add unnecessary parameters to their models. This might raise worries that game theoretical models are too simple to capture the sorts of complexity present in human behavior.

In response to this concern, I will first consider how far simple models can take us, and then move on to consider prospects for extending such models to accommodate more complexities.

Simplicity in explanation is not necessarily a bad thing. We may illustrate this point with an example from Hilary Putnam (1975). Suppose I have a square, wooden peg with two-inch sides, and a wooden board with a round hole two inches in diameter. How shall we explain the fact that the square peg will not fit through the round hole?

One proposed explanation might laboriously enumerate all the possible orientations of the peg and detail for each of these which parts of the peg would collide with which parts of the board surrounding the hole. This proposed explanation might convince us that the peg will not fit, but it will not afford us an intuitive understanding of why it will not fit, nor will it help us to approach similar cases.

An alternative explanation would point out that no rigid peg will be able to pass perpendicularly through a hole if it the peg contains two points in its cross section that are further apart than the diameter of the hole. Since the two corners of our peg are more than two inches apart, it therefore cannot fit through our two-inch-diameter hole. This alternative explanation makes no attempt to capture the full complexity of all the ways in which our peg might collide with our board. Instead, it simply shows how our peg is just one instance of a general pattern involving many pegs and many holes. This general pattern allows us to predict which pegs can pass through which holes, and it gives us recipes for intervention, telling us, for example, how much we would need to whittle our peg to get it to fit through our hole.

For very many purposes, we want our explanations to be like this simple alternative explanation, highlighting simple patterns that are predictively and pragmatically useful. Once we see this general pattern, a detailed list of potential collisions is quite beside the point—these extra complexities add little, if anything, to an explanation in terms of the simple general pattern cited above. Good explanations are supposed to help us see the forest, not just catalog all the trees.

Similar considerations apply to game theoretical modeling. Game theoretical models clearly make no attempt to capture all the psychological complexities that play a role in producing human decisions. Instead, game theoretical models attempt to capture some small number of parameters (e.g., features of the payoff structure, reputation effects, or heuristics for imitating neighbors' strategies), and show how these parameters can be used not only to predict what humans will do but also to give us recipes for intervention, telling us, for example, what sorts of incentives we would need to offer in order to get players to engage in cooperative ventures.

There is a lot to be said for an explanation that picks out a few highly relevant factors and shows how a pattern of results depends upon those factors. It is quite challenging to come up with a model that captures a great deal of data using only a few parameters, especially parameters that give us a handle for intervening to bring about particular outcomes. Models such as these are explanatorily valuable, often more so than models that use a large number of parameters to better fit the data.

But ultimately, we want both to see the forest and to know about all the particular trees. We want our explanations not just to highlight simple highly relevant patterns but also give us links to more detailed information, more accurate predictions, and more nuanced interventions. In Putnam's square peg example an ideal explanation wouldn't *just* tell us that the corners of the square are too far apart to fit through the hole, but it would also offer us links to more detailed information, for example, about how rigid objects maintain their geometrical structure during collisions. Similarly, ideal explanations of human behavior won't *just* highlight the simple patterns that current game theoretical models highlight. Instead, they must also offer links to more detailed information, for example, information about how humans manage to arrive at choices that are fairly rational.[7]

It is difficult to say how well game theory currently measures up in these regards. As a case in point, Almenberg and Dreber (Chapter 6) describe a number of games that experimental economists have watched human subjects play. There are two general ways in which economists' predictions have differed from observed behavior of human subjects. One difference is that human subjects apparently care more about other players' payoffs than economists originally expected. Almenberg and Dreber note that this might be accommodated in game theoretical models either by presuming that players have a different preference structure than was originally thought, or by presuming that subjects are choosing in something less than a fully rational manner. A second difference was that behavior varied significantly across human subjects. For game theory to accommodate this variation, it will likely need to introduce further parameters—perhaps different subjects have different beliefs about what consequences their actions would have, or perhaps they have different levels of other-regarding desires. Ultimately, one would hope that whatever modifications game theorists make to their models could be empirically motivated by psychological evidence involving how humans make decisions. An honest assessment of this work must acknowledge that much behavior of human subjects is not accounted for by current game theoretical models. So there is a great deal of room for improvement here, but also reason to hope that improvement will occur.

Critics of game theory will likely emphasize the gaps between the predictions of current game theoretical models and the wide variance in human behavior, as well as the gaps between simple game theoretical presumptions about psychology and the complex psychological story that surely underlies human decision making (see Hall's Chapter 12). Such criticisms are valid in that they highlight ways in which our current best explanations fall short of the sorts of explanations we would like to find. These criticisms are also valuable in that they help to point the way toward future models that will be better integrated with our understanding of human psychology, and will make better predictions.

But there are good grounds for optimism here. In the preceding sections we considered worries about dualism and libertarianism that threatened to show that game theory was the *wrong kind* of theory for describing human cognition. In contrast, the present worry involves not a matter of *kind* but just a matter of *degree:* can game theoretical models accommodate *enough* of the complexities of human cognition? This worry allows that game theory might at least be the *right kind* of theory for the task—it just insists that we need to find richer and more complex models of this sort.

This is good news for the game theorist. For if the bar is complexity, complexity is something that always can be added to game theoretical models. Even if current game theoretical models leave a great deal of room for improvement vis-à-vis the complexities of human psychology, these are improvements that game theorists can and surely will make.

Notes

1. I say "at first blush" because it may be questionable whether certain views of libertarian free will are actually coherent at all. If libertarianism contains logical contradictions within itself, it certainly will not be logically compatible with any other views.
2. For a good survey, see Chalmers 2003.
3. Some elaborate accounts of the afterlife are compatible with the view that living humans are purely physical beings. For example, one might hold that, at the moment of death, human bodies are transported away to the afterlife and replaced with look-alike corpses (see van Inwagen 1978), or some physicalists hold that there is a potential for uploading our minds into computers (Bostrom 2004) and one might imagine that a deity has arranged it so that something similar already happens at the time of bodily death.
4. The editors have encouraged me to explain why I myself am not attracted by dualism. My strongest reason is Occam's razor. We know that brains exist, that they are immensely complex, and that many physical changes in brains cause changes in behavior. Given all this, the default hypothesis is that our brains are in sole control of our behavior. We would need a compelling reason to posit some further thing helping to control our behavior. As I shall note below, I do not accept views of free will that would motivate dualism. Regarding consciousness, I am not so confident of my introspective abilities that I would adopt dualism merely on the basis of the fact that my conscious experiences do not seem physical (cf. Dennett 1991). Furthermore I hold no religious beliefs that require dualism.
5. Notice that, in this case, Princess Elisabeth's worry (how could mind and body possibly interact?) completely evaporates.
6. The editors have encouraged me to explain why I myself am not attracted to libertarianism. My main reasons for this are, first, that libertarianism strikes me as being either incoherent or else too bizarre to be plausible, and, second, that

the intuitive evidence that supposedly favors libertarianism instead seems, on reflection, to support compatibilism. Why would libertarianism be incoherent or bizarre? Because it is difficult to see how my actions could be sensitive to my antecedent knowledge, desires, and plans, and yet be free of antecedent causes. Even if I pretend it makes sense to posit some special sort of "agent causation" here, Occam's razor cautions against positing such things without good reason. Why do I not see a good reason? My actions do not seem to me to be uncaused by antecedent factors, and even if they did seem that way, I doubt I would trust such seemings. As for intuitions that free action and moral responsibility are incompatible with our actions' having antecedent causes, it seems to me that our systems of education, reward, and punishment presume that things we do *can* cause changes in what people will freely choose, and hence I think that, on reflection, we are intuitively committed to compatibilism at least as strongly as we are to libertarianism. Furthermore I have no religious beliefs that require libertarianism.

7. For an account regarding how explanations might offer such links, and regarding where these links should lead, see Fisher (2006, Chapter 5).

References

Ayer, A. J. 1954. "Freedom and Necessity." In *Philosophical Essays*. London: Macmillan, 271–84.

Blom, J. 1978. *Descartes, His Moral Philosophy and Psychology*. New York: New York University Press.

Bostrom, N. 2004. "The Future of Human Evolution." In *Death and Anti-Death: Two Hundred Years after Kant, Fifty Years after Turing*, ed. C. Tandy. Palo Alto, CA: Ria University Press, 339–71.

Chalmers, D. 2003. "Consciousness and Its Place in Nature." In *Blackwell Guide to the Philosophy of Mind*, ed. S. Stich and F. Warfield. Cambridge: Blackwell, 102–42. Also online at http://consc.net/papers/nature.html. Last accessed December 15, 2009.

Dennett, D. 1991. *Consciousness Explained*. Boston: Little, Brown.

Fisher, J. 2006. "Pragmatic Conceptual Analysis." PhD diss. University of Arizona.

Mele, A. 2006. *Free Will and Luck*. New York: Oxford University Press.

Murray, M. 1993. "Coercion and the Hiddenness of God." *American Philosophical Quarterly* 30: 27–38.

Putnam, H. 1975. "Philosophy and Our Mental Life." In *Mind, Language and Reality, Philosophical Papers*, vol. 2, ed. H. Putnam. Cambridge: Cambridge University Press, 291–303.

Van Inwagen, P. 1978. "The Possibility of Resurrection." In *Immortality,* ed. P. Edwards. Amherst, NY: Prometheus Books, 242–46.

Van Inwagen, P. 1983. *An Essay on Free Will.* Oxford: Clarendon Press.

Woodward, J. 2003. *Making Things Happen: An Account of Causal Explanation.* Oxford: Oxford University Press.

12

How Not to Fight about Cooperation

Ned Hall

Billy and Suzy hold distinctively opposed philosophical viewpoints. Suzy defends a version of *Sophisticated Christian Theism,* Billy a version of *Reductionist Atheist Physicalism.* Suzy believes in the Christian God—an omniscient, omnipotent, omnibenevolent being capable of loving personal relationships with its subjects, whose sole offspring was Jesus, and who was more or less directly responsible (through suitable human conduits) for much if not all of what is recorded in the Bible. Billy believes that nothing like such a being exists.[1] He further believes that physics can, at least in principle, provide a correct *and complete* description of reality.

Clearly, Suzy and Billy have plenty to disagree about. But of special interest here is whether, and in what ways, they should disagree about the nature and origin of *cooperative behavior,* particularly among humans. Just the other day, Billy cut short his vacation to help Suzy move: a sweaty and unpleasant business, vastly less enjoyable than time spent luxuriating on the beach. The two of them worked together in a way that imposed a cost on Billy (cultural, of course, and not genetic), while conferring a benefit upon Suzy—so, a clear case of cooperation, in the technical sense of that term used in this volume. Should the profound philosophical differences between

Suzy and Billy lead them to analyze this episode, and the larger patterns of cooperative human behavior of which it is a part, differently?

Maybe. Perhaps Suzy should consider this, and other manifestations of cooperative behavior, to be instances of genuine *altruism,* as we are understanding that category in this volume: cooperative behavior motivated, ultimately and at least in part, by unselfish concern for another.[2] Perhaps she should trace the capacity for altruistic behavior to the divine image in which we humans have been created, adding that *no further explanation* is called for—certainly not an evolutionary one. By contrast, perhaps Billy should insist on just such an explanation, adding that once in place, it will show that there is, in fact, *no such thing* as altruism. Cooperation, yes, but just as cooperation in the *biological* realm has a neatly Darwinian, genetically "selfish" source in the enhanced prospects for propagation of the hereditary material that gives rise to it, so too cooperation in the *cultural* realm invariably has as its ultimate motivation some expectation of future reward, or satisfaction of narrowly self-regarding desire—hence, motivation that disqualifies cooperative behavior from counting as altruistic.

If these are the correct positions for Billy and Suzy to take, indeed the battle lines are drawn here no less starkly than in any other arena where SCT and RAP might face off. But I do not think these positions are correct. On the contrary, if we are careful to construct the philosophically most sensible versions of each position, we will find that, with respect to human cooperation and altruism, it is not at all clear that a SCT-based investigation and a RAP-based investigation will look all that different.

What follows is a sketch of my reasons for thinking so. I will outline what I think SCT should look like, pause to acknowledge the very real points of conflict with RAP, and then spend a bit more time exploring what RAP is *not* committed to. That will set the stage for some simple closing lessons, pertinent to the exploration of cooperation and altruism.

What Makes SCT Sophisticated

Suzy is a *sophisticated* Christian theist. What makes her sophisticated? There are two things in particular.

First, there is her epistemology. Unsurprisingly, she thinks we humans can have epistemically special experiences of the divine. But she respects enlightenment values enough to recognize the dispassionate, reasoned evaluation of publicly accessible evidence as a nonnegotiable feature of the good epistemic life. Accordingly, she has no doubts about evolution, and finds all forms of creationism, including its most recent incarnation as "Intelligent Design," contemptible. Suzy is, in short, a scientist and scholar in good standing. Now, how *precisely* she reconciles scientific and scholarly norms of good epistemic behavior with her Christianity, and with her view of the Bible as a divinely inspired document, is not a trivial matter.[3] But we will take it that she is committed to working out the most amicable reconciliation possible—one that will provide scientific and scholarly inquiry with a central and secure role in guiding her belief.

Second, there is her metaphysics. She thinks God exists, but not as some kind of large, male humanoid, with a flowing beard and occasionally irascible temperament, and possessing, in the ordinary human sense, beliefs, desires, intentions, and so on. Her God is omniscient, omnipotent, omnibenevolent—yes, but *what it is* for God to have these attributes cannot be understood merely by imagining a good, knowledgeable, and powerful *person* and then "ramping up" the intensity of these characteristics to their logical maximum. In fact, one important reason why theologians have, according to Suzy, legitimate employment is precisely that it is so difficult to convey in words a conception of God that is at once illuminating and generally intelligible.

She also thinks the mind or soul is distinct from the body. Still, she sees a clear need for a conception of the mind/body relation that leaves ample room for neurophysiology as a discipline that can provide genuine insights into the mind. She has no patience for the metaphysical simpleton's view that, at some point during gestation, a bundle of cells in a mother's womb suddenly gets a fully formed soul somehow "attached" to it, whereupon the soul begins to develop in psychological complexity in a manner and at a pace that just happen to neatly track the neurophysiological development of the ensouled brain.

Unavoidable Conflict

The sophisticated character of Suzy's theism makes her easy company for someone like Billy—and we shall see that he returns the favor. Still, there are deep points of disagreement. Billy believes, roughly, that "it's all atoms in the void," and that the best and deepest understanding of this "it" comes *solely* by way of the reasoned evaluation of publicly accessible evidence—in other words, by scientific inquiry. Suzy believes that reality has much more to it than atoms in the void, and understanding remains profoundly incomplete if unaccompanied by some element of revelation.

Still, one might suppose that with respect to *empirical science,* Billy and Suzy can get along just fine. Are we not often reminded that science and religion belong to different "domains," and provided that we respect the boundaries (e.g., in the way Suzy does), we need never fear genuine conflict? Poppycock. *Of course* conflict will extend into the scientific domain, regardless of how sophisticated Suzy makes her theism, or how moderate Billy makes his scientifically minded atheism. Avoid the kind of wishful thinking that seems to infect the view that science and religion belong to safely separate domains, and you'll find plenty of examples.

One deserves special mention. That is the inquiry related to neurophysiology, and the phenomenon of revelation: Billy thinks it is inevitable that experiences that Suzy takes to involve a direct acquaintance with the divine will be exposed, by the developing science of the brain, as *delusions.* Suzy thinks that our growing understanding of the mind–body connection will *vindicate* this essential component of her epistemology. Granted, brain science is not far enough along for this battle to be joined in a serious way—yet—but once it is, it will make the wars over creation and evolution look like schoolyard squabbles, and no amount of blather about how science and religion "address different questions" will keep the peace.[4]

All the same, for those who relish a fight, work *on cooperation and altruism* provides not so promising a venue. To see why, we need to consider how Billy's endorsement of RAP ought—and more importantly, ought *not*—to constrain his judgments as to what constitutes legitimate scientific inquiry.

Sophisticated RAP

Let us unpack "reductionist atheist physicalism" so that we can see what its various components contribute, and how they hang together.

The "A" in "RAP" is easy: it is just the denial of a certain class of metaphysical-cum-epistemological views, of which Suzy's SCT is a paradigm instance.

As for the "P," here is the sort of thing Billy has in mind: physics has as its basic job description the task of explaining *motion,* or more generally *change of configuration over time,* or more generally still, the *spatiotemporal structure of the world.*[5] Billy is a *physicalist* because he places a fairly specific bet as to how physics must accomplish this task: he thinks that a complete and correct physics will follow a pattern clearly visible since Newton of positing some small set of fundamental physical magnitudes that characterize the very tiniest bits of the world (be they particles, or strings, or points of spacetime itself), and positing laws governing the way these magnitudes are distributed.[6] And he holds, crucially, that however different these magnitudes turn out to be from those that figure in present-day physics, *they will not include irreducibly representational or phenomenal magnitudes.* For Billy, the two phenomena distinctive of conscious mental life—namely, that it has a phenomenology (i.e., there is "something it is like" to be in a given conscious mental state) and a representational content (i.e., conscious mental states are typically "about" or "directed at" something)—will not appear at the level of a correct fundamental physics.

Why this last constraint matters becomes clear when we turn to the "R" in "RAP". Billy is a thoroughgoing reductionist. He thinks that *physics,* and physics *alone,* is the discipline that aims to uncover the *fundamental structure of the world.*[7] It is not, for example, that there is chemistry, and there is physics, and these are two separate disciplines, each with its own proprietary domain. No, there is chemistry, and there is physics, and chemistry is *renting* its domain from physics: any successful chemical theories or explanations must be *reducible* to (a correct) physics. And biology must be reducible to chemistry, and thence to physics, and psychology must be reducible to biology, thence to chemistry, thence to physics, and so on.

What is "reduction"? Partly, supervenience. For example, two things cannot differ in chemical respects without differing in physical respects.[8] But reductions must also answer certain characteristic questions about the theory being reduced. Suppose some theory posits a kind of entity, process, property, or relation X. We can ask, "What is an X?" In the best cases, a reducing theory will answer this question fully, in its own proprietary terminology. "What is a methane molecule?" Answer: "An aggregate consisting of one carbon atom with four hydrogen atoms covalently bonded to it." But a theory infected with too much inaccuracy may not permit this sort of analysis but may still be reducible, for all that. For example, suppose that the term "mass," as it figures in Newtonian gravitational theory, simply fails to refer to anything. Then we cannot give an illuminating answer to the question, "What is mass?" We cannot even give illuminating truth conditions to such claims as "The mass of the Earth is about 6×10^{24} kg."—after all, all such claims are *false.* But we can *still* give—using, say, the resources of Einsteinian gravitational theory—a perfectly illuminating explanation of *why Newtonian gravitational theory is as empirically successful as it is* (and why claims about Earth's mass are, even if strictly false, nevertheless *appropriate*).

This last kind of explanation is key: for one theory to reduce to another, the second theory must possess the resources to explain, at least in principle (i.e., setting aside issues of computational intractability), why and to what extent the first theory is empirically successful. Billy's reductionism thus consists in his conviction that there is a hierarchical structure to the possible theories and explanations of natural phenomena such that those lower down have the resources to account for the success of those higher up—and physics is right at the bottom.

Billy's reductionism has bite. Consider homeopathy: according to Wikipedia, "Homeopathic practitioners contend that an ill person can be treated using a substance that can produce, in a healthy person, symptoms similar to those of the illness. According to homeopaths, serial dilution, with shaking between each dilution, removes the toxic effects of the remedy while the qualities of the substance are retained by the diluent (water, sugar, or alcohol)." It is well known—on chemical and physical grounds—that, after a typical series of dilutions, the only chemical remaining in the diluent is the diluent itself. For Billy, that is case closed against homeopathy. For *there could be no* reductive explanation (in terms of chemical theory, say) of the empirical success of homeopathy, given that the "remedy" is chemically indistinguishable from diluent that never contained any toxin to begin with. That is, if homeopathy *were* empirically successful—at *all*—that would show that our more fundamental chemical and/or biochemical theories—the theories to which homeopathy would have to reduce—were *radically mistaken.* Since they are *not,* homeopathy is a crock.

Reductionism does real work here. Consider Billy's friend Nancy, an arch antireductionist. She might likewise hold that homeopathy is a crock. But she *also* holds that we need to wait on *direct empirical tests* of that theory before deciding so. Billy, thanks to his reductionist convictions, does not think we need to wait at all: the relevant empirical data has already been gathered in the form of the overwhelming evidence in favor of those theories to which homeopathy would have to—but cannot—reduce.

What RAP Is Not

It might have seemed perfectly obvious to you that reductionism has bite. But the point was worth emphasizing, since I'm about to argue that reductionism is, in fact, *much more* permissive than, I think, is commonly supposed.

To begin with, nothing in reductionist physicalism suggests that the concepts and distinctions most useful to any given inquiry must be those of fundamental physics.[9] Indeed, they need not even be *definable* in terms of those of fundamental physics, and not just because higher-level categories are often "multiply realizable": capable of having instances, under very different fundamental physical regimes.[10] As the example of Newtonian gravitational theory demonstrates, a perfectly successful account, in terms of a reducing theory, of the extent of another theory's empirical success need not hinge on the availability of such definitions. Billy is committed to viewing the empirical success of, say, a biological theory of cooperation as ultimately explicable in fundamental physical terms. However, for all that, he does not doubt that such a theory might achieve a stunningly rich *conceptual* autonomy.

What is more, Billy considers this autonomy a *very good thing,* given the overarching explanatory aims of science. We could not possibly discern many of our world's most important organizing patterns if we limited ourselves to the concepts and distinctions supplied by fundamental physics. (Imagine trying to understand complex multicellular organisms without the concept of the *cell.*) This observation in no way contradicts reductionism. For it is no part of that thesis that all we *really* need to understand the world is physics, or that explanations framed using the conceptual resources of some other discipline are necessarily *worse.*

Everything reduces to physics. So Billy believes. But his understanding of what this comes to allows him to avoid serious mistakes that too often plague discussion of reduction and science. That everything reduces to physics places *some* constraints on the sorts of concepts that can be legitimately introduced, and explanatory strategies legitimately deployed, in other areas of inquiry—witness the scientific disaster that is homeopathy. But these constraints are hardly straightforward, and for that reason Billy is, despite his fervent commitment to RAP, *exceedingly* wary of passing judgment on the scientific standing of any given inquiry merely on the basis of concerns about reduction.

Consider acupuncture. Traditional acupuncture theory posits a vital energy called "qi" that flows along bodily "meridians." You might expect that Billy, knowing as he does that the body is a complex biochemical machine, would dismiss this theory as so much quackery. He does not—at least, not on *that* basis. For he has a good sense of just *how* complex the body is, and so it is far from obvious to him a priori that biochemistry will not permit a reduction of qi theory that *vindicates* it, at least partially (e.g., as much as Einsteinian gravitational theory vindicates Newtonian gravitational theory).

More generally, consider this schema:

1. If higher-level theory X is legitimately scientific, then it must be reducible to lower-level theories (and ultimately to physics).
2. X is not so reducible.

Therefore,

3. X is not legitimately scientific.

Billy believes every instance of (1), on account of his reductionism. His physicalism is, of course, relevant to how he will assess any instance of (2). All the same, he considers a typical instance of (2) to be quite difficult to establish. There are exceptions: in the case of homeopathy, for example, rudimentary knowledge of chemistry and physics shows that reduction is impossible. But homeopathy is an outlier; most cases of interest will not be nearly so easy to assess.

Altruism, Folk Psychology, and RAP

Let us consider the case central to the project of this volume: altruism, or cooperative behavior motivated by concern for another's well-being. The thesis that humans are capable of altruism comes as part of a package, a certain *folk-psychological theory of human behavior.* According to this theory, humans sometimes exhibit behavior that has, among its immediate causes, *contentful motivational states.* More colloquially, humans sometimes act *for reasons.* Furthermore, the reasons on which a human acts can exhibit a *hierarchical structure.* I might go to the store because I want some cream. We can ask: why do I want the cream? Answer: I want it so that I can put it in my coffee. Why do I want to do that? Answer: because I enjoy the taste of coffee with cream—and there the inquiry into *motivation* ends. For while we can certainly ask why I enjoy the taste of coffee with cream, the answer *will not* advert to some further *end* or *goal* of mine that this enjoyment *serves.*

So this theory posits enough psychological structure to allow for a distinction between *ultimate* and merely *instrumental* motivations. In so doing, it allows us to say more carefully what we mean by altruism. Consider once more Billy, who has cut short his vacation. Why does he do this? What are his reasons? A first-pass answer might be this: his reason for cutting short his vacation is that he wants to move Suzy's furniture. Okay, but why does he want to do this? Because he knows that he will thereby contribute to her well-being (even if only in a small way), and he wants to contribute to her well-being. Now comes a crucial question: why does he want to contribute to her well-being? It *might* be that another answer *of the same psychological kind* is forthcoming. For example, it might be that he wants to contribute to her well-being *only because* he anticipates that if she is happy, she will do nice things for him, and he wants her to do nice things for him. Perhaps *that,* in fact, is his *ultimate* motivation, in which case contributing to her well-being is only an *instrumental* reason for Billy to move the furniture. If so, we should not count his cooperative behavior as altruistic. Perhaps, on the other hand, *no* such answer is available. We can give *some* answer: for instance, it might be that Billy was raised to be the kind of person who naturally cares about his friends. But that is not an answer that exposes his concern for Suzy as merely instrumental. A reason for acting does not become instrumental *merely* because we can give an account of its origins. No, it loses its status as ultimate only if that account shows that there is some *other goal* to which the given reason is subservient, in the sense that the agent acts on the reason *solely in order to* fulfill this other goal.[11]

Should Billy's commitment to RAP give him *any* reason to be suspicious of folk psychology, and the distinctions and explanatory strategies it deploys? Of course not. The brain is an enormously complicated organ, capable of astonishingly subtle behavior. It is *quite* capable of displaying the complexity it would need to, in order to support a reductive account of folk psychology that vindicated, as opposed to debunked, folk psychology's pronouncements. And folk psychology, it is worth remembering, passes demanding empirical tests on a daily basis, all over the world. (Contrast, say, folk theories of *medicine,* which do not have such an impressive track record.) There is, finally, no good empirical evidence for the dark view that all ultimate motivations are—the obvious appearances to the contrary notwithstanding—at bottom selfish.[12]

It is a distressing feature of modern intellectual life that a position on human altruism tends to be deemed "hard-nosed" or "scientific" to the extent that it doubts its very possibility. I think that this is sheer fraud. There is nothing intellectually respectable about being an empty-headed cynic.

So Billy and Suzy can (and should!) both believe in real, genuine, honest-to-God (or at least honest-to-goodness) altruism. There is, to be sure, one lingering disagreement: being a fan of RAP, Billy is *open* to the possibility of a scientifically based debunking of the claim that humans can act altruistically, whereas Suzy is not (her theism precludes it). But we should not get too excited about this disagreement. It does not take much to breathe life into a mere possibility, after all—and for Billy, this particular possibility has, at this stage in our investigations of the nature and origins of human behavior, almost nothing to recommend it.

We should pause to debunk one particular reason for doubting folk psychology, which is that when we *think* we are acting, say, for the sake of someone else, what we are *really* doing is acting in a way designed to guarantee the propagation of our genes into future generations. Here, for example, is a famously purple passage from Dawkins's *The Selfish Gene:*

> Was there to be any end to the gradual improvement in the techniques and artifices used by the replicators to ensure their own continuation in the world? There would be plenty of time for improvement. What weird engines of self-preservation would the millennia bring forth? Four thousand million years

> on, what was to be the fate of the ancient replicators? They did not die out, for they are past masters of the survival arts. But do not look for them floating loose in the sea; they gave up that cavalier freedom long ago. Now they swarm in huge colonies, safe inside gigantic lumbering robots . . . (Dawkins 1989, 19)

Which is to say, us.

Does this picture threaten the folk psychological conception of humans as capable of genuine altruism? No. To make the point as clearly as possible, let us pretend that there is actually a *gene for altruism*, a bit of genetic material such that those and only those who have it develop, psychologically, in such a way as to be capable of acting on genuinely other-regarding reasons. Also let us suppose there is a nice Darwinian story of how this gene came to flourish among humans (maybe humans who have it tend to act altruistically only toward other humans who have it). That *still* would not make it the case that humans never act altruistically. The only thing that might lead you to think otherwise is if you anthropomorphize the gene for altruism in Dawkins's regrettable fashion, thinking of it as an agent that *manipulates* its host into acting on certain kinds of reasons, because doing so serves *its* selfish ends—so that humans end up suffering from a kind of pervasive false consciousness, and would rebel and act in a consistently *self-centered* fashion, if they could only realize how they are being duped by their genes. But however much this metaphor of "selfish genes" makes for readable pop science, it is just plain stupid. It should hardly be necessary to remind ourselves that genes are relatively uncomplicated collections of atoms, not nearly complex enough to sustain the kinds of psychological states required to "manipulate" anything.

Reductionism and Cross-Platform Explanations

So much for one serious confusion about reductionism, directly relevant to the investigation of cooperation and altruism. Here is another: it is all too common to find a confusion between *reductionism* and the availability of what I call "cross-platform explanations." Let me explain, by example.

Consider a simple spring. It has an equilibrium point. Perturb it a little from that point, and it will begin to oscillate in a characteristic sinusoidal fashion. Now, *many* systems in nature have this abstract structure: they feature some equilibrium state, and a kind of dynamical behavior that pushes them in the direction of this state with a "force" proportional to their deviation from it. All such systems exhibit the kind of harmonic motion that can be described, to a first approximation, by the mathematical models familiar from textbook treatments of springs—and this, even though some of the systems might be chemical, some biological, some economic, and so forth. One *explanatory model* works on many different kinds of *platforms.*

It is a very cool thing when you can find a powerful, widely applicable cross-platform explanation. But however cool it is, the discovery and deployment of a powerful cross-platform explanation should not be mistaken for *reduction;* the two have nothing to do with each other. Yet I detect (mostly in conversation) a tenacity to the idea that they are somehow linked, and this leads to various mistakes. For example, we are sometimes told, by overzealous biologists or philosophers of biology, that we should take seriously the idea that *cultures* evolve via Darwinian selection on "memes." We are sometimes told, by economists, that the model of decision making as utility maximization applies to every sphere of human deliberation. We are sometimes told (look at other selections in this volume) that evolutionary game theoretical models can explain, in a uniform fashion, the development of cooperative behavior in slime molds, and the development of cooperative behavior in human societies. All of these strike me as myths, made plausible only against a background view that sees a hard-nosed, scientific attitude as one that places the first importance on the development and application of cross-platform explanations.

That is foolish. If you want to be hard-nosed and scientific, follow Billy's example, and endorse RAP. But do not engage in the kind of wishful thinking that sees deep, powerful, cross-platform explanations where there are none.

As a helpful antidote, remember that different platforms are, well, *different,* in ways that can matter a great deal to the depth of applicability of a given explanatory model. Consider the abstract framework of a Darwinian model. It requires a set of entities that vary in some respects, where those variants are capable of persisting through time (perhaps in the same entities or in their offspring, and perhaps with some modification), and where the expected frequency of a given variant over time is sensitive to the nature of that variant. *Lots* of things can instantiate this basic, abstract framework—maybe even "memes"—but the *details* will matter tremendously. (For example, it was crucial to the success of a Darwinian account of biological evolution that inheritance works by a *particulate* mechanism, and not a *blending* mechanism.) So it is a serious mistake to sign on to an allegedly cross-platform explanatory model without carefully examining the ways in which differences in platforms matter to its implementation.

Here reductionism can actually help. Consider harmonic behavior again. Suppose you have found a spring that oscillates around some equilibrium position, and also a *population* that oscillates around some equilibrium value. If you are like Billy, you will believe that there is some explanation in more fundamental terms of these oscillatory behaviors. But the population and the spring are *built very differently*. So you have no good reason to expect that the reductive explanations will look the same. To the extent that they differ, they may well reveal that it is only in a *very shallow* sense, or only over a *narrow range*, that both phenomena can be modeled as instances of harmonic behavior.

There is a lesson of immediate relevance to us: when a biologist starts waxing lyrical about the prospects for using evolutionary game theoretical models to explain, say, both slime mold behavior and the development of complex institutions of cooperation in human societies, you should nod politely and then ask for the details. What is it, precisely, about slime molds that makes the model applicable to them (assuming it is)? And what is it, precisely, about human societies (and human psychology) that makes the model applicable to *them* (again, assuming it is)? It is crazy to think the answers will look anything but profoundly different. One can lose sight of this point if, in the human case, one indulges too complacently in talk of "genes for cooperation." I suspect that in the contexts in which such talk appears, it is tacitly understood to be a kind of *placeholder* for that feature of human psychology—whatever it is!—in virtue of which the evolutionary game theoretical model applies. (The alternative view—that the model applies, in virtue of the presence in human populations of a genuine gene for cooperation—seems too silly to attribute to any serious scientist.) And that fact merely exposes our *ignorance* (as yet, anyway) of the detailed psychological mechanism or mechanisms that would explain the applicability of the model.

One final point. Suppose it turns out that evolutionary game theoretical models *can* be usefully applied to the analysis of human cooperation. Would that discovery favor RAP over SCT? Probably not: for remember that SCT is already committed to standard evolutionary theory, and so committed to there being *some* kind of evolutionary account of the origins of our cognitive capacities. Granted, if we discover the "gene for altruism" mentioned earlier, it might be difficult to square this discovery with the idea that we are *all* created in the image of a loving God. (For why did He arrange things so that some of us got the gene, but not others?) But such an outcome seems unlikely—the crude genetic determinism behind this hypothesis being so laughable. Much more likely—again, on the assumption that evolutionary game theoretical models prove scientifically useful, in the analysis of human cooperation—is that the *way* in which they are deployed proves perfectly compatible *both* with RAP and with SCT.

Closing Lessons

Here is the take-home message: fans of RAP have no particular reason to doubt that there are, as a matter of human psychology, genuinely other-regarding attitudes that frequently influence humans in ways that generate cooperative behavior. "Billy helped Suzy simply because he cares about her—and *not* because he selfishly expected something in return." That can be a *fully accurate* explanation of Billy's behavior, even if it turns out that the full story about how cooperative—and indeed, altruistic—tendencies developed among humans is a Darwinian one. More to the point, a devotion to RAP provides no compelling reason to doubt such an explanation. RAP and SCT will, of course, disagree about what *ultimately grounds* the human capacity for selflessness: SCT will say that humans have this capacity because it was essential to God's designs for creation to bring into being creatures capable of genuine moral worth, which in turn requires the capacity to act selflessly; RAP will say something else, a full specification of which will have to await developments in cognitive neuroscience, but *part* of which may be that evolutionary pressures on our ancestors favored the development of such capacities. But with such a capacity in place, and with a proper recognition on the part of SCT that neuroscience has an important role to play in the explanation of human behavior, SCT and RAP can give remarkably similar accounts of the *cultural* evolution of more refined and fully developed norms of cooperation in human societies. To think otherwise is, I suspect, to confuse RAP with a Crass variant—and the difference in acronyms should serve as an adequate reminder of what a bad idea this is.

Notes

1. The "nothing like" matters: for one could certainly deny the specific tenets of SCT while believing in some sort of divine presence. Billy believes in no such thing.
2. "In part," because we are complicated creatures, whose actions typically trace back to a variety of motives; we shouldn't narrowly restrict the term "altruism" to those cooperative actions free of *any* hint of a selfish agenda. And "ultimately," because the deep structure of motivation matters to the assessment of altruism: Billy might want to help Suzy but *only* in order to secure a later favor from her. If so—if, that is, it is *no* part of the ultimate motivation behind his cooperation simply to help someone he cares about—then his action doesn't qualify as altruistic.
3. The problem of evil looms large, in this regard—for it is exactly the problem of reconciling belief in God with widespread, prima facie evidence against the existence of any such being (see Rota, Chapter 19).
4. This is not to say that deciding on a victor in this upcoming battle will be at all straightforward. It *could* be straightforward, if the data are kind to us. For example, suppose that controlled experiments were to reveal that subjects who claim to have revelatory experiences regularly come to believe, on the basis of these experiences, specific and independently verifiable claims about the world to which they had no normal epistemic access; if so, that would constitute modest scientific evidence in favor of Suzy's theism, and rather serious evidence *against* RAP. Or suppose that careful studies reveal that allegedly revelatory experiences are strongly correlated with, and readily inducible by, specific kinds of dietary intake; that would rather tell against the view that they are typically caused by some kind of direct contact with the divine. But, of course, the evidential situation may not prove nearly so clean. In particular, the mere ability to induce seemingly revelatory experiences would not itself debunk all of them, any more than the ability to induce visual hallucinations shows that we cannot see. (Thanks, here, to Sarah Coakley.)
5. This is not to say that physics does not have *other* sorts of explanatory tasks—for example, the task of accounting for the various fundamental forces. But these

tasks only arise en route to fulfilling the *basic* task, which is that of accounting for spatiotemporal structure.

6. Cognoscenti will recognize Billy as a fan of the thesis that the philosopher David Lewis so brilliantly defended and that he called "Humean supervenience" (see in particular the introduction to Lewis 1986), although Billy does not necessarily subscribe to Lewis's controversial addition to that thesis that the fundamental physical laws themselves are nothing more than certain distinctive patterns in the distribution of fundamental physical magnitudes over space and time. Other cognoscenti will also recognize a problem—potentially very serious—generated by quantum mechanics: for that cluster of theories appears to teach us that we cannot completely specify the physical state of a system simply by specifying the physical magnitudes that characterize each of its tiniest parts, together with the spatiotemporal configuration of those parts. The quantum mechanical wave function of a system appears to capture an aspect of its overall configuration that goes *beyond* this material. Billy grants that he will probably need to concede this point but hopes he can do so without in any way compromising the philosophically significant aspects of his physicalism.
7. At least, in its *natural* aspects. Thus, perhaps one aspect of the fundamental structure of reality concerns the existence of numbers, or other mathematical objects; perhaps the world has an objective *moral* structure as well. If so, it is presumably not up to *physics* to say so. And if it were not for Billy's atheism, he might be willing to grant that there are *divine* aspects to reality that are outside the purview of physics—though it is less clear that the natural/nonnatural distinction really holds up here. Note, in addition, that Billy is well aware that *physicists* are not necessarily the best people to consult about every question concerning the fundamental structure of reality. For example, they will not necessarily have much that is useful to say about the question of what laws of nature are. So philosophers—in particular, philosophers of science who are also metaphysicians—still have plenty of work to do. It is just that they are, at bottom, really in the same business as physicists.
8. See, for example, Stalnaker 1996.
9. Supposedly, some early geneticists were quite impressed at the thought that their "genes" might prove to be the biological analog of the physicists' atoms. This thought is depressing. Biologists should know better than to succumb to *that* kind of physics envy.
10. "Rigid body" is a good example: presumably, there are various different ways that the fundamental physical structure of the world could be compatible with the existence, on the macro-level, of more-or-less rigid bodies.

11. Of course, there might be *mixed* cases. It might be that Billy has *two* ultimate reasons for moving the furniture: to contribute to Suzy's well-being and to get her to do nice things for him. In that case—presumably, the normal case—his action is less than perfectly altruistic. But by the same token, it is less than perfectly selfish.
12. Here it is necessary to mention a depressingly common and utterly sophomoric suggestion, which is that whenever a person *seems* to be helping another merely for the sake of that other's well-being, the ultimate motivation is *really* to enjoy a pleasurable feeling of satisfaction, or to avoid an unpleasant feeling of guilt. See Blackburn (2003) for an especially able debunking of this suggestion.

References

Blackburn, S. 2003. *Being Good: A Short Introduction to Ethics.* Oxford: Oxford University Press.

Dawkins, R. 1989. *The Selfish Gene,* 2nd ed. Oxford: Oxford University Press.

Lewis, D. 1986. *Philosophical Papers.* Vol. 2. Oxford: Oxford University Press.

Stalnaker, R. 1996. "Varieties of Supervenience." *Philosophical Perspectives* 10: 221–41.

V

Cooperation, Ethics, and Metaethics

13

The Moral Organ

A Prophylaxis against the Whims of Culture

Marc D. Hauser

Visual illusions are of interest to cognitive scientists because they reveal two fundamental aspects of our mental architecture: specificity of processing mechanisms, and encapsulation or insularity of processing (Fodor 1983). That is, when we stare at an illusion such as the Müller-Lyer case (Figure 13.1), our visual system immediately tells us that one line is longer than the other, even though our belief systems, empowered by the veridicality of measuring each line and determining that they are identical, tell us otherwise.

Other systems of the mind operate similarly, at least in the sense that there is an automaticity to our response, a sense in which the mind obligatorily decides for us how we will perceive the world. The idea I pursue in this chapter is that some aspects of our moral psychology function similarly (Dwyer 1999; Hauser 2006; Mikhail 2007). That is, I will suggest that when we confront certain moral situations, our minds automatically and obligatorily respond, generating specific moral judgments of right and wrong. And, like our visual system's response to an illusion, the system underlying our moral sense is also insulated to some extent from cultural influences, and especially, culturally created belief systems that often empower our religious

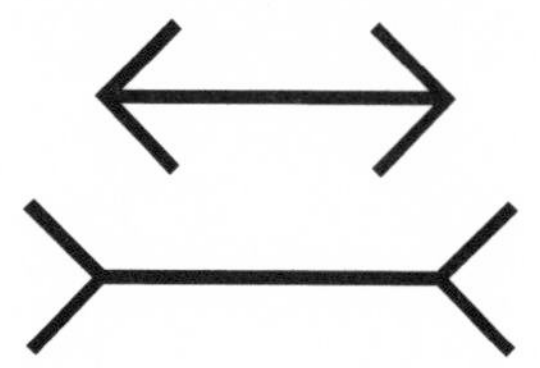

Figure 13.1 The Müller-Lyer illusion. Source: Müller-Lyer 1889.

and political institutions. As such, the intuitive moral judgments we generate operate on the basis of their own, internal logic, implementing abstract principles that gain their specific content from the highly local moral norms that specify the conditions permitting helping and harming others, as well as assigning labels of virtues and vices. What we typically think of as a rational process that moves from clearly justified principles to logical conclusions is often the outcome of an unconscious and involuntarily process, a point made forcefully by Haidt (2007, 2012). This suggestion in no way rejects rational deliberation as a mode of thought. Nor, as stated, does it reject cultural influences. Rather, it displaces the dominant role of rational deliberation from its presumed position as the engine of moral judgment to the interpreter of intuitive and unconscious moral judgments. And it displaces the dominant role of culture as the source of moral content and thus, specific judgments, to the role of filling in the details of biologically evolved but abstract rules. In this light, our evolved moral system constrains the form of each culture's specific moral system in much the same way that our evolved linguistic system constrains the specific language each culture speaks and comprehends.

Insomuch as our moral responses are evolved, functioning to coordinate cooperative activities among social group members, they may also fit at least generally into the concept of cooperation as it has been defined in this volume. That is, if cooperation, as articulated by Nowak and Coakley in the introduction and elsewhere, is a form of mutual benefit with a perceived potential cost, the ingrained moral responses I outline below would potentially subsume cooperation as an evolutionary phenomenon, especially in humans and at least some other social animals. This would indicate that cooperation is not only a stable evolutionary strategy (as potentially indicated by Nowak and Hauert in Chapters 4 and 5, respectively), but, critically, forms the basis for what we perceive as universal moral responses in humans and other social animals.

In what follows, after having described the theory of a moral grammar in Section 1, I turn in Section 2 to comparative evidence from animals that highlights potential building blocks to our moral faculty, cross-cultural data that speak to potentially universal principles, and neurobiological data that speak to some of the underlying mechanisms.

The Possibility of a Moral Grammar

The theoretical perspective I favor is, in many ways, closely aligned with the position that Haidt and other sentimentalists of the Humean tradition have proposed (Haidt 2001, 2007; Nichols 2002, 2004; Prinz 2004), but with a few crucial differences. First, whereas Haidt looks to the emotions as the cause of our intuitive moral judgments, I look to our perception of causality and intentionality. Importantly, however, in the same way that I do not deny the role of emotion in moral judgments, Haidt does not deny the role of our causal-intentional psychology. In some sense, where we differ is in emphasis, and, as a result, in the kinds of empirical evidence we seek. Thus, Haidt sees the phenomenon of moral dumbfounding (i.e., the inability to articulate a principled reason for a particular moral judgment) as mediated by unconscious emotions (e.g., the disgust we sense when we contemplate passionate love among siblings), whereas I see the same phenomenon as driven by an unconscious grammar of permissible action that relies on a cold calculation of intentionality, linking the causes of action to the ensuing consequences. Second, whereas Haidt places the emotions in a causally prior and necessary position for moral judgment, I see the emotions as *following* from our moral judgments, and, as such, will often be unnecessary for making moral judgments. These differences, though seemingly small, turn out to move our research programs in different directions, inspired by different theoretical traditions.

The inspiration for the theoretical framework I favor stems from the political philosopher John Rawls (1951 and 1971) who proposed that we think of our moral judgments in much the same way that linguists in the Chomskyan (1957 and 1986) tradition have thought about our grammaticality judgments. In brief, when we make grammaticality judgments, we do so spontaneously, without reflection, and, unless trained in formal linguistics, we do so without access to the underlying principles that govern syntactically legitimate sentences in the native, expressed language. From this general theoretical stance, the empirical agenda is largely set, requiring answers to five questions:

1. What are the descriptive principles that account for the mature speaker's knowledge of language?
2. How are the principles that are in place in the mature speaker acquired by the developing child?
3. How is our competence for language put into use during our communicative exchanges?
4. Are the mechanisms subserving our knowledge of language specific to language as opposed to being shared among domains?
5. How did the mechanisms subserving language evolve, and specifically, are these mechanisms uniquely human as opposed to being shared with other species, either by common descent or as the result of convergent evolution?

These questions have motivated several decades of empirical research in the various branches of linguistics, and continue productively to this day. By drawing upon an analogy to language, advanced in philosophy by Harman (1999), Dwyer (1999, 2004), and Mikhail (2000, 2011), and empirically by Mikhail (Mikhail 2007), and Hauser and colleagues (Cushman, Young, and Hauser 2006; Hauser et al. 2007; Young et al. 2007), the study of morality has picked up significant empirical steam. This chapter provides a sampler of what we have learned as of the writing of this chapter in 2007, and what is on the horizon.

The Empirical Landscape

By drawing on an analogy to language, studies of moral psychology have opened up a new set of research questions focusing on the folk psychological representations of cause and intention, the distinction between our competence to judge and our capacity to carry out morally relevant actions, the degree to which our moral judgments rely on domain-specific as opposed to domain-general processes, and the possibility that some aspects of our moral psychology are universal and immune to cultural influences because they reflect evolved specializations that, at one point, even if not today, provided significant solutions to adaptively relevant problems. Each of these very broad problems is affiliated with a suite of subsidiary problems that I myself and others have touched upon elsewhere (Dwyer 1999, 2004; Hauser 2006; Stevens, Cushman, and Hauser 2006; Hauser, Young, and Cushman 2007; Mikhail 2007; Hauser, Young, and Cushman in press).

Evolutionary Building Blocks

There has been much discussion about the evolution of morality. Most of this work has been discussed within two relatively distinctive perspectives. On the one hand are those who have focused on our emotions as the driving source of moral action, and thus have adopted a Darwinian approach to find insights into both potential precursors as well as adaptive function. Blair, Damasio, Greene, and Haidt (Blair 1995, 2007; Damasio 1994, 2003; Greene et al. 2001; Greene and Haidt 2002; Greene et al. 2004;) have pursued this angle from a cognitive neuroscience perspective, whereas Frank (1988), de Waal (1996, 2006), and Ridley (1996) have done so from an evolutionary-economic perspective. A second approach, one adopted in some form in this volume, has been to focus on cooperation as a general moral problem, looking at selection pressures to cooperate as well as mechanisms designed to detect and punish cheaters. This approach has been inspired by evolutionary game theory and behavioral economics, and, most recently, by neuroeconomics (Trivers 1971; Cosmides and Tooby 1994; Camerer 2003; Fehr and Fischbacher 2003; Glimcher 2004; Stevens, Cushman, and Hauser 2006; also see contributions in this volume from Nowak, and Almenberg and Dreber, Chapters 4 and 6, respectively). Here I focus on a few results that bear on the latter approach and, specifically, explore some of the psychological building blocks that were in play prior to the evolution of our own, distinctively human morality.

Imagine that, in the earliest forms of multicellular life, the only thing that mattered with respect to interacting with others is whether the *consequences* of action paid off in genetic fitness—that is, in the form of more offspring. At this stage, there was no capacity to make inferences about the psychological means or rules of action, no ability to detect why an individual engaged in one action as opposed to another. At some point in evolution, however, selection will have favored those individuals who picked up on the *means* underlying different actions and their consequences, as the means made a difference when predicting future interactions. Thus, although two individuals might act in such a way as to cause harm to another, one may have intended this harm, whereas the other may have brought it about accidentally. For a myopic consequentialist, the simplest strategy was to avoid all individuals associated with previously harmful actions. For organisms that could detect the means, however, a more sophisticated strategy was available: avoid the intentionally harmful agent but give the accidentally harmful agent another chance. Insofar as this distinction paid off in reproductive currency, selection would favor the capacity for mind reading, for making inferences about another's mental states. Once both attention to consequences and means were in place, internal psychological conflict emerged, a push-me-pull-you state where individuals might feel bad about intentionally harming another but not if a greater good emerged. Thus was born, loosely at least, the battles between utilitarian consequentialists arguing in favor of a myopic focus on consequences, and deontological nonconsequentialists arguing in favor of rules or means. Intriguingly, this distinction between consequences and means is played out in human development, with young children placing greater moral weight on the outcome of an action and only later in life realizing that what someone believed or intended at the time mattered (Piaget 1932/1965; Karniol 1978; Baird and Astington 2004). Similarly, recent work by Greene and colleagues (Greene et al. 2001; Greene et al. 2004) has shown where this battle is played out in the brain, and how, depending on the individual, different levels of neural conflict are experienced.

In the last few years, comparative research has revealed that nonhuman animals are also sensitive to the means underlying various actions, including morally relevant actions that arise in the context of cooperation. For example, it is now clear that a wide variety of nonhuman primates, including both distantly related monkeys (e.g., cotton-top tamarins, capuchins) and more closely related apes (e.g., chimpanzees), are sensitive to the distinction between intentional and accidental actions (Hauser et al. 2003; Call et al. 2004; Hare and Tomasello 2004; Herrmann and Tomasello 2006; Wood et al. 2007; Philips et. al. 2009). Thus, for example, chimpanzees show greater frustration when a human experimenter intentionally teases them by presenting and then pulling away a piece of food than when the same experimenter accidentally presents and then drops a piece of food. Similarly, tamarins, rhesus monkeys, and chimpanzees perceive a human experimenter's intentional grasp of a box as goal-directed but not if the same box is contacted by an accidental flop of the hand. In each of these cases, the consequence is the same, but the means differ.

The distinction between means and consequences also emerges in a more social and morally relevant context, specifically, cooperative games. Hauser and colleagues (2003) put cotton-top tamarins in a situation where reciprocally altruistic actions would generate the highest resource payoff for both players. In this game, one individual played the actor-1 role, where pulling a tool resulted in one piece of food for actor 1 and three pieces of food for actor 2. When the second individual playing the actor-2 role had access to the tool, pulling resulted in zero pieces of food for actor 2 and two pieces for actor 1. Each game entailed twenty-four trials, twelve alternating opportunities to pull for each actor. The interest of this game lies in actor 2's perception of actor 1's behavior, and specifically, whether actor 2 is focused on the consequences or intentions of actor 1. On the basis of both selfish and altruistic actions, we would expect actor 1 to pull on virtually every trial. If actor 2 perceives actor 1's pull as altruistic—biologically incurring a cost to benefit another—then actor 2 should pull to return the favor and maintain the reciprocally altruistic interaction. In contrast, if actor 2 attends to the means as opposed to the mere consequence of receiving food, then actor 2 should not pull, as actor 1's pull is purely selfish, with food delivered to actor 2 as a by-product of selfishly motivated behavior. The results were clear: individuals playing the actor-1 role pulled on virtually every trial, whereas individuals playing the actor-2 role, including those who had played the actor-1 role in other games, virtually never pulled. These results, together with those from other games played with this species, show that tamarins attend to both the means and the consequences of action. That is, they not only care about the benefits obtained but about the ways in which such benefits are provided.

In summary, research on event perception in nonhuman primates, as well as other animals, has begun to reveal some of the rudiments of the human capacity for mind reading, or what David Premack (Premack and Woodruff 1978; Premack and Premack 2002) famously called a "theory of mind." These capacities enable animals to read beneath the surface structure of an event, looking at an agent's intentions and goals. These capacities play into their ability to engage in cooperative interactions, handing off a suite of morally relevant building blocks (de Waal 1996, 2006; Hauser et al. 2003; Call et al. 2004; Hare and Tomasello 2004; Flombaum and Santos 2005; Hauser 2006; Herrmann and Tomasello 2006; Melis, Hare, and Tomasello 2006; Stevens, Cushman, and Hauser 2006; Warneken et al. 2007). And, from the game theoretical perspective emphasized in this book, a bourgeoning theory of mind would at least give a causal explanation to the role of reputation in iterative game-theoretic interactions that result in some form of cooperative behavior. It may also, on some crude level, help partially to explain the emergence of cooperative behavior, where the *perception* of mutual benefit despite cost is critical in maintaining cooperative dynamics.

Universal Principles and Parametric Cultural Variation

If the analogy to language is tenable, moral systems should show both signs of universality that cut across cultures as well as cultural variation that is constrained by the biology that permits each child to acquire a moral system. Moreover, the underlying biology should rule out, as impermissible, certain moral norms (i.e., even if cultures can invent them, they should have a short tenure), in the same way that a universal grammar represents a theory of both the permissible and impermissible expressed languages (i.e., cultures could, in theory and practice, invent all sorts of languages, but some will be unlearnable and thus have a short tenure). Though we are nowhere near a descriptive explanation of moral principles, here we can at least highlight the kind of evidence that would be relevant, and touch upon some findings that are consistent with this theoretical perspective.

To be clear, the idea that humans are endowed with a set of principles governing our moral judgments is not a theory about specific moral rules with specific moral content. Rather, it is a theory about abstract principles that operate over variables that have optional settings and a limited range of relevant content. For example, there is no principle, encoded in the brain, that states "killing is wrong." Rather, what I have in mind, and what other advocates of the linguistic analogy have proposed as well, is that there are principles dictating when harm is permissible (Dwyer 2004; Hauser 2006; Hauser, Young, and Cushman 2007; Mikhail 2011). As such, the computations that determine permissible harm evaluate whether the action was intended or accidental, whether another agent was harmed as a means to a greater good or as a mere side effect, whether the consequence of an action was directly or indirectly caused, whether one particular outcome can be said to supercede another, what the agent's primary goal was, and his options for attaining this goal.

In studies that I have carried out with my students, Fiery Cushman and Liane Young, as well as with the legal scholar John Mikhail, we have taken advantage of a web-based engine, the Moral Sense Test (http://moral.wjh.harvard.edu), to obtain large sample sizes enabling us to explore how variation in age, gender, religious background, and education impact upon some of the classic moral dilemmas engaged by philosophers to explore the nature of permissible harming and helping (Hauser et al. 2007; Cushman and Young 2011). In the case of harming actions, we have targeted three psychological distinctions, each associated with moral dilemmas that vary in their specific content:

1. Harming someone as the *means* to a greater good is morally worse than harming someone as a *side effect* to a greater good.
2. Harm caused by an *action* is morally worse than an equivalent harm caused by *omission* of an action.
3. Harm caused by *physically contacting* another is morally worse than an equivalent harm caused by means of *noncontact.*

Some of the classic trolley problems, created by the philosophers Foot (1967) and Thomson (1970), capture the first and third distinction. As the story begins, an out-of-control trolley is heading down a main track toward five people; if it continues, it will hit and kill these five. In one version of the story, if a bystander flips a switch, the trolley will turn onto a side track, thereby saving the five but hitting and killing one person on the side track. In a second version, if a bystander pushes a person off a bridge onto the tracks, the trolley will hit and kill this person, and because of the person's weight, the trolley will stop, thereby saving the five up ahead. Most people see the action of flipping a switch as more permissible than the action of pushing a person off the footbridge. We can explain this difference by saying that using a person as a *means* in the footbridge case is worse than the foreseen *side effect* of harming a person; in the footbridge case, the one person is the only means to saving the five (if no one is on the bridge, the only possible saving action is self-sacrifice), whereas in the switch case, if there is no one on the side track, flipping the switch is obligatory as cached out by the principle of cost-free rescue. Alternatively, we can invoke the third distinction, explaining these cases by saying that pushing the man (*contact* harm) is worse than flipping the switch (*noncontact* harm).

As recent work suggests, all three of the distinctions above, though morally relevant, recruit more general, nonmoral conceptual resources (Cushman, Young, and Hauser 2006). Specifically, the act/omission distinction is largely mediated by *causal* attributions whereas the means/side effect distinction is mediated by *intentional* attributions, conceptual representations that figure in a wide variety of social and even nonsocial situations.

Do these distinctions cut across cultural variation, showing up as universal parameters, a signature of our species? Though our web-based population is by no means representative of our species' potential or even actualized cultural variation, the sample consists of teenagers and retirees, men and women, atheists and the deeply religious, as well as primary school graduates, PhDs, JDs, and MDs. Thus far, such cultural variation has either very small effects on the patterns of moral judgment, or none at all. Thus, atheists and devout Catholics are equally likely to say that it is morally permissible to flip the switch and morally forbidden to push the man off the footbridge. And so too are men and women, young and old, bookish and not. And, such *moral prophylaxis,* which guards against cultural factors, carries over to other dilemmas as well, providing at least a first challenge to those who wish to see our moral psychology as open to the whims of culture (Banerjee et al., 2010).

For some, the fact that religious background seems to have little effect on the pattern of moral judgments is unsurprising. The basis for this response is usually the claim that Christianity has so deeply permeated the world that it alone is responsible for what is universal. But this cannot be right. Consider the fact that our sample of respondents consists of Jews,

Catholics, Protestants, Muslims, and Buddhists. In terms of doctrinal principles, these religions differ in many ways, and each claims that doctrine is handed down from a particular divine (or nondivine) source. But if one looks at the kinds of psychological distinctions people make when judging moral dilemmas, these distinctions often play no role within a given religious doctrine, or more importantly, do not play a role across all of the religions sampled. Further, even if religions were teaching such doctrine, we would have the further problem of explaining why, for many moral dilemmas, people provide robust judgments of right and wrong without being able to articulate why such actions are permissible or forbidden. That is, it would be odd if religiously tutored people were incapable of justifying their moral decisions with the lessons learned from their religious tutors. People with a religious background are thus left with a quandary: either stick with religious doctrine because this is what has been conveyed, independently of divine prescription, or abandon religious doctrine because it is neither connected to divine prescription nor of sufficient power to explain why certain cases feel intuitively right or wrong. I suppose there is another possibility: recognize that biology hands off, to all human beings, a set of principles for navigating within the moral domain, and these principles provide the building blocks for creating explicit moral systems.

We can look at the role of religious background, and thus a culturally created moral institution, from a different direction. First, consider a real world situation, that of abortion. Given current discussions and legal decisions, it should come as no surprise that people brought up within a religious tradition are more likely than atheists to judge abortion as forbidden. Based on a large internet sample, this shows up as a highly significant effect, even for some mildly apocryphal cases, such as:

> Sasha wakes up one morning and senses that she is pregnant. A man standing next to her bed informs her that overnight she has been impregnated with Princess Maria's two-month-old fetus. Princess Maria is incapable of carrying the fetus to term, and her next-born child will inherit the throne. Without Sasha's help, Princess Maria will not be able to deliver a child, and no one from her family will inherit the throne. Sasha decides to have an abortion.

Of relevance to the moral grammar thesis is whether the difference between religiously raised people and atheists would disappear if we enhanced the artificiality of the dilemma in order to make it less obviously about abortion. The thesis is that religion teaches explicit doctrine—namely, rules about highly specific moral situations. As soon as one departs from the specific case, however, presenting dilemmas that capture a core rule or principle within the moral grammar, one will find commonality across cultures—that is, shared judgments that cut across religious background.

To test this thesis, Ryan Boyko and I took advantage of a moral dilemma created by Judith Thomson (1971), designed to understand which aspect of abortion is specifically impermissible in the eyes of some. In particular, Thomson's dilemma attempts to isolate the issue of when life begins, focusing instead on whether the fetus has an obligatory right to the mother's body, or alternatively, whether the mother has an obligation to carry the fetus to term. The following is a minor modification of Thomson's case that Boyko and I used with our internet sample:

> Jill wakes up one morning and finds a strange man next to her in bed, plugged into her kidney. A man from the Society for Music Lovers introduces himself and explains to Jill that she has been plugged into a famous violinist who is dying of kidney failure. Without Jill's help, the violinist will die. The man from the society explains that Jill must stay plugged into the violinist for nine months in order for him to recover and survive. Jill pulls the plug and the violinist dies.

Results showed that most people judge Jill's decision as morally permissible. This case, one could argue, is not really like abortion because Jill never made a commitment to the violinist. For at least one aspect of Thomson's argument, this is irrelevant because what Jill's decision shows is that even in cases where someone else needs your body to survive, you are not obliged to share it, even if it would be nice if you did. More important, however: even if you change this case by having Jill agree to the request, staying plugged in but then unplugging two months later, you find an increase in the proportion of respondents who claim that Jill's action is forbidden, but no difference between atheists and people with a religious background. Although our work along these lines is still very preliminary, the observations thus far motivate the hypothesis that what religion and other culturally created moral institutions do is provide an explicit handbook for how to think about specific moral cases. *They do not, however, alter the operative force of our moral grammar's abstract principles.*

Moral Circuitry

A crucial question raised by the previous sections is whether the neural circuitry that guides moral judgments is specific to the moral domain or shared more generally with other domains of knowledge. A classic test of this kind of question is to look for evidence of selective breakdown—that is, cases where brain injury results in a deficit in moral thinking in the absence of other psychological deficits. For example, cases of psychopathy (Hare 1993; Blair 2007), together with studies of patients with frontal-lobe damage, such as the famous example of Phineas Gage (Damasio 1994), have strongly suggested that certain parts of the brain are not only necessary for normal moral functioning but perhaps specific to the moral domain. In particular, these studies, together with recent research in social psychology (Haidt 2001, 2007) and imaging-based cognitive neuroscience (Greene and Haidt 2002; Blair 2007; Moll et al. 2007) suggest that our emotions are fundamentally and causally intertwined with our moral judgments. When our emotional systems fail, especially the social emotions, so too do our moral distinctions. Thus, in Blair's classic work, psychopaths failed to make the moral-conventional distinction, a psychological contrast that, among normals, is universally perceived, present early in child development, and for some, at the core of what makes the moral domain different from other social situations (Turiel 1998, 2005). In particular, whereas psychopaths tend to judge violations of social and moral rules as equally impermissible, normals perceive moral transgressions as more severe. Since psychopaths have clear emotional deficits, the conclusion has been that a healthy moral sense depends upon a healthy emotional system.

Motivated in part by the linguistic analogy, I have argued that current work in moral psychology has often assumed, without test, an *emotion-first* view, one in which emotions are not only central to moral judgment but causally prior to such judgments (Hauser 2006; Hauser, Young, and Cushman

2007). At stake here is the temporal appearance of emotion in moral judgment as opposed to challenging the role of emotion altogether. That is, emotions may play a significant role in our moral thinking or behavior but more as a consequence of moral computation than as a cause. Specifically, we may compute what is morally right or wrong in the absence of emotional input but then, as a result of this computation, trigger an emotion that may, in turn, trigger specific kinds of behavior. Thus, an alternative explanation for the moral psychology of psychopaths is that their emotional deficits, along with their compromised capacity for certain kinds of reversal learning, lead to problems of moral behavior as opposed to judgment. That is, their moral knowledge is intact, but along the path from generating a judgment to carrying through on a particularly moral action, they fail to do the right thing—they fail to inhibit violence in a situation where most people would have quashed all violent tendencies (Blair 1995).

Although I do not believe any of the current data on psychopaths can resolve these alternative explanations, there is an ongoing effort to settle this issue with psychopathic populations. Of particular relevance to the distinction between normal and abnormal moral responses, as well as the causal role of emotional response, is recent work on patients with frontal-lobe damage, where, from clinical descriptions, patients have exhibited deficits in their ability to integrate the social emotions into socially relevant decisions, including moral decisions (Bechara et al. 1994; Damasio 1994; Bechara et al. 1997; Anderson et al. 1999;). Recently, this population has been tested on a battery of moral dilemmas, and this work provides new insights into the specifically causal role of social emotions on moral judgments (Koenigs et al. 2007). In particular, based on a study of six patients with bilateral damage to the ventromedial prefrontal cortex (VMPC), results showed that the pattern of judgments was indistinguishable from both normals and brain-damaged controls for nonmoral social dilemmas as well as for impersonal moral dilemmas; an example of the latter is the bystander trolley problem in which a person can flip a switch (impersonal) to turn the trolley onto a side track where it will hit and kill one, but save the five. For the impersonal moral dilemmas in particular, cases such as these are certainly emotionally evocative, and yet they trigger computations in the VMPC patients that must be sufficient, and sufficiently similar to normals, to create consistent moral judgments. For this class of dilemmas, therefore, emotions are not causally necessary.

For the final and third set of dilemmas—considered "personal" because the protagonist's act involves up-close-and personal contact with another agent—all were subjectively rated as more emotionally intense than the impersonal moral dilemmas. Here, the VMPC patients showed a different pattern of response from normals and brain-damaged controls, at least for a subset of the personal cases. Specifically, for personal dilemmas where the protagonist faced a self-serving but aversive act (e.g., abandoning a newborn infant to avoid the burden of caring for it), the VMPC patients showed the same pattern of response as normals. In contrast, for personal dilemmas where the protagonist was faced with the possibility of maximizing aggregate welfare but only by means of a harmful act associated with strong emotions (e.g., smothering a baby to save the lives of many), the VMPC patients were far more likely to endorse this harmful act than were normals. Said differently, damage to the VMPC apparently lessens the effectiveness of the emotions on this kind of moral decision, resulting in a greater tendency to focus on consequences. In these personal cases, therefore, emotions are causally necessary and operate prior to the moral judgment.

Results from patient studies are now joined by recent studies using imaging as well as transcranial magnetic stimulation (Saxe and Kanwisher 2003; Borg et al. 2006; Young et al. 2007). Though we are only at the beginning of this research enterprise, it seems clear that for a significant class of moral dilemmas, the computation relies more on the folk psychological or mental states of the protagonist than on the emotions that covary with their beliefs and actions.

Thank Goodness for Human Nature

The argument I have developed in this essay is that the linguistic analogy provides a useful way of framing questions about the nature of our moral psychology. In particular, I believe it opens the door to exploring the nature of our moral sense that is both unique, compatible with other lines of research, and has, within a relatively short time, generated novel data sets on nonhuman animals, cross-cultural samples, and neural mechanisms. I started out with the provocative suggestion that our moral organ may provide prophylaxis against cultural influences and could subsume some of the cooperative dynamics seen in recent work in game theory. Thus far, it looks as though certain moral distinctions such as that between means-based harms and side effects cut across genders, age, education, and religion. Further, some of the moral attitudes that appear to be shared within religious communities may turn out to have only local effects. That is, once people with a religious background are confronted with unfamiliar moral dilemmas, they judge these cases in the same way that atheists do. If this turns out to be the way our moral faculty works, it will not represent a denial of the role of culture but, rather, one of the ways in which our evolved capacities constrain the format of our cultures. In this sense, the forms of cooperation modeled and predicted in contemporary evolutionary game theory, especially for humans and primates, fits into a universal capacity for moral constraint based on inferential processing of agent intentionality. Cooperation is thus an expected outcome (at least for populations) of long-term interactions between individuals with similar moral and psychological dispositions.

Massive progress has been made in this area of research since this chapter was written, including important work by J. Blair, F. Cushman, A. Damasio, J. Greene, J. Haidt, J. Knobe, A. Raine, and L. Young.

References

Anderson, S. W., A. Bechara, H. Damasio, D. Tranel, and A. R. Damasio. 1999. "Impairment of Social and Moral Behavior Related to Early Damage in Human Prefrontal Cortex." *Nature Neuroscience* 2: 1032–37.

Baird, J. A., and J. W. Astington. 2004. "The Role of Mental State Understanding in the Development of Moral Cognition and Moral Action." *New Directions for Child and Adolescent Development* 103: 37–49.

Banerjee, K., B. Huebner, and M. Hauser. 2010. "Intuitive Moral Judgments are Robust across Variation in Gender, Education, Politics and Religion: A Large-Scale Web-Based Study." *Journal of Cognition and Culture* 10/3: 253–81.

Bechara, A., A. Damasio, H. Damasio, and S. W. Anderson. 1994. "Insensitivity to Future Consequences Following Damage to Human Prefrontal Cortex." *Cognition* 50: 7–15.

Bechara, A., H. Damasio, D. Tranel, and A. Damasio. 1997. "Deciding Advantageously before Knowing the Advantageous Strategy." *Science* 275: 1293–95.

Blair, R. J. R. 1995. "A Cognitive Developmental Approach to Morality: Investigating the Psychopath." *Cognition* 57: 1–29.

———. 2007. "The Amygdala and Ventromedial Prefrontal Cortex in Morality and Psychopathy." *Trends in Cognitive Science* 11: 387–92.

Borg, J. S, C. Hynes, J. Van Horn, S. Grafton, and W. Sinnott-Armstrong. 2006. "Consequences, Action, and Intention as Factors in Moral Judgments: An fMRI Investigation." *Journal of Cognitive Neuroscience* 18: 803–17.

Call, J., B. Hare, M. Carpenter, and M. Tomasello. 2004. " 'Unwilling' versus 'unable': Chimpanzees' Understanding of Human Intentional Action." *Developmental Science* 7: 488–98.

Camerer, C. 2003. *Behavioral Game Theory*. Princeton: Princeton University Press.

Chomsky, N. 1957. *Syntactic Structures*. The Hague: Mouton.

———. 1986. *Knowledge of Language: Its Nature, Origin, and Use*. New York, NY: Praeger.

Cosmides, L., and J. Tooby. 1994. "Origins of Domain Specificity: The Evolution of Functional Organization." In *Mapping the Mind: Domain Specificity in Cognition and Culture*, ed. L. A. Hirschfeld and S. A. Gelman. Cambridge: Cambridge University Press, 85–116.

Cushman, F., and L. Young. 2011. "Patterns of Moral Judgment Derive from Nonmoral Psychological Representations." *Cognitive Science* 35: 1052–75.

Cushman, F., L. Young, and M. D. Hauser. 2006. "The Role of Conscious Reasoning and Intuition in Moral Judgments: Testing Three Principles of Harm." *Psychological Science* 17: 1082–89.

Damasio, A. 1994. *Descartes' Error*. Boston, MA: Norton.

———. 2003. *Looking for Spinoza*. New York: Harcourt Brace.

de Waal, F. B. M. 1996. *Good Natured*. Cambridge, MA: Harvard University Press.

———. 2006. *Primates and Philosophers: How Morality Evolved*. Princeton: Princeton University Press.

Dwyer, S. 1999. "Moral Competence." In *Philosophy and Linguistics*, ed. K. Murasugi and R. Stainton. Boulder, CO: Westview Press.

———. 2004. *How Good Is the Linguistic Analogy?* 2004, www.umbc.edu/philosophy/dwyer. Last accessed February 25, 2004.

Fehr, E., and U Fischbacher. 2003. "The Nature of Human Altruism—Proximate and Evolutionary Origins." *Nature* 425: 785–91.

Flombaum, J., and L. Santos. 2005. "Rhesus Monkeys Attribute Perceptions to Others." *Current Biology* 15: 1–20.

Fodor, J. A. 1983. *The Modularity of Mind*. Cambridge, MA: The MIT Press.

Foot, P. 1967. "The Problem of Abortion and the Doctrine of Double Effect." *Oxford Review* 5: 5–15.

Frank, R. H. 1988. *Passion within Reason: The Strategic Role of the Emotions.* New York: Norton.

Glimcher, P. 2004. *Decisions, Uncertainty, and the Brain.* Cambridge, MA: The MIT Press.

Greene, J. D., and J. Haidt. 2002. "How (and Where) Does Moral Judgment Work?" *Trends in Cognitive Science* 6: 517–23.

Greene, J. D., L. E. Nystrom, A. D. Engell, J. M. Darley, and J. D. Cohen. 2004. "The Neural Bases of Cognitive Conflict and Control in Moral Judgment." *Neuron* 44: 389–400.

Greene, J. D., R. B. Sommerville, L. E., Nystrom, J. M. Darley, and J. D. Cohen. 2001. "An fMRI Investigation of Emotional Engagement in Moral Judgment." *Science* 293: 2105–8.

Haidt, J. 2001. "The Emotional Dog and Its Rational Tail: A Social Intuitionist Approach to Moral Judgment." *Psychological Review* 108: 814–34.

———. 2007. "The New Synthesis in Moral Psychology." *Science* 316: 998–1002.

———. 2012. *The Righteous Mind.* New York: Pantheon.

Happe, F. 1999. "Autism: Cognitive Deficit or Cognitive Style?" *Trends in Cognitive Science* 3: 216–22.

Hare, B., and M. Tomasello. 2004. "Chimpanzees Are More Skillful in Competitive than in Cooperative Cognitive Tasks." *Animal Behaviour* 68: 571–81.

Hare, R. D. 1993. *Without Conscience.* New York: Guilford Press.

Harman, G. 1999. "Moral Philosophy and Linguistics." In *Proceedings of the 20th World Congress of Philosophy.* Vol. 1, *Ethics*, ed. K. Brinkmann. Bowling Green, OH: Philosophy Documentation Center.

Hauser, M. D. 2006. *Moral Minds: How Nature Designed Our Sense of Right and Wrong.* New York: Ecco/Harper Collins.

Hauser, M. D., M. K. Chen, F. Chen, and E. Chuang. 2003. "Give unto Others: Genetically Unrelated Cotton-Top Tamarins Preferentially Give Food to Those Who Altruistically Give Food Back." *Proceedings of the Royal Society B* 270: 2363–70.

Hauser, M. D., F. Cushman, L. Young, R. K-X. Jin, and J. Mikhail. 2007. "A Dissociation between Moral Judgments and Justifications." *Mind and Language* 22: 1–21.

Hauser, M. D., L. Young, and F. Cushman. 2007. "Reviving Rawls' Linguistic Analogy: Operative Principles and the Causal Structure of Moral Actions." In *Moral Psychology.* Vol. 2, *The Cognitive Science of Morality,* ed. W. Sinnott-Armstrong. Cambridge, MA: The MIT Press.

———. In press. "Reviving Rawls' Linguistic Analogy: Operative Principles and the Causal Structure of Moral Actions." In *Moral Psychology.* Vol. 2, *The Cognitive Science of Morality, ed.* W. Sinnott-Armstrong. Cambridge, MA: The MIT Press.

Herrmann, E., and M. Tomasello. 2006. "Apes' and Children's Understanding of Cooperative and Competitive Motives in a Communicative Situation." *Developmental Science* 9: 518–29.

Karniol, R. 1978. "Use of Intention Cues in Evaluating Behavior." *Psychological Bulletin* 85: 76–85.

Koenigs, M., L. Young, R. Adolphs, D. Tranel, M. D. Hauser, F. Cushman, and A. Damasio. 2007. "Damage to the Prefrontal Cortex Increases Utilitarian Moral Judgments." *Nature* 446: 908–11.

Melis, A., B. Hare, and M. Tomasello. 2006. "Chimpanzees Recruit the Best Collaborators." *Science* 311: 1297–1300.

Mikhail, J. 2000. "Rawls' Linguistic Analogy: A Study of the 'Generative Grammar' Model of Moral Theory Described by John Rawls in 'A theory of justice.'" PhD.diss. Cornell University, Ithaca.

———. 2007. "Universal Moral Grammar: Theory, Evidence, and the Future." *Trends in Cognitive Science* 11: 143–52.

———. 2011. *Elements of Moral Cognition: Rawls' Linguistic Analogy and the Cognitive Science of Moral and Legal Judgment.* New York: Cambridge University Press.

Moll, J., R. Zahn, R. de Oliveira-Souza, F. Krueger, and J. Grafman. 2007. "The Neural Basis of Human Moral Cognition." *Nature Reviews Neuroscience* 6: 799–809.

Müller-Lyer, F. C. 1889. "Optische Urteilstäuschungen." *Dubois-Reymonds Archiv für Anatomie und Physiologie,* supplement vol., 263–270

Nichols, S. 2002. "Norms with Feeling: Toward a Psychological Account of Moral Judgment." *Cognition* 84: 221–36.

———. 2004. *Sentimental Rules.* New York: Oxford University Press.

Phillips, W., J. L. Barnes, N. Mahajan, M. Yamaguchi, and L. R. Santos. 2009. "'Unwilling' versus 'unable': Capuchin monkeys' (Cebus apella) Understanding of Human Intentional Action." *Developmental Science* 12: 938–45.

Piaget, J. [1932] 1965. *The Moral Judgment of the Child.* New York: Free Press.

Premack, D., and A. Premack. 2002. *Original Intelligence.* New York: McGraw Hill.

Premack, D., and G. Woodruff. 1978. "Does the Chimpanzee Have a Theory of Mind?" *Behavioral and Brain Sciences* 4: 515–26.

Prinz, J. J. 2004. *Gut Reactions.* New York: Oxford University Press.

Rawls, J. 1951. "Outline of a Decision Procedure for Ethics." *Philosophical Review* 60: 177–97.

———. 1971. *A Theory of Justice.* Cambridge, MA: Harvard University Press.

Ridley, M. 1996. *The Origins of Virtue.* New York: Viking Press/Penguin Books.

Saxe, R., and N. Kanwisher. 2003. "People Thinking about Thinking People: The Role of the Temporo-Parietal Junction in 'Theory of Mind.'" *NeuroImage* 19: 1835–42.

Stevens, J. R., F. A. Cushman, and M. D. Hauser. 2006. "Evolving the Psychological Mechanisms for Cooperation." *Annual Review of Ecology and Systematics* 36: 499–518.

Thomson, J. J. 1970. "Individuating Actions." *Journal of Philosophy* 68: 774–81.

———. 1971. "A Defense of Abortion." *Philosophy and Public Affairs* 1: 47–66.

Trivers, R. L. 1971. "The Evolution of Reciprocal Altruism." *Quarterly Review of Biology* 46: 35–57.

Turiel, E. 1998. "The Development of Morality." In *Handbook of Child Psychology,* ed. W. Damon. New York: Wiley Press.

———. 2005. "Thought, Emotions, and Social Interactional Processes in Moral Development." In *Handbook of Moral Development,* ed. M. Killen and J. G. Smetana. Mahwah, NJ: Lawrence Erlbaum Publishers.

Warneken, F., B. Hare, A. Melis, and M. Tomasello. 2007. "Spontaneous Altruism by Chimpanzees and Children." *Public Library of Science, Biology* 5: 1–7.

Wood, J., D. D. Glynn, B. C. Phillips, and M. D. Hauser. 2007. "The Perception of Rational, Goal-Directed Action in Nonhuman Primates." *Science* 317: 1402–5.

Young, L., F. Cushman, R. Adolphs, D. Tranel, and M. D. Hauser. 2007. "Does Emotion Mediate the Relationship between an Action's Moral Status and Its Intentional Status? Neuropsychological Evidence." *Cognition and Culture* 2006: 291–304.

Young, L., F. Cushman, M. D. Hauser, and R. Saxe. 2007. "The Neural Basis of the Interaction between Theory of Mind and Moral Judgment." *Proceedings of the National Academy of Sciences of the USA* 104: 8235–40.

14

A New Case for Kantianism

Evolution, Cooperation, and Deontological Claims in Human Society

Friedrich Lohmann

Immanuel Kant's reputation among some game theorists, socio-biologists, and behavioral scientists is rather bad.[1] Ken Binmore, for example, thinks that Kant's categorical imperative is nothing more than "simply a grandiloquent rendering of the folk wisdom which says that it is immoral to lay oneself open to my mother's favorite reproach: 'Suppose everybody behaved like that?' " (2005, VII). And Marc Hauser criticizes "Kant's cold, rational, and calculated morality" (2006, 28). Many such judgments as these are caused by misinterpretations of Kant. But most of the time, they lead back to what seems to be a fundamental gap between a Kantian understanding of morality and game theoretical or evolutionary approaches to human behavior. The title of my contribution makes it easy to guess on which side of the gap I am standing. I would not make a case for Kantianism if I were not convinced of the superiority of the Kantian approach.

Things are, however, not that easy. First of all, we must keep in mind that Kantianism goes far beyond moral theory: Kant also made important contributions to epistemology, politics, and aesthetics, for example. And even if we narrow our scope to moral theory, Kantianism has several aspects, and

I am far from subscribing to all of them. You do not need to be a faithful disciple of Kant to make a case for Kantianism in the face of recent approaches to morality by behavioral scientists. Actually, what I will do in what follows is to take just three aspects of morality on which Kant has focused and show shortcomings of game-theoretical and evolutionary approaches with respect to them. The most important point is already highlighted in the title of my essay: for Kant, morality is based on deontological claims—that is, on obligations that have to be fulfilled for their own sake without any regard for inclinations or results in terms of success. Kant has a *normative* vision of human conduct, whereas most behavioral scientists take a *descriptive* approach. This is more than just a methodological difference. It is one of Kant's insights that the human mind *is* normative, whereas behavioral scientists usually do not pay attention to the fact that moral obligations might be an integral part of human consciousness as it *is*. Normative judgments are part of our daily life—also in their most rigid, deontological form—and thus behavioral scientists conceiving of morality and cooperation only in terms of (evolutionary) "success" already fall short of simply describing human moral behavior.

I am going to focus my discussion on two cases of human conduct that have been reported recently. Once the discussion is done, the fundamental gap I have mentioned will be evident. I will, however, argue in a concluding section that this gap is not the end of the story. When it comes to morality, Kant and behavioral scientists have purposes that are completely different. They speak about two different aspects of human consciousness and conduct that every serious theory of morality must include. That is why the gap between the two approaches can be made productive. There are shortcomings on both sides, and I will argue, at the end of my contribution, for an integrative approach to morality.

Sean Penn and Liviu Librescu

One day in September 2005, my attention was caught by a news headline on the Internet. It was a few days after the south of the United States had been struck by Hurricane Katrina, and the headline that captivated me said something like: "Actor Sean Penn criticized for selfish rescue effort." What had happened? Well, here is the story. In reaction to the destruction caused by

Katrina, Sean Penn had come to New Orleans and chartered a boat to save inhabitants from their flooded houses. Soon after the photo documenting his efforts had been released, suspicions were raised: were there not other things to do than taking a photo? Had he brought his personal photographer with him? What about the trendy and expensive jacket he was wearing? Was the whole thing a huge propaganda effort? Even after Sean Penn had declared his solely altruistic motives on television, the discussion did not stop.

I am not much interested in the real motives of this rescue action, which are difficult to explore anyway. What strikes me is the reaction of those who criticized it. The result was not in question: several dozens of people were saved, some of them praising the actor publicly for his philanthropy. What was in question were the motives behind the rescue effort, and for many people the positive result was completely tainted, if not erased, by the suspicion that a self-interested strategy was the leading motive behind all that apparent charity. For them, the suspicion of hidden selfishness was sufficient to change the moral worth of Sean Penn's action into the negative. No doubt the case in question was a case of cooperation, even altruism, given that help is just a special kind of cooperation with an urgent need and no direct reciprocity involved. But many doubts about the moral worth of this act of cooperation were generated, once the spirit of self-interested motivation began to linger around.

I shall come back later to the deeper reasons for the contempt Sean Penn was confronted with. For the moment, I would like to turn your attention to the opposite case: to an action where no self-interest at all seems to be involved.

On the morning of April 16, 2007, Seung-Hui Cho crossed the campus of Virginia Tech. The unassuming student had changed into a gunman, dressed to kill. He entered the Norris Hall building and started a shooting rampage, known as the Virginia Tech massacre. When he approached one of the classrooms, he was hindered from entering because the door was blocked from the inside. It was blocked by Professor Liviu Librescu, who at the same time ordered his students to leave the classroom through a window. All but one had escaped when Cho finally managed to enter, having shot through the door and killed the professor whose body was struck by five bullets.

In Librescu's case, the public reaction was unambiguous. Soon after his students had reported what had happened, news services around the world began to give tribute to his action as an example of unselfish and quiet heroism. He was awarded posthumously the Order of the Star of Romania, the Romanian government praising the fact that he had saved others by sacrificing his own life.

Morality and Intentions

A comparison of the two cases, including the respective public reaction, reveals a frequently made moral distinction between the intentions behind an action and the results of an action, which seems to be deeply rooted in our common sense. The result of Penn's and Librescu's rescue efforts is almost identical: a considerable number of people were saved who otherwise would probably have died. Therefore, the different appreciation they received from the broad public cannot be related to the result of their actions. It must depend on the different intentions for action. Whereas many comments on Librescu focused on his unselfishness, linked to his modest and responsible character as a teacher that enabled him to act as he did, Penn was suspected of following a hidden agenda and to be motivated *also* by selfish interests.

Intentions matter for morality. This is not just a vague common-sense sentiment that shapes our daily moral evaluations and our written law. It has also been an important strand in moral theory since its inception in early Greek antiquity. Plato and the Christian tradition can be mentioned here, but the philosopher who has made this claim most rigorously is Immanuel Kant, who states, for example, in the sentence that introduces the first part of his *Groundwork in the Metaphysics of Morals:* "Nothing can be thought in- and outside the world which could be considered good without any restriction, if it is not a good will" (Kant 1997, my translation). According to Kant, a good will has worth in itself, independently of the final result. The intention behind an action mirrors a deeply rooted attitude *(Gesinnung)* that alone makes the action good or bad. To understand Kant's point, it is important to place a stress on the "without any restriction" in the quoted sentence. One might say that there are other things than the will that can be considered good. There are, for example, virtues and good actions. But virtues can be abused by a bad will. Take the persevering criminal. Or take the officer in Kafka's *Penal Colony* who is a perverted aficionado of justice. And actions in the round (which are composed of the intention, the execution, and the result of the action) contain contingent elements not to be taken into consideration for a moral judgment. The result can be good, but with a restriction stating that this result was not intended (example: you stand accidentally in the way of a criminal who is pursued by the police). Conversely, the result can be accidentally bad in spite of a good intention (example: you warn someone of an imminent danger and cause him a fatal heart attack). Moral judgments attribute responsibility for what is in our hands. And what is in our hands is preeminently the intention, not the contingent execution and result of an action. Everything depends, as stated by Kant, on a good *will*.

Intentions matter for morality. This is stressed by Kant, but it is strangely forgotten by sociobiological or game-theoretical approaches to morality and cooperation. Actually, in terms of moral analyses, sociobiology and game theory have, since their inception, been almost purely outcome-oriented. The distinction between evolutionary and psychological altruism introduced by Sober and Wilson (1998) is the exception that confirms this rule. Whereas their concept of "psychological altruism" pays attention to the importance of intentions (only an action with "the welfare of others" as at least one goal of the acting person can be called "altruist"), the concept of *evolutionary* altruism counts only for "the effects of behavior on survival and reproduction" (Sober and Wilson 1998, 6).[2] The primate researcher Frans de Waal's criticism is another witness to this implicit shortcoming in sociobiological analyses of human action and intentionality: "Human moral judgment always looks for the intention behind behavior . . . These distinctions [i.e., between intended and unintended results] are largely irrelevant within a sociobiology exclusively interested in the effects of behavior . . . Having thus denied the most important handle on ethical issues, some sociobiologists have given up on explaining morality" (de Waal 1996, 15).

Indeed, by focusing alone on the outcome of actions with regard to fitness, sociobiologists have left out a big part of the story. What serves evolutionary fitness, for them, is good. In this outcome-oriented perspective, Librescu's self-sacrifice must appear as a bad action, whereas the voluntary creation of a good reputation—the intention Sean Penn was accused of—is irreproachable. A good example of such an evaluation of human altruism can be found in Richard Alexander's influential book, *The Biology of Moral Systems* (1987). Alexander's evaluation of self-sacrifice is clear-cut: either it is done in favor of one's own offspring (and is, therefore, actually no act of altruism but of nepotism) or it implies no such benefit—a rare exception and an "evolutionary mistake" (1987, 88; also see Timothy Jackson's discussion of this point, Chapter 16). Librescu, then, would have committed a "mistake" (unless one takes into consideration the positive effects of his sacrifice for his surviving children; this argument lies in the spirit of Alexander,[3] but it fails because it is not very likely that Librescu did a cost-benefit analysis with regard to his children before rushing to the classroom door). Sean Penn, on the other hand, would have done everything right in documenting his rescue effort by a photographer: in Alexander's eyes, morality comes down to indirect reciprocity that makes *reputation* its key theme. Morality is, then, about *strategy* (1987, 108: "strategies of beneficence") and *appearance* (1987, 104: "Everyone will wish to appear more beneficent than he is."). The moral importance of the intentions *behind* seemingly altruistic actions is neglected.

The same shortcoming characterizes game theoretical approaches to human behavior. "Strategy," "payoff," "fitness," and "success" are the keywords in Martin Nowak's recent summary of "evolutionary game theory" (Nowak 2006, 46). Cooperation is seen as a sole affair of cost and benefit, and the prisoner's dilemma—in which all is about strategy and "payoff values" (and nothing about moral values)—is considered to be "a game that captures the essence of cooperation" (2006, 90). The same focus on strategy can be found in John Maynard Smith's work, which introduced game theory into evolutionary biology.

Such references to game theory in terms of an outcome-oriented "strategy" are entirely faithful to the origins of the game theoretical approach. R. B. Braithwaite's seminal lecture on "Theory of Games as a Tool for the Moral Philosopher" reveals the limited sense in which game theory can serve as such a moral tool – in calculating the best strategy to arrive at a given aim, notwithstanding the moral value of that aim: "Theory of Games is concerned with calculating what is the 'best' strategy for playing such a game [i.e. a game of intellectual skill], where a precise meaning is given to the notion of *best strategy*" (Braithwaite 1955, 16).

This term "best" is, in this context, to be understood as "most prudent" (1955, 20), and the task of the "philosophical moralist" is reoriented toward giving "advice to people with different aims as to how they may collaborate in common tasks so as *to obtain maximum satisfaction* compatible with fair distribution" (1955, 4; italics added). The aims themselves are not judged by the philosopher: "The recommendations which I shall make for sharing fairly the proceeds of collaboration will therefore be *amoral* in the sense that they will not be based upon any first-order moral principles" (Braithwaite 1955, 5).

Braithwaite is pleased to escape by this reorientation of moral philosophy from the challenge of presupposing a *common* moral value system as had been done by welfare economics (1955, 14), and he indeed allows also for malevolent satisfactions (1955, 14: "Or, if he is a disagreeable man, the prospect of thwarting Luke may increase his satisfaction."). In the last case too, the philosopher is supposed to help the malevolent competitor "to obtain maximum satisfaction compatible with fair distribution."

This strategic, outcome-oriented understanding of morality as a tool for the fair maximization of satisfaction in social situations of competition and cooperation is in fundamental accord with the conception of game theory that had been presented a decade before by von Neumann and Morgenstern: game theory as a mathematical tool to establish a rational "strategy" (von Neumann and Morgenstern 1944, 79) in order "to obtain a maximum of utility" (1944, 1) in different social situations. But von Neumann and Morgenstern had restricted themselves to a description of *economic behavior* when they claimed "that the typical problems of economic behavior become strictly identical with the mathematical notions of suitable games of strategy" (1944, 2). Contrary to that restriction, Braithwaite and others such as Binmore extend the game theoretical approach to *all* human behavior, including to morality.

This step was well prepared for by the tradition of philosophical utilitarianism. Game theory and utilitarianism share the same basic conviction that human beings are primarily interested in maximizing utility in terms of satisfaction and well-being. Bentham, the first utilitarian in the proper sense of the word, wanted to conceive of a universal theory of human action constructed only upon the common striving for happiness. The aim of utility-maximization was understood in terms of quantity, not quality, as stated by Bentham's famous pun: "Push pin is as good as poetry." The subsequent history of utilitarianism has shown, however, that such a theory encounters massive problems. Utilitarian philosophers have been forced again and again to integrate qualitative distinctions in terms of intentions (as in

John Stuart Mill's distinction between higher and lower pleasures) and deontological claims (as in rule-utilitarian approaches). Not all satisfactions are allowed. Quality matters, in the sense of a *social* utility. It is not surprising, then, that Harsanyi's recent utilitarian theory of morality includes not only obligations but also evaluates intentions (1998, 290: Harsanyi excludes malevolent preferences—an interesting step beyond Braithwaite).

Utilitarianism has, in the course of its history, abandoned the sole interest in the calculable outcome of actions that had been characteristic of Bentham's theory. In that, this philosophical ally of game theoretical and sociobiological approaches to human behavior confirms that they fall short of an important aspect of morality that has been stated most clearly by Kant: the relevance of intentions.

Morality and Self-Interest

The lack of reflection on intentions for acting that can be diagnosed in almost all theories of morality based on evolutionary biology can be explained easily: there is no reflection because everything seems to be clear! It is a common presupposition of these theories that everything that happens over the course of evolution can be derived from self-interest. "Evolution doesn't favour mutations that sacrifice themselves to help their neighbours. It favours mutations that help themselves" (Binmore 1998, 279). Alexander's claim that true altruism—in the rare case that it is not hidden selfishness or nepotism—would be an "evolutionary mistake" is based on this presupposition: like all other human actions and aspirations, morality wants to promote the well-being of an individual or his or her own in-group. Whence follows the importance of the notion of reciprocity: moral actions are considered to be an "investment," given after a cost-benefit analysis and in anticipation of a comparable "return" (Alexander 1987, 82). Game theory is an important tool in performing the "cost-benefit analyses" (1987, 87) that have to be done in order to find the best among all the possible "strategies of beneficence" (1987, 108).

From a Kantian standpoint, such a conception of human behavior is not completely wrong. Kant recognizes that self-interest plays an important role in human thinking and acting. But Kant would sharply dismiss the claim that it is the proper basis for morality. His vision of the human self and what is good for it is much more complex than the vision presupposed by most sociobiologists and game theorists. Actually, Kant distinguishes three different types of goodness.[4] The first one means, very broadly, that something is good for the purpose it has been designed for: *technical* goodness. The second type of goodness corresponds to a strategic, self-interested perspective (often the subject of game theoretical and sociobiological analyses): something is good *for me*. Kant calls this type *pragmatic* goodness. Yet human thinking and acting does not stop at the pragmatic level. There is a way to speak of goodness that transcends all strategic considerations. And it is only here that morality comes into play: *moral* goodness. Its hallmark is the lack of any pursuit of selfish interests. The moral good is done—as Kant says in accordance with Plato, Aristotle, and the Christian tradition—for its own sake. We do not perform a *moral* act because it is to our own advantage but because we feel in our consciousness an obligation to do it.

Is there no purpose or interest at all in play when someone forgets his or her immediate interests and strives for the moral good? John Stuart Mill has pointed out that Kant claims, when he wants to show the importance of promise keeping, that the whole society would fall apart when the trust in given promises would be undermined by unkept promises. For Mill this means that Kant is, in fact, a utilitarian because he makes what is good for the future of humanity the criterion of good and bad. Faithful Kantians have tried to refute this utilitarian appropriation of Kant. I am not sure if this is necessary. There *is*, of course, a benefit in sticking to moral rules. If not, morality would be comparable to an annihilating forced labor or at least to a senseless *art pour l'art*. The point is that this benefit is the good for all (not the pragmatic good for me) and that it is the *direct* aim of moral actions. Rather, such a benefit is the consequence of the obedience to a rule that is not negotiable, and therefore categorical: the categorical imperative. Kant can thus be called a partisan of an extreme version of rule utilitarianism insomuch as the realization of the good for all in the kingdom of ends *is* the desired outcome of all moral action.[5]

Even in this rule-utilitarian interpretation, Kant's conception of morality does not fit at all with the anthropological premise of Alexander's ethics, which states that human thinking and acting are—and should be—always selfish. For Alexander, all ethicists who try to establish a moral code beyond self-interest are, in fact, selfish themselves because they want to exercise power by their moral considerations: "I see this form of utilitarianism as a concept useful to those with *more* power and influence in getting those with *less* power and influence to behave in fashions more useful to the former than to the latter" (Alexander 1987, 136).

This argument (which is very close to criticisms of morality that have been put forward by the Antique Sophists and by Nietzsche) is coherent within Alexander's presuppositions, but at the same time it reveals clearly the extent to which Alexander's underlying vision of human consciousness narrows the scope of his own inquiry and the normative judgments it produces. Is human morality really nothing else than an attempt to increase power?

For me, Kant's complex anthropology, including the constant fight between pragmatic and moral inclinations, is much closer to reality. Morality and human cooperation cannot be reduced to strategies of selfishness. This is confirmed by the common moral sense, which tends to praise actions reversely proportionate to the amount of selfishness involved, as is shown by the Penn and Librescu cases (also see Johnson, Chapter 8, and Hauser, Chapter 13 in this regard). A strategy of selfishness hidden under an altruistic surface, as advocated by Alexander, would be received with particular contempt, just as it was with Penn: people were suspicious about Penn's rescue effort because, as a strategy, it would have intended to misuse the misery of other people for his own selfish publicity interests. If there was any strategy to be publicly recognized, it failed and backfired completely, due to the moral consciousness of the Internet community, sensitive to authenticity. Alexander and others who advocate a morality of pure appearance, then, simply miss the core of standard moral judgments.

Morality and Empirical Facts

The discussion of the relationship between morality and self-interest takes me to another, more methodological point. What is the source of our statements on morality? The behavioral sciences—psychology, sociology, economics, biology—usually start their work with empirical facts. Their image of human attitudes and behavior is therefore shaped by observations of how human beings *actually* behave and what intentions they *actually* pursue when performing actions. It is not very surprising, then, that self-interest comes out as the most important human intention. Morality, however, does not deal with what usually happens but with what *ought* to happen. If this difference between moral claims and empirical facts is not observed, the result will be the so-called "naturalistic fallacy": descriptive statements are taken to have normative value. Jeremy Bentham arguably committed the most evident naturalistic fallacy in the history of ethics when he set out from the fact that the avoidance of pain and the pursuit of pleasure *are* the most important springs of action and deduced the normative conclusion that we *ought* to seek the greatest happiness of the greatest number when acting.

Bentham was proud of the empirical foundation of his ethics. He actually wanted to become the Bacon of moral science, the one giving empirical observation its due (Bentham 1983, 294–95). This was an important point in a period of speculative rationalism. Bentham failed, however, when he narrowed down the perspective of his own ethics and declared that empirical facts are the *only* source for a science of morality. The same reductionist perspective, including the risk of a naturalistic fallacy, threatens current accounts of morality coming from the behavioral sciences. Marc Hauser's *Moral Minds,* for instance, is full of empirical material gained either in the laboratory or via questionnaires. By reading Hauser's book, one gets an admirable picture of what people actually think about morality or moral dilemmas, and what the neuronal bases of those thoughts on moral questions might be. But is this the whole picture of morality? Morality is not a question of how the majority may, or may not, judge—a fact Hauser is well aware of (see, e.g., Hauser 2006, 96). Librescu's act is depicted by popular opinion as exemplary just because it is extraordinary (it is "heroism"). One does not need to subscribe to the old Stoic *fiat iustitia, et pereat mundus* (Let there be justice, even if this will make the world perish) in order to realize that there is—at least sometimes—a huge difference between the current state of affairs and what is morally required.

Kant was very clear-cut in pointing out this difference. His distinction between obligation *(Pflicht)* and inclination *(Neigung)* leads us right to the point. Kant was conscious of the fact that people are inclined to pursue their self-interest. It was, for him, the task of morality to limit this natural inclination and to take humanity up to the higher, morally valuable goals it was created for. This gives his ethical theory an antiempirical drive, avoiding any naturalistic fallacy. Kant's ethics is, however, less dualistic than some of his followers—up to Habermas and his stressing of the counterfactual *(kontrafaktisch)* character of morality—have stated. Kant was eager to claim that the consciousness of the categorical imperative is a *fact* of pure practical reason. Everyone has some true moral intuitions when reflecting upon good and bad. With this claim, Kant was in continuity with the doctrine of natural law, which is based on the idea of an innate knowledge of basic moral rules in the sense of a "natural inclination" (Aquinas). The more dominant inclination toward immoral selfishness overshadows this natural knowledge of moral obligation (in his *Religion within the Boundaries of Mere Reason,* Kant interpreted this overshadowing as an effect of original sin).

Nevertheless, Kant acknowledges that the *two* inclinations are still present in human consciousness. They are fighting against each other, giving the human character all its fundamental ambiguity. Once again, Kant is much more aware of the complex struggle of competing interests that is continually going on in each human's moral consciousness than most of the behavioral scientists who have written theories on morality. To quote Frans de Waal as a notable exception: "By denying the existence of genuine kindness, however, these theories miss out on the greater truth emerging from a juxtaposition of genetic self-interest and the intense sociality and conviviality of many animals, including ourselves. Instead of human nature's being either fundamentally brutish or fundamentally noble, it is both—a more complex picture perhaps, but an infinitely more inspiring one" (de Waal 1996, 5).

It has only been recently that behavioral scientists have become interested in experiments showing the presence of unselfish impulsions in human beings: "inequity aversion" (Fehr and Schmidt 1999), "altruistic punishment" (Gintis et al. 2005), and empathy in infants (Warneken and Tomasello 2006). Those researchers usually remind us that the presence of selfishness should not be neglected (Fehr and Fischbacher 2005, 155)—but there is at least *some* empirical evidence for the existence of truly moral impulses. This result is in remarkable agreement not only with Kant's interpretation of the human character but also with an empirical observation Kant mentions in his *Metaphysics of Morals*. It has already been said that Kant stresses considerably the *difference* between the empirical and the moral self. However, even Kant acknowledges from time to time the empirical reality of moral sentiments, as he does in his treatise on friendship, a topic that is close to the question of cooperation. Kant distinguishes different types of friendship. The most valuable of them, "moral friendship," is characterized by a relationship of complete trust and the absence of any selfish calculations about mutual benefit.[6] Such a friendship is, Kant admits, a rare thing on earth, something like a black swan. However, Kant continues by insisting: "This (merely moral friendship) is not just an ideal but (the black swan) actually exists here and there in its perfection." (Kant 1996, 217; translation amended)

Kant thus reminds us that morality is not a question of majority rule or of matters of fact. True moral friendship is, to be sure, "a rare bird." Therefore, empirical facts cannot be the base for moral judgments. The moral perspective is characterized by an "ought," seen as a state of affairs that has not yet been reached. On the other hand, Kant is aware of the necessity that moral claims (as the one stating what is a true, uninterested friendship) must have a reference point in empirical reality. That is why he insists on the *fact* of a moral consciousness and the presence of at least some realizations of moral friendship. Behavioral scientists usually reverse this order of reasoning: they begin with empirical facts and *then* construct a theory of morality around these facts. In proceeding this way, they risk not only committing naturalistic fallacies; they also fall short of the *ought* dimension of morality, in which human action transcends each simple observation of "what is the case."

Deontological Claims in a Society of Human Beings: Toward an Integrative Theory of Morality

Kant's example of true friendship as a black swan represents a good bridge to my concluding remarks. Why is it important that moral claims have a reference point in empirical reality? Kant is clear on that point: it is important because it shows that the ideal of moral friendship "is not just an ideal." An ideal is, in Kant's language, a product of rational imagination, containing a degree of perfection that cannot be found in empirical reality. It is, therefore, open to ideology. In addition, its relevance for life can easily be rejected. But morality must avoid all ideological assumptions. It must also stress the relevance of its claims for daily life. In other words: morality has to be implemented.

The question of implementation is important for Kant. He spends some long time on it in the last part of his *Critique of Practical Reason*, speaking mainly about questions of methods of moral education. In this part, Kant pays attention to the human empirical self: moral consciousness, even if innate, has to be elevated above, and liberated from, the selfish impulsions that are its natural enemies. Therefore, morality cannot just be taught in its pure rational form; it must be vulgarized, for example, by the use of biographical material showing the moral pupil that this or that moral attitude *has* been realized in a concrete life.

Even from a Kantian standpoint, then, a good knowledge of the empirical human character is crucial for applying moral rules to reality, either in teaching about morality or in making laws that are in accordance with the categorical imperative. Morality has to be implemented, adapted to the conditions of real humans living in a real human society. And it is here, in the process of implementation, that anthropological findings from biology and psychology or mathematical models of game theory become relevant for moral theory. Business ethics, for example, has to take into account the empirical fact that competition is an important factor in human life, notwithstanding its questionable moral value. Mathematical strategies of game theory can show the importance of cooperation even for those who are interested only in personal advantages.

Kant's ethics admits of such realities in human behavior, if only to transcend them normatively. For Kant, the empirical description of actual drives, actions, and inclinations remains critical. On the other hand, a solely descriptive ethics falls short of the *ought* dimension of deontology, which is decisive for a moral theory worth its name. Therefore, in spite of the gap between Kant's theory of morality and game theoretical or evolutionary approaches to moral behavior, both have their right. They are complementary. In adapting a well-known formula from Kant's *Critique of Pure Reason,* one can say: moral ideas without empirical clues are empty; empirical facts without orientation to moral categories are blind.

Notes

1. By a game theoretical approach to human behavior, I understand, here and in the following, all attempts to interpret human behavior by means of mathematical game theory; "sociobiology" summarizes all attempts to interpret human behavior by means of evolutionary biology. For roughly thirty years, game theory and sociobiology have merged into "evolutionary game theory" to explain coopera-

tive behavior (Maynard Smith 1982; Nowak 2006). When I speak in the following of "behavioral sciences," I mean by that *all* sciences that deal with actual human behavior, usually from an empirical starting point: psychology, sociology, economics, biology.

2. As explained in the Introduction, this current volume also makes a clear distinction between "cooperation" and (true) "altruism," the latter being inherently motivational.
3. See Alexander 1987, 84: heroism as indirect nepotism.
4. The following lines are inspired by the distinction between the technical, pragmatic, and moral imperatives Kant introduces in his "Groundwork" (Kant 1997, 24–27).
5. Cf. Kant 1996, 216: "Hence friendship cannot be a union aimed at mutual advantage but must rather be a purely moral one. . . ."
6. If I have, in the last section, spoken of deontological claims in utilitarianism, I am stressing here the teleological aspects of Kant's ethics. Actually, the whole distinction between deontological and teleological ethics is only of heuristic value, pointing out two poles each serious conception of morality must integrate.

References

Alexander, R. D. 1987. *The Biology of Moral Systems.* New York: Aldine de Gruyter.

Bentham, J. 1983. "Article on Utilitarianism: Long Version." In *Deontology together with a Table of the Springs of Action and the Article on Utilitarianism,* ed. A. Goldworth. Oxford: Clarendon Press, 285–318.

Binmore, K. 1998. "Evolutionary Ethics." In *Game Theory, Experience, Rationality: Foundations of Social Sciences, Econonomics and Ethics. In Honor of John C. Harsanyi,* ed. W. Leinfellner and E. Köhler. Dordrecht/Boston/London: Kluwer Academic Publishers, 277–83.

———. 2005. *Natural Justice.* Oxford/New York: Oxford University Press.

Braithwaite, R. B. 1955. *Theory of Games as a Tool for the Moral Philosopher.* Reprint 1969. Cambridge: Cambridge University Press.

de Waal, F. 1996. *Good Natured: The Origins of Right and Wrong in Humans and Other Animals.* Cambridge, MA: Harvard University Press.

Fehr, E., and U. Fischbacher. 2005. "The Economics of Strong Reciprocity." In *Moral Sentiments and Material Interests: The Foundations of Cooperation in Economic Life,* ed. H. Gintis, S. Bowles, R. Boyd, and E. Fehr. Cambridge, MA: The MIT Press, 151–91.

Fehr, E., and K. M. Schmidt. 1999. "A Theory of Fairness, Competition, and Cooperation." *The Quarterly Journal of Economics* 114: 817–68.

Gintis, H., S. Bowles, R. Boyd, and E. Fehr, eds. 2005. *Moral Sentiments and Material Interests: The Foundations of Cooperation in Economic Life.* Cambridge, MA: The MIT Press.

Harsanyi, J. C. 1998. "A Preference-Based Theory of Well-Being and a Rule-Utilitarian Theory of Morality." In *Game Theory, Experience, Rationality: Foundations of Social Sciences, Economics and Ethics. In Honor of John C. Harsanyi,* ed W. Leinfellner and E. Köhler. Dordrecht: Kluwer Academic Publishers, 285–300.

Hauser, M. D. 2006. *Moral Minds: The Nature of Right and Wrong.* New York: HarperCollins.

Kant, I. 1996. *The Metaphysics of Morals,* trans. and ed. Mary Gregor. Cambridge: Cambridge University Press.

———. 1997. *Groundwork of the Metaphysics of Morals,* trans. and ed. Mary Gregor. Cambridge: Cambridge University Press.

———. 1998a. *Critique of Practical Reason,* trans. and ed. Mary Gregor. Cambridge: Cambridge University Press.

———. 1998b. *Religion within the Boundaries of Mere Reason,* trans. and ed. Allen Wood and George di Giovanni. Cambridge: Cambridge University Press.

Maynard Smith, J. 1982. *Evolution and the Theory of Games.* Cambridge: Cambridge University Press.

Nowak, M. A. 2006. *Evolutionary Dynamics: Exploring the Equations of Life.* Cambridge, MA: Belknap Press of Harvard University Press.

Sober, E., and D. S. Wilson. 1998. *Unto Others: The Evolution and Psychology of Unselfish Behavior.* Cambridge, MA: Harvard University Press.

von Neumann, J., and O. Morgenstern. 1944. *Theory of Games and Economic Behavior.* 6th printing, 1955. Princeton, NJ: Princeton University Press.

Warneken, F., and M. Tomasello. 2006. "Altruistic Helping in Human Infants and Young Chimpanzees." *Science* 311: 1301–3.

15

Nature, Normative Grammars, and Moral Judgments

JEAN PORTER

For several years now, my research and writing have focused on the concept of natural law developed by jurists and theologians in the Latin West during the twelfth and thirteenth centuries.[1] This may seem to be an unpromising basis for contemporary ethical reflection, in part because the scholastic approach to natural law depends on a broadly Aristotelian philosophy of nature, within which widely shared human inclinations are rendered intelligible and given shape and direction in the light of their contributions to the overall well-being of individuals and the community. For many of our contemporaries, this aspect of medieval, natural-law thinking invalidates it from the outset. Have not Hume and Darwin between them shown that teleological analysis has no place in either scientific or ethical thought? Clearly, I would not agree with this assessment. On the contrary, I would argue that a broadly Aristotelian approach has much to offer to both science and ethics, and moreover, it has something to offer to the current dialogue regarding cooperation in particular. Obviously, I cannot develop this argument in any detail here, but I do want to say enough to explain what I mean, and to suggest that an Aristotelian approach can still offer fruitful lines of inquiry for both scientific analysis and ethical reflection.

In order to achieve even these modest aims, I need to make a case for a kind of natural teleology—that is to say, I need to show that at least some judgments about what is good, advantageous, and worthwhile are rooted in objective reality rather than being reflections of contingent human desires. I suspect that many will find this implausible—and I would agree with my critics to this extent that there are many kinds of teleological analysis that really are implausible, including both modern and contemporary versions of the argument from design, or the claim that the overall process of evolution reflects guidance in view of some divine plan. Current debates over the naturalness of the good and the goodness of nature generally do not get outside the terms set by these assumptions.

Yet, the Aristotelian approach that I want to commend does take us outside the terms of these debates—and that, in itself, represents a real contribution to contemporary scientific and ethical reflection. This approach is indeed teleological, but the kind of teleology in question does not presuppose conscious design or the imposition of purposes extrinsic to the functioning of the creature. It does, however, presuppose a close link between intelligibility and normativity, at least in the sense that the characteristic features of human life—or of any other kind of living creature—are explained in terms of their contributions to a species-specific way of life. This presupposes, in turn, a kind of realism about natural kinds, according to which well-formulated concepts of natural kinds are grounded in real qualities or aspects of the things in question, in virtue of which they count as the kinds of things that they are. It presupposes, in other words, that a living creature instantiates a specific *form,* in and through which it exists as an individual entity of this or that determinate kind. By the same token, this view presupposes that the forms of things can be grasped through concepts, albeit imperfectly, and moreover that these concepts are genuinely explanatory. That is to say, they help us to make sense of the operations of natural things in ways that would otherwise be inaccessible to us.

The kinds of explanations in question are styled by Aristotelians as appeals to a formal cause, but it is important not to be misled by this expression. In this context, the language of "cause" does not necessarily imply efficient causality but refers more broadly to any principle of intelligibility and explanation. More specifically, the "formal cause" indicates the description under which something can be understood in terms of the characteristic modalities of existence and operation proper to a specific kind of organism or entity. As James Lennox puts it, referring specifically to Aristotle's conception of form, "the form of a living thing is its soul, and Aristotle considers soul to be a unified set of goal-ordered capacities—nutritive, reproductive, locomotive, and cognitive."[2] It is thus a kind of category mistake to think of the forms of living creatures as among the kinds of things that can have independent existence, prior to specification in some individual; they are instantiated, or they are not, and it is these instantiations, not the forms themselves, that come into being and pass away. By the same token, the forms of living creatures do not exist over against the creatures in such a way as to exercise efficient causality on them. Rather, the form is manifested in and through the modalities of efficient causality proper to the kind of creature in question.

I hope I have said enough to indicate that nothing in the Aristotelian account of formal causation implies that the processes of evolution in particular, or biological processes more generally, require special divine intervention or other kinds of external causation. On the contrary, the analysis just developed is entirely consistent with a view according to which biological processes can be wholly and completely explained through natural causes. Indeed, there is good reason to believe that we cannot fully account for the emergence of species, or the development and functioning of living individuals, without drawing at some points on considerations stemming from the kinds of creatures with which we are dealing. And that means that at some points we must appeal to formal causes—that is to say, explanations that irreducibly refer to the kind of creature that is in question.

This brings me back to the point with which we began. For Aristotle himself and for those who follow his lead on these issues, formal causes and teleological explanation, or in other words, explanations in terms of final causes, are inextricably linked. The proper form of a given kind of living creature can only be adequately understood by reference to some idea of a paradigmatic instance of the form—that is to say, a healthy and mature individual of the kind in question. It is only by reference to this paradigm that we are able to identify immature, sick, or defective individuals of this kind. By the same token, appeals to formal and final causality provide a framework within which one may develop and test hypotheses about the functions of the characteristic organs and operations of a given kind of creature and to explain them in terms of their contributions to the functioning of the organism as a whole.

Yet surely even this kind of teleological analysis, naturalistic and modest though it is, is ruled out by contemporary biological sciences? As the biologist and philosopher Lenny Moss remarks, the incompatibility between teleological claims and contemporary evolutionary biology is a shibboleth, so widely accepted that it has become difficult even to canvass the relevant issues (Moss 2003, 4). However, over the past few years, two different strands of thought have converged to undermine this assumption. The first follows historical and philosophical reappraisals of Aristotle's understanding of teleology and its afterlife in later scientific thought. The second draws on efforts by biologists themselves to make sense of the proper subject matter of their field. As Michael Ruse points out, "Design language reigns triumphant in evolutionary biology," (Ruse 1998, 16) and while no serious biologist interprets this language in terms of externally imposed design, it is clear that it does carry connotations of purpose and function.

But whose purposes, and which functions? Here we come to the crux of the matter. For Aristotle, the language of purpose and function is legitimate and indeed necessary in order to speak of the organs, structures, and recurrent activities of living creatures, *within* a context set by the well-being of the creature itself. In other words, for him, teleological language finds its appropriate context by reference to the kind of well-being characteristic of the creature itself, and, correlatively, to the ordered activities through which the creature pursues its well-being—that is to say, to the creature's own goals, broadly construed. Thus understood, teleological language does not imply any reference to externally imposed purposes, whether ours, those of some demiurge or creator, or a generalized nature or evolutionary process. Rather, in this context the language of purpose functions in such a way as to render the different components of a living creature intelligible in terms of their contributions to the life processes of the creature. As Lennox puts it, "the 'package' that constitutes a distinct bird form precisely adapts it to its way of life—specifying 'what it is to be a Spoonbill' is explaining how its differentiae are precisely those needed to live as it does" (Lennox 2001, 129). Interpreted in this way, the observation that the eye is meant for seeing does not presuppose a story about the way in which the physical mechanism of the eye reflects the contrivance of an omnipotent designer (as it would do on most modern understandings of design). Instead, it presupposes a more straightforward account of the role that the eye plays in promoting the overall well-being of the creature by means of providing an organ for sight, which will include a further account of what sight itself is good for, given an overall orientation toward life, growth, and reproduction shared (in some way) by all living creatures.

An Aristotelian approach to teleology may seem on this interpretation to be trivial—do we not all know that one sees through the eyes, and do we not all agree that sight is a good thing? And it is certainly true that this kind of explanation is not generally developed in such a way as to provide new information. However, what it does offer is an interpretation in terms of which the different components of living creatures—including both organs and activities—can be rendered intelligible by means of their relation to an overall pattern of functions and activities, all directed toward the aims shared by all living creatures. While this kind of explanation may seem trivial when applied to familiar kinds of creatures, it is far from trivial when applied to less familiar kinds of creatures, or to puzzling aspects of the bodily structure and activities of familiar kinds. By the same token, teleological explanations provide biological science with fruitful agenda for research—not so much by supplying new information but by suggesting interpretations of data and generating hypotheses.

Mark Hauser's *Moral Minds* offers one example of an important body of biological and behavioral research that can be fruitfully illuminated by an Aristotelian teleological perspective (see Hauser 2006). In this book Hauser offers a remarkably comprehensive case for the claim that normative judgments are grounded in, and necessarily shaped by, the innate structures of our natural cognitive faculties. He marshals an impressive body of evidence, drawn from a wide range of disciplines, for persistent, cross-cultural features of normative judgment and practice. What is more, he offers a cogent theoretical explanation for these constants and shows how they themselves, properly understood, have explanatory power—mostly, by illuminating the consistent structural patterns informing wide-ranging cultural variables. This latter aspect of Hauser's work is particularly valuable because it transforms what might have been a very long list of data into something more like a classical concept of human nature, through which our knowledge about ourselves is rendered intelligible and given explanatory force. Whatever one may think about the details of Hauser's arguments, it is an important achievement simply to show that we do know quite a bit about human normative judgments, and, moreover, that this body of knowledge can be interpreted in light of a plausible explanatory framework.

But just what is it, in Hauser's view, that we know and understand about ourselves as moral agents? The title, and even more, the subtitle of his book—"How Nature Designed Our Universal Sense of Right and Wrong"—are misleading, because they lead the reader to expect a scientific defense of the universal force of specific principles, for example, that murder is always wrong, or that it is always good to rescue someone in distress. Hauser apparently does believe that there are moral universals of this sort—he refers to universal rules, qualified by culturally specific exceptions—but it quickly becomes apparent that these kinds of rules are not the focus of his argument. Rather, drawing on an analogy with the universal deep structures of grammar defended by the linguist Noam Chomsky, he argues that our moral judgments reflect a deep structure, a normative grammar, as it were, that is normally inaccessible to consciousness but that ineluctably shapes normative reasoning in demonstrable ways.[3] Like Chomsky's universal grammar, this normative grammar sets parameters for intelligibility, but it does not in itself determine any particular set of moral principles. Just as the universal grammar innate in every developmentally normal human being must be specified through acquiring a particular language, so our innate normative grammar must be specified through formation in a particular moral system in order to become practically effective. By the same token, divergence at the level of specific norms can be interpreted as diverse expressions of what is recognizably an underlying normative grammar. Thus, paradoxically, the evident reality of moral pluralism actually supports Hauser's analysis: we can only recognize diverse normative systems as different expressions of one kind of human practice because we can identify normative judgments in terms that are not tied too closely to a specific set of moral judgments.

It will be apparent that Hauser's work is important for anyone who is interested in retrieving classical naturalistic or natural law theories of morality. By identifying the deep structures informing our practical judgments, he opens up the possibility that human nature may possess an intrinsic normative structure, while granting that it underdetermines the specific moral norms that give it concrete expression in any given society. On this view, considerations of naturalness are morally relevant without being, by themselves, morally determinative. By the same token, however, Hauser's work raises a further question. If we grant that human nature incorporates normative structures of some kind, just how are these structures morally significant? Defenders of a universally binding, natural-law morality have one kind of answer to this question: moral judgments track naturalness in very specific ways, whereas defenders of the autonomy of morality offer another kind of answer: moral judgments have nothing to do with considerations of naturalness. If we reject both of these alternatives, what other options do we have? Clearly, if we are to draw out the implications of Hauser's work in a persuasive way, we will need to think more carefully about the relation between the normative structures of human nature, and moral judgments properly so-called. And by the same token, his work offers a kind of case study for considering just what it might mean to say that human nature can be understood in terms of an Aristotelian teleological approach, thus uncovering its intrinsic normative structure. This approach, in turn, may perhaps offer something to Hauser's project—namely, a way of thinking about the relation between the normative grammar that Hauser uncovers, and the practices, principles, and ideals of virtue that constitute morality as we understand it. Or so I will try to show in what follows.

In key part, if not entirely, the normative grammar that Hauser uncovers is a grammar for the identification and evaluation of kinds of actions, distinguished by reference to the causal relations between what the agent does and the outcomes that she produces, whether intentionally or not. Generally speaking, this grammar tracks familiar distinctions between doing something and allowing something to happen, between bringing something about directly or indirectly, and the like—distinctions that are further formulated in terms of broad categories of permissible, obligatory, and forbidden kinds of acts. These categories, in turn, set parameters that are further specified through culturally specific moralities, that specify what counts as fairness in exchange and for permissible exercises of violent force. For example, they inform familiar distinctions between killing someone and allowing him to die. (Considerations of fairness and the legitimate uses of force comprise two domains of substantive principles that Hauser identifies, in my view quite rightly, as central to the life of complex social primates such as ourselves.) At this level, we find a wide range of cultural variability, so much so that the same distinctions are sometimes given diametrically opposed significance by different societies. Once the parameters for permissible, and especially obligatory and forbidden acts have been set, however, they determine our moral judgments, even against the grain of explicitly held beliefs and felt sentiments across a wide spectrum of cases.

Hauser remarks that the deep structures of this normative grammar are inaccessible. But clearly, as his own work illustrates, they are less inaccessible to some individuals—and to some periods—than to others. While our contemporaries are aware of traditional distinctions among kinds of actions, following the lines Hauser draws, there is a widespread tendency to reject these as irrational anachronisms, or, at best, as approximations to judgments that could better be made on other—usually, Kantian or utilitarian—grounds. One influential school of moral philosophy, taking the consequentialist logic of utilitarian theories to its ultimate conclusions, argues that these distinctions have no rational force at all.[4] On this view, there is no difference, morally, between (for example) killing someone and failing to avert her death—and this is so, no matter how remote our actions are from the foreseen consequences of doing one thing rather than another. Of course, this line of argument would do away with familiar distinctions between, for example, withholding medical treatment and euthanasia, and for that very reason, many find it to be attractive. But taken to its conclusions, this argument implies that we are equally responsible for any and every outcome of everything that we do, no matter how remote or indirect the link may seem to be—thus, we have murdered everyone whom we might possibly have saved throughout our lives—all those who starved because we did not write a check to Oxfam, who froze to death because we did not take them into our homes. This, it seems to me, is the point at which Hauser's work is most directly relevant to the project on evolution and cooperation because, as will be apparent by now, this radical consequentialism would seem to call for a very stringent version of altruism indeed, one in which the needs, interests, and desires of oneself and one's nearest associates carry no distinctive weight at all over against the (always overwhelmingly greater) needs, interests, and desires of human beings, or sentient creatures, taken in the aggregate.

But is this kind of altruism something that we would really commend? Most moral theorists find radical consequentialism of this sort to be appalling, and yet they are hard-pressed to respond to it beyond a curt dismissal in the name of our fundamental moral intuitions. The philosopher Jonathan Bennett, more candidly, says of this claim that "the proposed morality is too demanding (not to be *plausible,* but) to be *acceptable:* I am unwilling to hold myself to such a standard" (Bennet 1995, 16). Bennett goes on to admit that this is an awkward conclusion, but "awkward" scarcely seems to meet the case. If the radical consequentialists are correct, our most deeply entrenched normative intuitions are not only unjustified but profoundly wrong-headed, directing us to actions that are at odds with our central rational principles. It may well be that on this view, the demands of morality are simply beyond us—and in this way, what begins as an attempt to purify and redirect our moral intuitions results in a kind of moral nihilism, or, at best, a series of very poor approximations of a radically altruistic ideal.

Hauser at least offers a plausible explanation about why it is that our moral judgments track considerations of causality in the ways that they do. On his account, the relevant distinctions are so deeply entrenched that we simply cannot act—or much less reflect on how we ought to act—without invoking them in some way any more than we can speak without employing the deep grammatical structures identified by Chomsky and others. At the very least, this implies that natural cognitive structures set parameters that must be respected if we are to act at all—and in this way, it provides some justification for drawing the distinctions that we draw between direct and indirect, proximate and remote, ways of bringing something about or allowing something to happen. We might also suspect that these structures comprise preconditions for effective cooperation, considered simply as a set of mutually beneficial interactions—after all, we cannot cooperate unless we can in some meaningful way act in concert, and this would seem to imply that we can identify the actions of others as such and intend to act in meaningful ways ourselves. Yet this line of justification is limited at best; we may be compelled to accept these parameters as limitations on our capacities for judgment and action, but that does not mean that there is anything positive about them, anything that should be prized in its own right as a positive component of our best moral ideals.

Is that really what we want to say? To continue with our earlier example, it might seem obvious that it is better to save a life than allow someone to die, all other things being equal. But would we really want to pursue the kinds of activities that we would need to undertake in order to save as many lives as we could, even assuming that such a project represents a real possibility? In order to do so, every possible course of action that did not have some kind of "rescue payoff"—either directly, or indirectly, by generating income that could be used for aid—would have to be foregone. (Can it be a coincidence that this kind of radical consequentialism has emerged at a time and place in which the logic of the market, and capitalist exchange, has begun to take over every aspect of human life?) We could not in good conscience take time out for nonremunerative artistic activities or scholarly research. Gratuitous expressions of affection, together with every form of preferential treatment for our nearest and dearest, would be ruled out on this view, to say nothing of long hot baths and other such expressions of self-regard. Not only is it impossible for us to live in this way, I would suggest that we should not even try. Nor should we expect these kinds of sacrifices from others who might conceivably aid us in distress. In order to pursue this kind of life, we would have to sacrifice almost all the goods and values that are characteristic of the only forms of human flourishing and neighbor-love that we actually know—almost everything that makes human life worthwhile, praiseworthy, or genuinely loving.

I suspect that Hauser would share this judgment, but in any case the normative grammar that he uncovers lends it support. If we are indeed "hard-wired" to draw certain distinctions among different kinds of actions, we should not be surprised that these distinctions are deeply interwoven with a range of other judgments regarding what is genuinely praiseworthy and worthwhile, what we owe (and do not owe) to others, what constitutes authentic human love and friendship, and the like. Even so, Hauser's analysis by itself cannot do more than suggest a line of defense. In order to cash this suggestion, we need to go further. It is at this point that an Aristotelian teleological approach, particularly as developed by the early scholastic jurists and theologians reflecting on a natural law, can provide an unexpected complement to Hauser's analysis.

Many readers will wonder why I introduce early scholastic reflection on the natural law at this point, given my reliance so far on an Aristotelian teleological approach. Yet the scholastics' concept of the natural law was developed within a broadly Aristotelian philosophy of nature, within which widely shared human inclinations are rendered intelligible and given shape and direction in the light of their contributions to the overall well-being of individuals and the community. Given widespread assumptions about natural-law arguments, this may suggest that, for them, the broad normative structures of human existence are morally binding in the sense that we have some obligation to pursue typically natural patterns of activity—to pursue reproduction, for example. Yet whatever Aristotle's own views on this question may have been, this was (perhaps surprisingly) not the view taken by the early scholastics. On the contrary, they were well aware that the broad structures of human existence must be further qualified through rational reflection and specification in order to be translated into properly moral norms.[5] We noted above that on Hauser's showing, certain kinds of distinctions are always marked as normatively significant, but the precise significance that they have varies from culture to culture—and this is just what the scholastic appropriation of Aristotelian teleology would lead us to expect. Nonetheless, even prior to its specification in terms of some particular moral system, the broad patterns that Hauser identifies should be intelligible in the light of their contribution to an overall, distinctively human way of life—and to be good and valuable for that very reason, considered simply as a characteristic way of functioning.

When we turn to the beginnings of systematic normative analysis, as these emerged in the work of twelfth- and thirteenth-century lawyers and theologians, we find that the kinds of distinctions among actions that Hauser suggests, so far from being suspect, are foundational and central to their work. These authors are at great pains to work out what we would call the grammar of normative analysis through a close conceptual analysis of the categories of permissible, obligatory, and forbidden, paying special attention to the components of human choice and behavior informing these categories and reflecting on the ways in which these components come together to determine the normative character of a particular act.[6] What is more, there is nothing in their debates over these matters to suggest that they find this kind of close-grained analysis of the structures of human action to be in any way incongruous, much less irrational. On the contrary, if anything they are too quick to assume that this is the natural and proper form for normative analysis. Not only do they spell out traditional moral prohibitions in these terms through painstaking analysis of the characteristics of an act of murder, for example, seen in relation to other kinds of homicide, they also spell out ideals of virtue and counterideals of vice in terms of the kinds of acts characteristic of each. There are reasons why the moral significance of the structure of the human act is so salient for them, and once we take those reasons into account, we can better appreciate why a comparison between Hauser and these unexpected allies might prove to be fruitful.

The most immediate such reason is suggested by the fact that it was scholastic jurists and theologians, rather than philosophers, who took the lead in this line of analysis. In order to appreciate the significance of this fact, it is necessary to realize that the twelfth and thirteenth centuries comprised a period of far-reaching reforms in legal and sacramental practice, together with corresponding attempts to develop comprehensive accounts of jurisprudence and theology within which these reforms could be defended and carried forward. In this context, they tended to frame normative analysis within broadly juridical categories, defining kinds of actions as blameworthy or meritorious. This tendency was as marked among the theologians as the legal scholars—indeed, it was if anything more marked among the theologians because of their focus on the concept of sin, which, again, prompted a close analysis of the components of human action. Yet this period was also marked by intense attention to the inner life of the individual and to the value and appropriate expressions of inner freedom. In many respects, these tendencies arose out of the same matrix of causes and were mutually reinforcing. With respect to the questions we are considering, however, these tendencies stood in tension with one another in such a way as to shape what became the defining issue for debates over merit and sin—namely, what is the relation between the exterior act, so carefully defined in institutional and legal contexts, and the inner intention, so vitally important to the life of the individual?

It would take us very far afield to trace the various ways in which the scholastics attempted to address this question. At this point, I simply want to call attention to the question itself, and to suggest that it will inevitably arise in some form in every human society, although it will have more salience in some communities than in others. And this, I want further to suggest, reflects one of the intrinsic features of the form of life proper to us as social primates of a certain kind. As such, we have a stake in setting boundaries for permissible behavior so as to maintain group cohesion and to protect the vulnerable. At the same time, we also have a stake in maintaining a conceptual framework within which we can assess our intentions, attitudes, and stances *vis-a-vis* one another—a framework that can be applied readily to others and also reflexively toward oneself. We might be tempted to think that this framework, and the concerns that it represents, are simply a further reflection of the need to draw boundaries, but the relation between these two sets of concerns is more complex than that. We could draw boundaries through simple punishment or exclusion without reference to considerations of intention or much less culpability. Correlatively, our concerns about the stances of others toward us, and our own (not obvious!) disposition toward them, go beyond the requirements of boundary maintenance, to include most of the questions that arise when we reflect on our relations with others and our sense of ourselves as worthy, deserving of respect, persons of integrity—or not, as the case may be.

If this line of analysis is at all valid, it implies that any viable normative tradition will find a way to bring together its boundary setting and identity mediating components in a complementary and mutually re-enforcing way. And this, I would argue, is precisely why the scholastics' analysis of normative principles and human agency in terms of a normative grammar of action was so felicitous—because it illuminated the ways in which external normative considerations and the agent's dispositions and overall stance come together in mutually conditioning ways. The key insight here is that we cannot distinguish so sharply between normative considerations and attributions of intention, responsibility, and the like—our judgments about each will necessarily and properly reflect some consideration of the other. What is more, these two sets of considerations can best be held together through sustained attention to the human act—to our conceptions of morally salient kinds of actions, on the one hand, and to the grammar of intention and responsibility expressed in individual acts, on the other. By implication, this line of analysis lends support to the earlier suggestion that these kinds of distinctions are essential to any kind of human cooperative behavior. By the same token, it suggests further that any genuine ideal or virtue of altruism (as precisely driven by love or other regard) must incorporate and respect them as well, if only because our virtues and ideals must be tethered in the natural inclinations that they perfect, while remaining consonant with our best ideals of human life taken as a whole.

The scholastics offer us an example of what this kind of sustained attention would look like, and what it might have to offer in the way of normative insight. Hauser has given us good grounds to believe that the scholastic approach reflects deep structures of human cognition—in other words, that they identified and drew on an aspect of human life that is genuinely natural and not conventional, even though their analysis was couched in more or less conventional terms. When we bring these two perspectives together within the framework of the classical approach to natural law sketched above, we begin to see how the grammar that Hauser uncovers can be defended as an intelligible and broadly normative component of human nature, good in virtue of its indispensable place in the form of life proper to us as rational, social primates. In other words, the grammar of action—which is a grammar of intention and accountability, paradigmatically located in assessments of equity and harm—serves the critical function of enabling us to place ourselves and others within a framework of communal expectations and individual choices and ideals in such a way that these are not disjoined or set at odds with one another but brought together in sustainable and mutually reinforcing ways. Of course, the specifics of this framework, and especially the relative weights given to individual integrity and communal solidarity, will vary widely, most probably in incommensurable ways. Yet any such framework will necessarily leave room for both considerations and will offer ways of combining the two. Only in this way can human beings act in principled ways—that is, on the basis of norms that they grasp and endorse, while at the same time playing their diverse parts in a cohesive community and locating themselves in relation to that community and to other people. Whatever the mechanisms that led to the emergence of the deep structure that Hauser identifies, it was appropriated, through a synthesis of cultural and biological evolution characteristic of us as a species, to enable our further development as creatures of the specific kind that we are.

This brings me finally to the broader question of the relevance of an Aristotelian teleology to the subjects of cooperation and altruism tackled in this book. Admittedly, an Aristotelian analysis does not provide us with an explanation for the emergence of cooperation, which would stand as an alternative to the careful reconstruction of possible evolutionary mechanisms developed by Martin Nowak and his colleagues. I hope I have said enough by now to indicate that it would be a mistake to expect Aristotle and Nowak to compete in these terms. What an Aristotelian analysis does offer, I would suggest, is a helpful way of thinking about and formulating the issues that Nowak and his colleagues are addressing.

As I understand it, this project begins with a puzzle: how can we account for the emergence and persistence of cooperative behavior, given that cooperation ought to detract from, rather than promote, the reproductive success of the individual? Note that in identifying this as a problem, we already make certain assumptions about cooperative behavior. We assume, first of all, that cooperative behavior is, in fact, a recurrent and identifiable component of the functioning of many different kinds of living creatures, difficult though it has proven to be to define exactly what cooperation means! This implies, in turn, that cooperative behavior is not just a recurrent feature of living creatures but a kind of functioning that contributes to the overall way of life of the organism in some way or other. There are many aspects of the typical course of any living creature that are clearly not positive—everything, sooner or later, wears out or gets sick or just decays—and we do not look for evolutionary explanations for these processes, except perhaps to try to explain why the state of optimal fitness just breaks down sooner or later. Cooperative behavior, in contrast, does appear to play a positive role in the life of individuals, and still more, in the corporate life of populations. It just does not play the (direct) role of optimizing the reproductive success of the individual, and that is precisely why it raises so many interesting questions. What is more, reflection on the positive aspects of cooperative behavior suggests hypotheses for the ways in which cooperation might have embedded itself through evolutionary processes—through promoting the well-being of one's kin and their shared genetic inheritance, for example, or through mechanisms of group maintenance and selection. Once these hypotheses are formulated, it falls to Nowak and others to test them through the creation of mathematical models or through whatever other methodologies suggest themselves. Note, by the way, that through this process of reflection, our concept of reproductive fitness is itself transformed in subtle but important ways, reformulated in terms of individual promotion of the reproductive fitness of a group or a kind rather than narrowly understood in terms of the generation of one's own kin. And this, in turn, suggests a more comprehensive teleological account of what it is for a population of living creatures to flourish and to perpetuate itself, an account that allows us to formulate well-being and flourishing in terms of a collective form of life in which individuals play diverse and complementary roles and achieve their individual perfection in these terms.

This line of analysis can work in the opposite direction, as the example of punishment of noncooperators suggests. Here we find an example of a pervasive form of behavior that is demonstrably disadvantageous to the individual even when the above kinds of qualifications are taken into account. Yet, a tendency to punish appears to be deeply ingrained, not only among humans but among many other kinds of social animals as well. This suggests, to Nowak and his colleagues, that punishment contributes to the overall well-being of populations in yet unspecified ways. Note that in this case, the assumption is that such a pervasive feature of social life, human and nonhuman—not obviously a malfunction, or an expression of decline and decay—probably does have a part to play in the overall well-being of the kind, if not of the individual itself. And once the function of the behavior has been identified, or plausibly reconstructed—once we have a better sense of the point of punishment—we will be in a position to test our reconstruction through appropriate mathematical or other methodologies. Again, a teleological analysis suggests lines of inquiry and helps us to formulate the relevant kinds of questions to ask and identify what would count as evidence for, or disproof of, our hypotheses.

I have not attempted to say much about the relation between cooperation and altruism, and given the limitations of space I won't attempt to do so now. I do want to suggest, however, that this issue represents a particular case of a broader problem—namely, the relation between the well-being proper to the human creature considered as a living organism and the properly human form of perfection constituted by a virtuous life. The relation between these, as I have argued elsewhere, is complex. We cannot simply equate natural well-being with the fully human form of flourishing attained in and through the practice of the virtues—but we cannot detach the first level of analysis from the second, either. Rather, the virtues represent reflective reformulations and knowing appropriations of the kinds of values and functioning that structure human life at its most basic levels. Temperance is not just a matter of good, healthy eating and clean sex, but it does reflect an intelligent pursuit of health, reproduction, and yes, sexual pleasure, mediated through a reflective sense of the place that these activities should have in a human life that is enjoyable and admirable, taken as a whole. By the same token, altruism is not just cooperative behavior, but it does represent a reflective pursuit of the values inherent in cooperative behavior, reformulated in the light of a broader and more comprehensive sense of the scope of concern, going beyond individual or communal well-being to include the widest possible range of others with whom we can cooperate in shared projects, and whom we can realistically benefit. Yet any such account of altruism, if it is to remain both humane and realistic, must stay grounded in a realistic appreciation for the fundamental natural values promoted by what we might call the first level of cooperative behavior. The ideal remains grounded in the values and limitations inherent in our lives as creatures and as animals, in other words—or so I would argue. But this is clearly a topic for another day.

Notes

1. See Porter (1999, 63–97; 2005, 82–125). There, I defend an Aristotelian philosophy of nature and argue for its compatibility with contemporary biological science.
2. Lennox 2001, 128. See more generally 127–30.
3. The expression, "normative grammar," is mine and not Hauser's, but I hope it captures his intent in an economical way.
4. The *locus classicus* for this approach is Singer (1972); for more recent statements of his views, see Singer (1994). Other influential examples include Parfit (1984), Kagan (1989), and Garth L. Hallett (1995).
5. For further details and relevant texts, see Porter (1999, 63–97).
6. For a comprehensive account of scholastic act analysis, including an exhaustive survey of texts, see Lottin (1948).

References

Bennett, J. 1995. *The Act Itself.* Oxford: Oxford University Press.
Hallett, G. L. 1995. *Greater Good: The Case for Proportionalism.* Washington, D.C.: Georgetown University Press.
Hauser, M. 2006. *Moral Minds: How Nature Designed Our Universal Sense of Right and Wrong.* New York: HarperCollins.
Kagan, S. 1989. *The Limits of Morality.* Oxford: Oxford University Press.
Lennox, J. 2001. *Aristotle's Philosophy of Biology.* Cambridge: Cambridge University Press.

Lottin, O. 1948. "Le problème de la moralité intrinsèque d'Abélard à saint Thomas d'Aquin." In *Psychologie et morale aux XIIe et XIIIe siècles*, 21–465. Louvain: Abbaye du Mont César.

Moss, L. 2003. *What Genes Can't Do.* Cambridge, MA: The MIT Press.

Parfit, D. 1984. *Reasons and Persons.* Oxford: Oxford University Press.

Porter, J. 1999. *Natural and Divine Law: Reclaiming the Tradition for Christian Ethics.* Grand Rapids, MI: Eerdmans.

———. 2005. *Nature as Reason: A Thomistic Theory of the Natural Law.* Grand Rapids, MI: Eerdmans.

Ruse, M. 1998. *Philosophy of Biology.* Amherst, MA: Prometheus Books.

Singer, P. 1972. "Famine, Affluence, and Morality." *Philosophy and Public Affairs* 1: 229–43.

———. 1994. *Rethinking Life and Death: The Collapse of Our Traditional Ethics.* New York: St. Martin's Press.

16

The Christian Love Ethic and Evolutionary "Cooperation"

The Lessons and Limits of Eudaimonism and Game Theory

TIMOTHY P. JACKSON

> When you enjoy a human being in God, you are enjoying God rather than that human being. For you enjoy the one by whom you are made happy.
>
> —Saint Augustine, *On Christian Teaching*

> Natural selection will never produce in a being anything injurious to itself, for natural selection acts solely by and for the good of each.
>
> —Charles Darwin, *On the Origin of Species*

One of the longest-standing debates within Christian moral theology is that between eudaimonism and agapism. Eudaimonism holds that all human actions aim at personal thriving or "happiness"—*eudaimonia,* in philosophical Greek—with wisdom being a matter of properly appraising the objects and enterprises that contribute to this fulfillment. Agapism holds, in contrast, that Jewish and Christian ethics begin with theology rather than anthropology, with the real holiness of God rather than the ideal happiness of humanity. For the agapist, morality is founded on the divine bestowal of worth on others—agape, in New Testament Greek—rather than on the accurate appraisal of worth for oneself (see Nygren 1969). My main thesis in this essay is that evolutionary game theory's efforts to account for "altruism" and "cooperation" are akin to eudaimonism's efforts to account for neighbor love and social justice. Both game theory and eudaimonism start out with natural self-interest and the instinct for self-preservation and seek to analyze robust other-regarding virtues on that basis. Yet both fail, and for similar reasons.

I do not claim in these pages to disprove all possible naturalistic accounts of the genesis of self-sacrificial love, but I do contend that Darwinian evolution, defined as mutation and natural selection, does not provide an adequate explanation for altruism and neighbor love in the rich and demanding sense intended here. More specifically, I reject eudaimonism and evolutionary game theory as erroneous: they both conflate doing good with doing well, and past causal etiology with present moral purpose. These mistakes are the result of a single common failing: overestimating the power of eros, variously characterized as the pursuit of happiness or, in the case of all species, natural selection. I do not doubt the existence of the pursuit of happiness or natural selection, nor do I deny that they bear derivatively on ethics, but I do contend that they are not the *teloi* or normative bases of all ethics.

Defining Key Terms

Cooperation. Sarah Coakley and Martin Nowak write: "For the purposes of this current book, we propose the following working definition of the word 'cooperation' in its evolutionary sense: 'Cooperation is a form of working together in which one individual pays a cost (in terms of fitness, whether genetic or cultural) and another gains a benefit.'" This definition is straightforward but problematic, since it appears to rule in parasitism and agapism as forms of "cooperation," even as it rules out mutualism and commensalism. In mutualism, both parties benefit from an interaction; in commensalism, one party benefits while the other is unaffected; but surely it is odd to decline to call these instances of "cooperation." Moreover, it seems downright contradictory to call (coercive) predation and (willing) self-sacrifice instances of "cooperation." The problem here is not in Coakley and Nowak but rather in the linguistic custom of many evolutionary biologists and game theorists.

What is needed for true cooperation, I believe, even in evolutionary contexts, is the stipulation that the "cost" paid by one individual and the "benefit" gained by another must add up to a greater aggregate benefit for both than could be achieved by either independently. In the iterated prisoner's dilemma, for instance, both parties benefit overall and in the long run by each taking less than he or she might acquire alone and in the short run. With this alternate definition, one sees that neither parasitism nor agapism counts as cooperation, since both of these interactions involve real and permanent costs and benefits that are not compensated for or overridden across time. A parasite gains while its host loses, even as an agapist loses while her neighbor gains—the difference being that the parasite forces a loss on the other for the parasite's own benefit, while the agapist voluntarily serves the other at the agapist's own cost. (Here the question of motive fully emerges.)

Altruism. The word *altruism* (French, *altruisme,* from *autrui:* "other people," derived from Latin *alter:* "other") was coined by Auguste Comte (1798–1857), the French founder of positivism (see Dixon, Chapter 2). *The Oxford English Dictionary* (1971, 65) defines "altruism" as "devotion to the welfare of others, regard for others, as a principle of action; opposed to egotism or selfishness." *Merriam-Webster Online* (2006) defines the term as follows: "(1) unselfish regard for or devotion to the welfare of others; (2) behavior by an animal that is not beneficial to or may be harmful to itself but that benefits others of its species." What Coakley and Nowak call "cooperation"—loss by one for another's gain—actually seems synonymous with "altruism," as defined directly above. But again, *pace* the editors of this book, surely "cooperation" entails mutual benefit, or at least no intentional profiting from the woe of another, whereas "altruism" involves *x*'s willingly paying a cost (without the possibility for compensation) for *y*'s benefit.

Agape. I take agape to be an elaborated form of altruism. Drawing on the Judeo-Christian tradition, I identify agape with three interpersonal features: (1) unconditional willing of the good for the other, (2) equal regard for the well-being of the other, and (3) passionate service open to self-sacrifice for the sake of the other (Jackson 2003, 10). To the extent that agape involves disinterested concern for others, including openness to self-sacrifice even for a stranger or enemy, it would seem inevitably to be weeded out by natural selection, understood genetically. Agapists willingly reduce, or risk reducing, their reproductive success for the sake of others.

Eros. Eros is the Greek philosophical term for preferential desire, a love that loves by virtue of the excellence of the loved object, especially when that excellence benefits the loving subject. As Diotima says in *The Symposium,* eros is "desire for the perpetual possession of the good"—its "function" being "procreation in what is beautiful," with procreation being "either physical or spiritual" (Plato 1951, 206a–b, 86). So understood, eros drives both nature and culture toward personal advantage and reproductive success.

A Brief Critique of Eudaimonism

There is a burgeoning literature seeking to document the myriad ways in which altruistic behavior redounds to the agent's benefit. The frequent thesis is that behaving benevolently makes one happier and healthier and/or secures a public reputation that makes one better off in the long run. Endorphins, if not Eden, are the payoff for other-regarding ethics. I remain skeptical of such medical and sociological optimism (or skepticism, for that matter); Jesus Christ went to the cross a despised and physically broken man after all, and Gandhi and King were assassinated. But one must distinguish, in any event, between the beneficial *consequences* of altruism and its primary moral *motivation.* Otherwise one will inevitably be tempted by a eudaimonistic ethic, inspired by the Greeks, in which I "love" others primarily because this secures my own flourishing. Self-interested flourishing has its proper place, but it is the fragile flower rather than the vigorous root of charity. It leads inevitably to self-justification, rather than to gratitude to God.

Jesus, at any rate, was neither a eudaimonist (focusing on personal benefit) nor a nepotist (focusing on direct or inclusive fitness):

> Do not think that I have come to bring peace to the earth; I have not come to bring peace, but a sword. For I have come to set a man against his father, and a daughter against her mother, and a

> daughter-in-law against her mother-in-law; and one's foes will be members of one's own household. Whoever loves father or mother more than me is not worthy of me; and whoever loves son or daughter more than me is not worthy of me; and whoever does not take up the cross and follow me is not worthy of me. Those who find their life will lose it, and those who lose their life for my sake will find it. (Matt. 10:34–39)

Jesus sets family members against each other precisely to the extent that he commands them to love all neighbors regardless of biological (genetic) or ethnic (cultural) relatedness. He commands his followers, in effect, to overcome both the selfish gene and the selfish meme, even going so far in the Sermon on the Mount (Matt. 5–7) to instruct them to love their enemies. Some speculate that this universalism was motivated, in part, by Jesus's knowing himself to be the bastard son of Mary and an unknown human father and his having grown up in a context in which Hebraic, Greek, and Roman cultures interacted, however tensely. Be that as it may, Jesus clearly held, and knew, that his radical obedience to God and his inclusive attitude toward humanity would be costly in this life, far from a gospel of worldly prosperity or reproductive advantage.

Undoubtedly, practicing charity can awaken one to a fuller or higher sense of what makes for genuine happiness and well-being. In ordinary temporal terms, however, Christlike love/altruism does not always redound to the benefit of the lover or his or her family. Sometimes (perhaps often) such love means real loss and sorrow in this life, and to try to explain that away or lessen its offense is to preach another gospel. For Christians, at any rate, the Passion of Christ only "makes sense" if agape is its own reward, a participation in God's own holiness. That is to say, the Passion does *not* "make sense" in strictly adaptational, rational terms. Presumably this is why the tradition has called agape a "supernatural" virtue.

The most basic criticism of "Christian eudaimonism" is that it seeks to arrive at love of God and neighbor (agape) by beginning with self-interested desire (eros). The problem is not that eros is illegitimate as such but rather that it cannot be the starting point for biblical ethics. Jewish, Christian, and Muslim morality begins, not with immanence but transcendence, not with the goal of human flourishing but the reality of divine holiness.

An Analogous and More Extended Critique of Evolutionary Game Theory: Consequences, Motives, and Means

Much of evolutionary game theory recapitulates the problems with eudaimonism, including confusions concerning consequences, motives, and means. In a recent review essay, Martin Nowak and Karl Sigmund (2005, 1291) observe that "in the terminology based on Hamilton, Trivers, and Wilson, an act is said to be altruistic if it is costly to perform but confers a benefit on another individual. In evolutionary biology, costs and benefits are measured in Darwinian fitness, which means reproductive success." Such reproductive utilitarianism is characteristic of evolutionary psychology in general and game theory in particular,[1] and the consequentialism at work is unaffected by whether the "reproductive success" is measured in genetic or cultural terms. To focus on replicating genes or on proliferating memes is not to ask about the motives of agents or the forms of their actions.

One might grant that, to a degree, the authors mentioned can define technical terms as they see fit, but is it really edifying to omit the motive of the act as well as its form? Is an act really altruistic if it is done expecting, or even demanding, a return, and what if the act is coerced rather than voluntary? Would not we say that someone who helped another only because he anticipated a reward was merely shamming altruism? We might not fault him for such behavior; we might even concede that few human motivations are altogether free of self-seeking; but if the predominant or sole aim is self-interest, would we praise the agent *morally?* I think not. Moreover, if the benefit to the other is extracted unwillingly, would not we both fail to admire the actor and (probably) indict the one who compelled him?

The song of a rose-breasted grosbeak or a yellow warbler may be beautiful to us in spite of the lack of any avian intention to create that beauty, even as the Grand Canyon is awe-inspiring without its being self-aware. Unlike with aesthetic effects or natural states, however, motives and means of production matter in ethics, as well as end results. This is because human beings have inwardness, and this inwardness is partly constitutive of their experiences and actions (see Nagel 1974). If I understand English and sincerely announce to myself and others, "I am in pain," for instance, then, in fact, I am in pain. Similarly, if I give food and shelter to a hungry and homeless person because I honestly feel compassion for him and concern for his well-being, there is at least strong reason to believe that I am genuinely altruistic (a Good Samaritan). Conversely, if I offer the food and shelter but do so with a malicious or manipulative intent, I fail to be altruistic. This is the case even if the homeless person is actually benefited and even if my action is actually costly to me.

A certain motivational structure is a necessary condition for real altruism, then, though it is not a sufficient one. Again, actions and effects matter too.

The Perils of Panselectionism

Much evolutionary game theory makes two unquestioned assumptions: (1) that competitive advantage in the context of scarcity is basic to explaining the origin and durability of all traits, and (2) that competitive advantage is enjoyed by individuals rather than groups. A trait (like cooperation or altruism) emerges and endures, that is, only because it has utility for the agent that bears it, to the relative disadvantage of other agents vying for limited resources. These assumptions model, in game theory, what Stephen Jay Gould (1997a, 34–37, and 2002, 197–201) calls "Darwinian fundamentalism" or "organismal panselectionism" in biology: the view that all morphological and behavioral features of an organism are a function of adaptation for reproductive success for that organism. The problem with evolutionary game theory, however, is that it is blind to the levels of selection and overrates natural selection itself. Let me now substantiate this charge.

Many evolutionary biologists and game theorists presume that, given the two assumptions above, they can answer two pressing questions: How could natural selection have first generated an altruistic impulse, given its seeming reproductive disadvantages? If natural selection is behind altruism after all, should we continue to practice or applaud it once we know its evolutionary explanation? Sophisticated discussions of the ultimatum game and the prisoner's dilemma—with titles such as "Chaos and the Evolution of Cooperation," "The Arithmetics of Mutual Help," "The Economics of Fair Play," "Empathy Leads to Fairness," and so forth—often claim to explain how naturalistic evolution and traditional virtue are compatible, even mutually implying. The common refrain, however, is that what looks like selflessness actually pays dividends, if only indirectly, either to oneself or to one's close biological relatives ("kin altruism" and "inclusive fitness"). In effect, most biologists and game theorists assume the reality of evolution as panselectionism and then seek to show how agape, despite appearances, has adaptive value for the agent or his or her relatives. But, if successful, that exercise would *explain away* agape, as usually defined (see above). Nothing with unrequited reproductive cost can be tolerated by natural selection alone, but without openness to such cost, there is no agape.

Now what is someone like myself, a theistic evolutionist, to do? I affirm *both* the reality of robust altruism *and* the fact of biological evolution, but how can we reconcile cultural and genetic imperatives? If ideas are analogous to genes, with "survival of the fittest" being applicable to moral traditions as well as physical bodies, what are the mechanisms of, and standards for, such fitness? And can game theory capture these mechanisms and standards?

The brilliance of many of the game theoretical models is beyond question, but there is often a creeping conflation of mutual advantage or prudence with agapic love or compassion.[2] In addition, causal explanations of impersonal processes (efficient causes) are at times confused with the moral motivations of personal agents (final causes). On the first score, it does not matter whether we talk about "general utility" or "rational cooperation" or "indirect reciprocity." For it is one thing to endorse fairness because it pays or because you fear that you or your relatives will otherwise get the short end of the evolutionary stick, but it is something else to affirm another's well-being even though you and yours need never fear injustice or barrenness yourselves—indeed, even though it involves your voluntarily surrendering properly accessible values (such as progeny or good reputation).

Two Objections

Let me clearly state my two main objections to game theoretical and other naturalistic accounts of "altruism" that, as noted, I treat as synonymous with what Christians call "agape" or "love of neighbor." (I am not alone in my misgivings.) First, there is *the biological problem:* standard game theoretical models of the "evolution" of altruism—for example, those based on the iterated prisoner's dilemma—trade on a confusion of Lamarckian and Darwinian mechanisms, treating or seeming to treat cultural evolution as synonymous with genetic evolution. As Kirschner and Gerhart (2005, 27–28) point out: "Modern molecular and genetic analysis has revealed no hint of directed genetic change in response to physiological need or experience. No mechanism is known to direct a specific environmental stress toward the alteration of a specific gene or set of genes, as a way to ameliorate that stress. Hence there is no evidence for 'facilitated genotypic change.' Genetic variation and selection are completely uncoupled." The prisoner's dilemma explicitly takes an "environmental stress" (the need to defect or cooperate) and construes a cultural strategy to cope with it (e.g., generous tit for tat) as a heritable trait that has been or can be selected for, something that is ruled out by the modern Darwinian synthesis itself. According to that synthesis, there can be no environmentally induced variation, since "only the genotype is inherited" and "only the phenotype is selected" (Kirschner and Gerhart 2005, 29). But doesn't the prisoner's dilemma, in effect, treat phenotypic change as if it were genotypic? If "experience, learned behavior, or physiological adaptation to the environment [can] not be inherited" (Kirschner and Gerhart 2005, 29–31), does not game theory fall apart as an explanation for how altruism evolved?

If it is claimed that a genetic disposition to be altruistic just happens (random mutation) and *then* is selected for, due to its promoting competitive reproductive advantage, the confusion I am pointing to is avoided. But in that case agape does indeed seem like a miracle. Even if it could survive (or even thrive) as a behavioral trait, once it appears, how it originally emerged from blind chance amid the "struggle for existence" is, to say the least, a mystery.

In any event, there is a second and still more major stumbling block to game theory: *the ethical problem*. I have noted that much evolutionary game theory either totally neglects moral motivation for action or conflates motives with consequences, identifying reproductive advantage as the reason why people are sometimes altruistic. This is a bad mistake, but there is an even deeper ethical issue here: if agape unconditionally wills the good for the other, for the other's and/or God's sake, and if agape is open to a self-sacrifice that outstrips reciprocity (much less personal advantage), then talk of "tit for tat," "win-stay/lose-shift," "kin altruism," "indirect reciprocity," and the like is largely beside the point. Such phrases may capture aspects of social *cooperation*, understood in the ordinary language sense as action coordinated for mutual benefit. They may even expand the normal bounds of our concern in some respects, but they still basically reduce love of neighbor to a form of personal prudence. It may be that apparently or reputedly altruistic behavior is, in reality, motivated by fundamentally self-regarding agendas. But then it is better simply to concede that Nietzsche and Freud were right: love of neighbor is unnatural and self-deceived.

Importantly, the judgment that evolutionary theorists often reduce altruism to rational self-interest holds for both biological and cultural evolution. Both genes and memes are selfish, it seems, with universal love being an ideal cultural construct that makes up for genetic reproductive failure. For Richard Dawkins (1989, 2), for example, we are "machines created by our genes" to help them replicate themselves. Thus are body and soul made epiphenomenal to the genome, even as old-school materialists made the soul epiphenomenal to the body. Agape, in turn, is but a false consolation for the agony of agamy—a kind of evolutionary sour grapes. The rub is that once human beings and their moral minds are judged, in effect, to be mere vehicles of genes and their reproductive success, there is no higher value than such success—indeed, there are no purposeful personal agents anymore, only biological urges operating under the guise of ethical imperatives.

The Scale of Evolution and Widening the Circle of Concern

How, to reiterate the key question asked by Darwin himself, could natural selection have first generated an altruistic impulse, given its seeming reproductive disadvantages for individuals? There are currently three (putative) answers to this puzzle, each offering a slightly different interpretation of the scale or timing of selection. The first is that a helping tendency evolved as an adaptation bestowing benefits on close relatives, via the "inclusive fitness" and "kin selection" of William Hamilton (1963 and 1964). One labors and suffers not merely for one's own genes but for those of relatives who share extensively similar genes. The problem with this response, however, is that it describes nepotism, not true altruism (in an agapeistic sense). The second would-be answer is the "reciprocal altruism" of Robert Trivers (1971), in which one aids unrelated others at cost to oneself in the expectation that they (or others) will aid you (or yours) in return. This is not really altruism either, but prudence. The third proposal is that, over and above kin selection, a deme or an entire species may be a unit of selection and thus the beneficiary of altruism. So-called group selection, as defended by Elliott Sober and David Sloan Wilson (1998), maintains that a tendency to help others at cost to oneself might have evolved because it allowed one collective to outcompete another. This explanation widens the circle of concern, but even it does not get us to full-blown neighbor love in Jesus's sense. If one sacrifices for a limited group, even one including persons unrelated to you, this is tribalism, not altruism. In each of the three cases, one has not escaped invidious contrasts between "us" and "them" in favor of genuine agape.

There is yet a fourth, more radical, alternative that is seldom recognized: that altruism is not an adaptation selected for at all but, rather, a structural part of the *Bauplan* of human beings and the environment in which they live. It is crucial to realize that past adaptive history in no way determines the current meaning or intent of helping others at cost to oneself. (To take a standard example, even if sexual dimorphism evolved for the sake of reproduction, its use can now be "shifted" and directed at mutual pleasure.) It is equally crucial to realize that, in the fourth alternative, other factors are at work in evolution besides natural selection.

Spandrels and Exaptations

If natural selection alone were behind altruism/agape, we could not continue to practice or applaud it once we knew its evolutionary explanation, for then it would be for the sake of self-interest. How, then, do we reconcile causal biological explanations with purposeful design of *any* kind at *any* level? Unless that can be done, cultural evolution seems in thrall to, not merely to supervene on, biological evolution—even as the person appears a mere tool of her genes. The thoroughgoing naturalist Michael Ruse (1986, 253), for one, has maintained that morality is "an invention of the genes rather than humans"; indeed, he concludes that "morality is a collective illusion foisted on us by our genes." The end of winning at a competitive game can cast up parasitism as a strategy, and even cooperation as I have defined it, but it cannot grant agapism. Hence either agapism is delusory—as Ruse, Richard Joyce (2006), and others argue—or there are elements of ethics that cannot be captured by game theory.

Either agape is evolutionarily "front-loaded" in some sense—for example, as an echo of God's kenotic act of creation—or it is unreal. Competition can no more generate charity than nature can create God, even though charity can be present amid competition even as God is omnipresent in nature. In fact, altruism: competition :: natural selection: random mutation.

Rather than being caused by competition, altruism sifts and limits competition as raw material, even as random mutation is the raw material acted on, but not caused by, natural selection. Think of altruism as a kind of moral gravity that acts on, and in, bodies with spiritual mass.

As Gould and Lewontin have argued, we are not limited biologically or culturally to what can be accounted for by random mutation and natural selection. Some attributes are structural spandrels rather than functional adaptations. In architecture, a spandrel is the space between arches in the design of a building; in biology, a spandrel is an internal feature of an organism's basic form. The space between arches might subsequently be decorated, but it is not initially planned for its aesthetic worth. Similarly, a phenotypic or behavioral feature of an animal might come along with physical and/or biological constraints, the "laws of nature," and/or "nature's God," rather than selective advantage (Gould 2002, 43–49).[3] To call a trait like altruism a "spandrel" is to affirm the trait's reality but concede that it cannot be explained as arising due to adaptive value. I myself am inclined to construe the human potential for agape to be a function of being made in the image of God, a feature that is a limit on, rather than a consequence of, natural selection. If God is agapic love (1 John 4:8), to bear the image of God is to have the need or ability to give or receive such love. As a spandrel, agape would be "a positive director and channeler of evolutionary change," to quote Gould's general description of such organizational features (2002, 47).[4]

Alternatively, one might judge altruism to be an "exaptation." If an adaptation is a character trait that arises directly for its current utility, an exaptation, in contrast, involves "the cooptation of a preexisting character for an altered current utility" (Gould 2002, 671). A spandrel, to repeat, exists for no adaptive reason, but as part of an organism's internal engineering; an exaptation, on the other hand, is an adaptation functionally shifted from its original use. In the latter case, the biochemical and social-psychological raw materials for agape would have evolved over time and then been put to a different purpose by a self-conscious agent. Parental affection, for instance, might well be exapted for more general employment by the will of human agents or the grace of the Holy Spirit. But, again, this is not to say that agape itself can be directly caused or explained by reproductive success. Reductionist contentions to the contrary are usually the result of an implausible drive to make morality "scientific."

A Revision of Hamilton's Rule

A vigorous effort to make morality scientifically rigorous, even quantifiable, comes again from William Hamilton. For Hamilton, an "altruistic" act should be expected only if $rb > c$, where r is the degree of genetic relatedness of the recipient to the actor, b is the benefit to the recipient, and c is the cost to the actor. (Alternatively, the inequality may be written as: $r > c/b$.) This rule is manifestly violated every day, not just by Good Samaritans and Mother Teresas but by ordinary folks touched by empathy. If Gould and Lewontin are correct, however, this should not trouble or surprise us. If the capacity for altruism is a spandrel or exaptation, then, *pace* "Darwinian fundamentalism," it need not be subject to explanation in terms of reproductive advantage/natural selection. This fact contemporary game theory simply cannot accommodate.

Evolutionary game theorists point out that, in the iterated prisoner's dilemma, strategies of "always defect" and "always cooperate" lose out first to "tit for tat" and then to "generous tit for tat." Theorists also note that when a strategy of "always cooperate" is reintroduced into the stable system, "always defect" makes a comeback. Obviously, if "always cooperate" is meant to be a "winning strategy" for competitive advantage, it is a failure. Indeed, in destabilizing "generous tit-for-tat" and allowing defection to re-emerge, it opens the door to "war" where there had been a tolerable "peace." The point, however, is that "always cooperate" is not intended to be a competitive winner to begin with. Just as pacifism is not an alternative way to wage war, so agape is not just another path to erotic conquest. Both nonviolence and altruism[5] break out of the survivalist system and its economies of exchange and point to a different and better "game"—a noncompetitive game. This is radical cultural critique, a repudiation of much that counts as politics, to which current game theory is largely oblivious. Putting the point another way, to assume that evolution is a zero-sum game is to deny, a priori, the possibility of a general principle of emergent complexity that imparts a less Malthusian *telos* to life (see, e.g., Davies 2004).

If one sees agape as a selective adaptation, this, in effect, conflates it with eros, which leads to all kinds of mischief. Because eros aims at perpetual possession of what is good for the agent, I maintained above that it is the engine behind eudaimonism. Because eros seeks reproductive advantage, especially in the context of scarcity, it is also the engine behind natural selection. In both contexts, optimizing individual utility is the traditional summum bonum, and this is certainly the common assumption of game theory. Even if group selection is acknowledged, however, the tendency to corrupt ethics and theology remains. Eros's quest for differential survival funds an invidious contrast, for example, between us the elect and them the damned.

If, in contrast, one recognizes agape as a spandrel or an exaptation that freely departs from competition, Hamilton's Rule must be set aside in favor of: $rb > c/a$, where a (agape or altruism) is an independent variable, a passive potential built into us ab initio. This would mean amending Hamilton to read: "A self-sacrificial action will be performed only if the benefit to a recipient times relatedness of recipient is greater than cost to actor divided by the actor's altruism." The straightforward implication is, as it should be, that the higher the level of altruism, the more biological relatedness will not matter. Altruism can be "built up" or "schooled" over time, but it is not adaptive for the individual; in fact, it *guides and constrains* natural selection as a primal sympathy with others sunk into our very beings. It limits what one is willing to do to promote one's own happiness or to prejudice the happiness of others. Altruism, so understood, *may* be advantageous for collectives—that is, groups with members willing to sacrifice for fellow members may be able to outcompete other groups with fewer heroes—but the stronger the altruism, the weaker the motivational connection to nepotism or tribalism. However rarely realized, the Christlike ideal is to transcend *all* chauvinisms such that *anyone* who bears the image of God is recognized as brother or sister.

With reference to $rb > c/a$, Martin Nowak (2008) writes that "you would need to prove that such a formula is the crucial condition for evolution of cooperation in a particular process. Thus the question becomes: what is the process that leads to that formula?" My answer is: evolution as a structured unfolding is the process rather than pure selectionism. Nowak helpfully notes that "altruism could be modeled in the following way: my utility function is not only my own payoff, but also the payoff of the other person. But . . . this is not an evolutionary answer, because the crucial question is: how do I compete with someone who does not have such a utility function and only maximizes his own payoff ? In other words: how could natural selection lead to a utility function that cares about the other person?" I can only reply that $rb > c/a$, with a utility function that looks to the payoff of the other person, does not aim to be a *selectionist* answer. *We must not equate evolution with natural selection.* The purpose of the revised rule is to show how certain human traits limit, control, or outstrip the process of competitive adaptation.[6]

Conclusion

Natural selection does not exhaust the meaning and mechanisms of evolution, even as erotic virtue does not exhaust the meaning and methods of ethics. Evolutionary theory need not talk exclusively in terms of adaptation, just as ethical theory need not talk exclusively in terms of happiness. Neither fitness nor advantage is an omnicompetent concept. More forcefully put, both adaptation*ism* and eudaimon*ism* are impoverished and impoverishing, forgetful of our origins and oblivious to our aspirations. If the argument of this paper is correct, altruism can and should be accounted for in nonselectionist terms—namely, in terms not amenable to game theoretical models such as the iterated prisoner's dilemma. Either altruism is a spandrel, a currently telling feature with a nonadaptive origin, or an intentional exaptation of a previous adaptation, to use Gould's terms (2002, 1258–63), or it is a delusion. For reasons given, I take agapic love/altruism to be real but the product of neither the pursuit of happiness nor competitive reproduction.

Evolution, defined as descent with modification or speciation across time, is not a delusion but a well-established theory. The question remains, however, how evolution occurred. Evolution does not create agape ex nihilo on the theistic hypothesis, but, rather, divine Agape creates evolution, which in turn makes human agape possible. This is not a matter of God's injecting some alien power or substance into dead chemicals, or of God's intermittently intervening *ab extra* to improve biological designs, but rather of God's "making creatures make themselves."[7] Why do this so indirectly and with so much attendant suffering? The Christian agapist can only affirm: "so that we might eventually love one another, freely and self-sacrificially." To sacrifice for each other is to imitate the kenotic God[8] who created us and died for us on the cross, and we can only presume that there is no other way to make such agape than via the agony and ecstasy of nature. As Peacocke (2004, 63) avers, "God has been creating all the time through eliciting all the possibilities of the matter which he brought into existence endowed with certain potentialities and governed by the laws of its transformations."[9] Or so Christian faith believes and evolutionary biology, properly conceived, gives us no compelling reason to deny. In fact, one of the paradoxical merits of evolutionary game theory is that, in taking eudaimonism to its dubious logical conclusion, it helps Christians resolve an ancient debate in favor of agapism.

Notes

1. Richard Dawkins (1989, 4) also explicitly states that his "definitions of altruism and selfishness are *behavioral,* not subjective." Robert Wright (1994, 186) observes that "behavior, not thought or emotion, is what natural selection passes judgment on."
2. Robert Trivers (1971, 189) observes that "models that attempt to explain altruistic behavior in terms of natural selection are models designed to take the altruism out of altruism." Trivers cites in this connection W. D. Hamilton's landmark 1964 work and its principle that "altruistic" sacrifice must benefit, in reproductive terms, those closely genetically related to the sacrificer. As I note below, Hamiltonian "kin selection" or "inclusive fitness" is, in effect, expanded nepotism. Trivers's own account of "reciprocal altruism," however, makes it a kind of delayed symbiosis rather than genuine charity—namely, rather than genuine self-sacrifice uninterested in or uncaused by payback.
3. Kauffman (1993, xiii) makes a related point: "Darwin's answer to the sources of the order we see around us is overwhelmingly an appeal to a single singular force: natural selection. It is this single-force view that I believe to be inadequate, for it fails to notice, fails to stress, fails to incorporate the possibility that simple and complex systems exhibit order spontaneously."
4. I focus here on spandrels at the organismal level, but, in light of population genetics, there is no theoretical reason why equivalent features could not also

apply to groups. If so, the dynamics of a system could exercise what has been called top-down causal influence on the members of a species. In this instance, in Arthur Peacocke's (2004, 126) words, "the system as a whole has emergent properties not obvious from those of the constituents and in many cases not strictly predictable from them." Peacocke takes such top-down influence to be suggestive of how God is creative in and through the world.

5. In my lexicon, "(generous) tit for tat" is plausibly called "cooperation" (in the ordinary language sense) while "always cooperate" is a kind of unconditional love, aka altruism (in my sense).
6. Although Hamilton sometimes treats his principle as probabilistic (see 1964, 31, 44)—that is, as a predictor of likely behavior—he more typically sees it as a firm transmission rule. Only if $rb > c$, could the relevant behavior have been selected for, on this reading, and if the behavior cannot be selected for, it cannot exist over time. Evidently, this judgment is what moves Hamilton (1964, 16, 49) to aver that "a tendency to simple altruistic transfers ($k = -1$) will never be evolved by natural selection" and that apparent instances of such are presumably "biological error." (In this context, k is the ratio of gain involved in an "altruistic" action: the number of units of fitness received by others for every one lost by the "altruist." Simple altruism would entail recipients' gaining one unit of fitness for every two lost by the agent, such that $k = 1 - 2$.) I, in contrast, make a (agape) an independent variable precisely to emphasize that not all behaviors can or need be accounted for in terms of natural selection alone.
7. The quoted phrase is a slightly altered version of Frederick Temple's famous line (1885, 115), cited in Peacocke (2004, 79).
8. That is, a "self-emptying" God. See Phillipians 2:5–7: "Let the same mind be in you that was in Christ Jesus, who, though he was in the form of God, did not regard equality with God as something to be exploited, but emptied himself, taking the form of a slave, being born in human likeness."
9. On these matters, see also John Polkinghorne (2004). Peacocke (2004, 46–7) notes that "God as creator we now see as somewhat like a composer who, beginning with an arrangement of notes in an apparently simple tune, elaborates and expands it into a fugue by a variety of devices."

References

Saint Augustine. 1997. *On Christian Teaching* [395–426], trans. R. P. H. Green. Oxford: Oxford University Press.

Darwin, C. 1964. *On the Origin of Species* [1859]. Cambridge, MA: Harvard University Press.

Davies, P. 2004. "Emergent Complexity, Teleology, and the Arrow of Time." In *Debating Design: From Darwin to DNA,* ed. William A. Dembski and Michael Ruse. Cambridge: Cambridge University Press: 191–209.

Dawkins, R. 1989. *The Selfish Gene.* Oxford: Oxford University Press.

Gould, S. J. 1997a. "Darwinian Fundamentalism." *The New York Review of Books,* 44(10), June 12, 1997: 34–37.

———. 1997b. "Evolution: The Pleasures of Pluralism." *The New York Review of Books,* 44(12), June 26, 1997: 47–52.

———. 2002. *The Structure of Evolutionary Theory.* Cambridge, MA: Belknap and Harvard University Press.

Hamilton, W. 1963. "The Evolution of Altruistic Behavior." *American Naturalist* 97: 354.

———. 1964. "The Genetical Evolution of Social Behaviour I and II." *Journal of Theoretical Biology* 7: 1–16 and 17–52.

Jackson, T. P. 2003. *The Priority of Love: Christian Charity and Social Justice.* Princeton, NJ: Princeton University Press.

Joyce, R. 2006. *The Evolution of Morality.* Cambridge, MA: The MIT Press.

Kauffman, S. 1993. *The Origins of Order: Self-organization and Selection in Evolution.* Oxford: Oxford University Press.

Kirschner, M. W., and J. C. Gerhart. 2005. *The Plausibility of Life: Resolving Darwin's Dilemma.* New Haven, CT: Yale University Press.

Lewontin, R. 2000. *It Ain't Necessarily So: The Dream of the Human Genome and Other Illusions.* London: Granta Books.

Merriam-Webster's Dictionary, online, 2006, http://www.m-w.com/dictionary/altruism. Last accessed on March 5, 2006.

Nagel, T. 1974. "What Is It Like to Be a Bat?" *The Philosophical Review* 83: 435–50.

Nowak, M. 2008. Personal e-mail correspondence on January 24, 2008.

Nowak, M., and K. Sigmund. 2005. "Evolution of Indirect Reciprocity." *Nature,* 437 (7063): 1291–98.

Nygren, A. 1969. *Agape and Eros,* trans. P. S. Watson. New York: Harper and Row.

Oxford English Dictionary, compact edition. 1971. Vol. 1. Oxford: Oxford University Press.

Peacocke, A. 2004. *Evolution: The Disguised Friend of Faith.* Philadelphia and London: Templeton Foundation Press.

Plato. 1951. *The Symposium,* trans. Walter Hamilton. New York: Penguin Books.

Polkinghorne, J. 2004. "The Inbuilt Potentiality of Creation." In *Debating Design: From Darwin to DNA,* ed. William A. Dembski and Michael Ruse. Cambridge: Cambridge University Press, 246–60.

Ruse, M. 1986. *Taking Darwin Seriously.* Oxford: Blackwell.

Sober, E., and D. S. Wilson. 1998. *Unto Others: The Evolution and Psychology of Unselfish Behavior.* Cambridge, MA: Harvard University Press.

Trivers, R. 1971. "The Evolution of Reciprocal Altruism." *Quarterly Review of Biology* 46: 189–226.

Wright, R. 1994. *The Moral Animal: Why We Are the Way We Are.* New York: Vintage Books.

∴ VI ∴

Cooperation, Metaphysics, and God

17

Altruism, Normalcy, and God

Alexander Pruss

There are two kinds of facts. Normative facts contain a claim about how things, in some sense, ought to be or function: "We *should not* harm others without good reason." "Her car is *broken*." "That was a *mean* thing to do." "Sheep *are four legged*." "Smithers committed a *foul*." "It's not *normal* to put food in one's ear instead of one's mouth when hungry." Nonnormative facts tell us only about how things actually *are* in the world. Some apparent examples: "2 + 2 = 4." "Electrons have spin and electric charge." "He caused her severe pain without believing himself to have a justification for it." "Her car is unable to start." "She told him that she always secretly laughed at him." "Most sheep have four legs." "Jones said that Smithers committed a foul." "One's hunger will not go away if one puts food in one's ear instead of in one's mouth." (I said these examples are "apparent" because it may be that in the end some of the concepts here also have hidden normativity in them—maybe "saying" or "hunger" are in part normative.)

Modern science specializes in explaining nonnormative facts in terms of further nonnormative facts. In doing this, it is a departure from an earlier Aristotelian science that would sometimes explain nonnormative facts in terms of nonnormative facts, just as modern science does, but would also sometimes explain facts, whether normative or not, in normative terms. Thus, an Aristotelian explanation of why I have sharp teeth in the front would involve the normative fact that it is *normal* for members of my species to have sharp teeth in the front. And in turn this normative fact would be explained by talking of a further normative fact about the *function* of these teeth—these teeth are for *cutting*, and cutting is promoted by the sharpness of the teeth.

Modern science has thus both constricted the range of facts to be explained and the range of admissible explanations. This constriction should not be seen in a negative way: it is one of the specializations that have made many of the great successes of science possible. The biologist looks for biological explanations of biological phenomena. This does not mean that she believes the absurd thesis that all phenomena are biological, or even that all biological phenomena have biological explanations (presumably the origin of life cannot have a biological explanation, on pain of vicious regress). Rather, she heuristically restricts the range of her inquiries, thereby making her task more manageable. Nonetheless, her inquiries may end up crossing disciplinary boundaries. Somewhat similarly, science in general looks for nonnormative explanations of nonnormative phenomena, but the scientist need neither believe the absurd thesis that there are no normative phenomena, nor even the thesis that normative facts never appear in explanations of nonnormative phenomena.

What I said so far is not without controversy. For one might, instead, think that normative facts can be *reduced* to nonnormative ones so that ultimately the distinction is of no avail. For instance, we might claim that to say that Patricia's car is broken is nothing mysterious. To say that her car is broken is just to say that it does not do what the manufacturer, in fact, intended it to do. This seems to be a reduction of a normative fact about her car—that it is not fulfilling its function—to a nonnormative fact about the intentions of the manufacturer.

In cases where a normative fact can be reduced to a nonnormative one, a largely scientific explanation of the normative fact might be available. Thus, after reducing the claim about the brokenness of the car to a claim about its lack of conformity to the manufacturer's intentions, I might go on to give a scientific explanation of why it fails to conform to these intentions. Such an explanation involves two steps: first, a conceptual one—that of reduction—and second, a scientific one—that of explaining the nonnormative claim.

A particular reductive claim may be highly controversial. Thus, one person might attempt to reduce the normative fact that slavery is wrong to the nonnormative fact that our society disapproves of slavery, while another might try to reduce it to the (nonnormative?) fact that God disapproves of slavery. In the next sections, I will argue based on specific cases related to cooperation that it is unlikely that all normative facts can be reduced to nonnormative scientific facts.

As the present book shows, there is excellent scientific work on how human tendencies to cooperation might have evolved. This work generally makes no reference to God or other supernatural influences. It appears, then, that we might eventually be able to give a thoroughly naturalistic explanation of how we came to have these tendencies, an explanation that makes no reference to anything but the nonnormative kinds of things that science deals with.

However, observe two things. First of all, in this scientific work, "cooperation" has to have a precise, scientifically amenable and thus nonnormative meaning. Second, what the scientific explanation explains is not why cooperation is *normal* or *appropriate* for humans but what makes humans, *in fact,* have a tendency toward it. I will show that attending to these issues shows that even after the scientist is done explaining cooperation, there will be work left for the philosopher or theologian, and moreover doing some of this work may provide us with evidence for the existence of God.

Biological Cooperation and Moral Altruism

Take this book's definition of cooperation in terms of losing a fitness cost in such a way that another derives a fitness benefit. On its face, this definition is not normative. The terms "cost" and "benefit" in ordinary English are often normative, but here seem to be defined simply in terms of what conduces to the propagation of genes or cultural artifacts.[1] One might question whether propagation is *really* a nonnormative matter, but I shall not press this worry. Instead, I want to distinguish this biological analysis of cooperation from what I will call "moral altruism" or "morally valuable altruism."

Suppose that Patricia is uninterested in having children, and so, instead of mating, she takes up the development of low-cost water purifiers as just a fun hobby. She develops such a purifier, and her invention saves countless lives in the third world. This is an instance of biological cooperation: Patricia's channeling of energies that could have been spent on childbearing carried a fitness cost for her, but a fitness benefit for people elsewhere.

Yet Patricia is a selfish person—she took up her hobby just for fun. There is no *moral* altruism there, and while we could honor her work in order to encourage other such inventors, we should not look up to her as a moral hero.

Nor will it do to tack a "motivation" condition onto "cooperation" in order to get morally valuable altruism, for instance, insisting that morally valuable altruism is cooperation *motivated by love for another.* It is not morally valuable altruism if, to ensure that a beloved friend get a job, I calumniate a better-qualified competitor, thereby suffering a fitness cost to myself through the potential of being eventually found out.

Furthermore, an activity can exhibit morally valuable altruism even if it enhances nobody's fitness. Thus, to give solace to the dying need not increase anybody's reproductive or cultural fitness. Nor need the thing one sacrifices be fitness. If you suffer pain in my place, you may be acting altruistically even if the pain does nothing to your reproductive or cultural fitness.

The above remarks indicate that exhibiting *moral* altruism is a normative property in at least two ways. First, the action has to be *morally praiseworthy.* Second, one is sacrificing a private *good* for the sake of the *good* of another. Reproductive fitness may not be a normative concept, but the good *is* normative, through and through.

In having explained the origins of the tendency to cooperation, or even cooperation motivated by love of the other, then, we still have not explained the origins of the tendency to *moral* altruism. The explanation there cannot be entirely scientific, at least not in the modern post-Aristotelian sense of "scientific," since moral altruism is a normative feature, dependent upon both what is good and on what is morally praiseworthy.

That is not to say that scientific work cannot be helpful here. For instance, it might turn out that in most situations biologically cooperative and morally altruistic actions are, in fact, identical. If so, a scientific explanation of why we tend toward biologically cooperative actions may partially explain why we tend toward morally altruistic ones (but see Fodor 2008). However, there would still remain a fascinating question: *why* is it that the biologically and morally altruistic things to do are, in fact, identical in most cases?

We could, after all, imagine selective pressures and situations that would drive the two apart. For instance, in a species where older individuals are a repository of experience, caring for an elderly member of the community may be biologically altruistic: it increases the reproductive fitness of other members of the community who benefit from the elderly member's wisdom, and so such care may be a sacrifice of one's own fitness for that of another. But one might also have a species where *no one's* fitness is improved by caring for elderly members of the community, for instance, because senility sets in early. In that case, the care would be morally altruistic but not a form of biological cooperation.

If our species is one where what is biologically cooperative and what is morally altruistic by and large coincide, this interesting coincidence calls out for an explanation. This explanation cannot be wholly scientific in the modern sense, because what is to be explained is in part normative.

A theist can offer the following explanation: God wanted there to be a species where body and soul would be united, with biological and spiritual goals harmonizing and interpenetrating to a significant extent, and thus where biological cooperation and moral altruism would be closely united. Furthermore, it may be that God wanted the world to participate in creation, and thus set things up so that various aspects of us would be developed through natural means, such as evolutionary ones, rather than created *ex nihilo,* and so God arranged for our evolutionary history to take place in an environment where the biologically cooperative and morally altruistic would usually coincide.

We can understand the kinds of motives that could make God desire this kind of arrangement, and so what I stated *is* a putative theistic explanation of the phenomenon, a species of the general form of explanation where we explain a phenomenon in terms of the intelligible motives that drove a person to produce it (e.g., explaining an arrangement of rocks in terms of an artist's agency). Now each new puzzling phenomenon that a theory can help explain provides additional evidence for the theory—that a physical theory that posits electrons can explain features of the Brownian motion of dust particles gives some evidence for the existence of electrons. The above theistic explanation for the convergence of biological cooperation and moral altruism thus provides some evidence for theism. How much evidence does it provide for the existence of God? Well, that depends on whether good nontheistic explanations for the phenomenon could be given instead.

If, on the other hand, it is false that most of the time biological cooperation and moral altruism coincide, the scientific story about the origins of biological cooperation is not by itself sufficient to tell us why there is moral altruism. And then a theist might again offer an explanation in terms of God's creative activity.

Normalcy

Let us now consider a second science-transcending aspect of altruism, starting with this seemingly irrelevant observation: "It is *normal* for an adult sheep to have four legs." What does this "normalcy" claim mean? Maybe it just means most sheep do, in fact, have four legs. This would let us reduce the normative claim about its being normal that adult sheep have four legs to the nonnormative statistical claim that most sheep have four legs. However, even if a mad sheep mutilator went around and cut one leg from every four-legged sheep, it would *still* be normal for an adult sheep to have four legs, even though, in fact, *no* sheep had four legs. The three-legged sheep would be in an *abnormal* state. There would be something *wrong* with such sheep: they would have a defect, and if a veterinarian managed to get one of these three-legged sheep to regrow the amputated leg, the sheep would thus be restored to its normal state.

So its being *normal* for an adult sheep to have four legs is not just a matter of statistics. We cannot reduce the normative claim that sheep *should* have four legs to a statistical claim about sheep, *in fact,* having four legs.

Maybe where statistics fails, DNA will help: Holly's DNA codes for four legs, and this makes four-leggedness normal for Holly. However this answer does not work either, for if Holly were a sheep whose DNA did *not* code for four legs, it would not follow that Holly's number of legs is normal. Rather, it would follow that Holly has a *genetic* defect. In the case of a genetic defect, we have an abnormal feature that, nonetheless, is the expression of the individual's DNA. So we cannot reduce what is normal for Holly to what her

DNA says, since sometimes the DNA is wrong. This argument is exactly parallel to the observation that we cannot reduce morality to our society's moral customs because sometimes our society's customs are simply wrong.

At this point it is tempting to try for something a bit more sophisticated. One way to proceed is through the notion of proper function. While I could extract my teeth and make a rattle out of them, that is not their proper function. Rather, the proper function of my teeth is to cut and grind food, and to help me fight enemies or maybe rivals. If a bodily system falls short of fulfilling its proper function, it is *defective.* A mammalian eye with no lens is defective, while a throat with a lens would be defective: the lack of a lens in the eye impedes vision while a lens in the throat impedes eating. One of the proper functions of the growth and self-maintenance processes of a sheep is to develop and maintain four legs. A sheep that did not develop and maintain four legs is a sheep that failed at that function.

We may be able to reduce a system's being normal to that system's functioning properly. But proper function is still a normative concept: it refers to what a system *ought* to do, rather than to what it does, since otherwise there would be no such thing as a defective system.

Now let us return to altruism. Suppose that we have given a full scientific account of how biological cooperation and maybe even moral altruism came about—the other chapters of the book give hope that such an explanation may one day be found. This will still leave unexplained a crucial fact. Currently, it is *normal* for a human to exhibit cooperation. A human who almost never cooperates is not just rare, but abnormal. On the other hand, if we go back far enough in our evolutionary history, we will get to species where most individuals did *not* exhibit cooperation and where, furthermore, the exhibition of cooperation would not have been normal. We may have to go *very* far back to get to such a species, given the extent of biological cooperation in the nonhuman world, but it seems likely that some of our evolutionary ancestors did not exhibit cooperation in the way that it has subsequently emerged.

We thus have at least *two* transitions to explain: (a) the transition from a species where most individuals do not exhibit cooperation to one where most do, and (b) the transition (perhaps simultaneous, perhaps not) from a species where it is not normal to exhibit cooperation to one where it is abnormal not to exhibit cooperation.[2] The first transition is one that, I have assumed, science can explain in nonnormative terms. However, the transition from cooperation not being normal to its being normal is a *normative* transition, because the *normal* is a normative concept. And unless we can reduce claims about normalcy to nonnormative claims, we cannot give a scientific explanation (at least in the post-Aristotelian sense of science) of the transition from cooperation not being a normal feature of an organism to its being a normal feature.

We have already seen that reductions of normalcy to statistics or DNA fail. I suggested that normalcy could be understood in terms of proper function. Maybe what makes cooperation be normal in our species, perhaps unlike in some protoreptilian species, is that we have systems (e.g., neuropsychological ones) whose proper function it is to impel us to act cooperatively, while some of our evolutionary forebears either lacked these systems, or these systems did not have cooperation as their proper function. However, unless we can explain how the change in the *proper* function of systems happened, this is of little use here. And since "proper function" is a normative concept, we have a problem here for a purely scientific approach. Again, I want to emphasize that the philosophical problem here is not the transition from a species where these systems did not produce cooperation to ones where the systems did, but the transition from a species where it was not the *proper function* of the systems to produce cooperation to ones where it was. The proper function of a system is not the same as what the system, *in fact,* causes, as cases of defect show.

There have, to be sure, been some serious attempts to reduce the concept of proper function to nonnormative facts of natural selection. Thus, Wright (1976) says that a feature *F* has *G* as its proper goal provided that *F* exists *because* it tends to produce *G*. In evolutionary cases, this would mean that some system has a particular goal if the system was evolutionarily selected for *because of* its tending to achieve that goal. This would reduce the notion of a proper goal or function to the nonnormative question of why some feature *F* exists. Thus, eyes do tend to produce visual representations, and we have eyes *because* they produce visual representations—it is because eyes produce visual representations that the genes coding for them were selected for. Hence, on Wright's account, visual representation is the proper function of the eyes.

However, Wright's simple account fails, as Alvin Plantinga (1993) has shown. For an example of the failure of Wright's account, suppose that it turns out that psychopathy has a strong genetic basis (this is at least imaginable) and a dictator kills everyone who lacks this disorder (she has it herself, of course). Then it will be true that the survivors have DNA that tends to produce the disorder, and moreover this DNA exists only *because* it tends to produce psychopathy. However, surely, the psychopath is abnormal, and will not become any more normal just because she has killed off every normal person.

Valiant attempts have been made to fix up Wright's account. For instance, Bedau (1992) adds conditions about the goodness of the function, while Koons (2000) has tried to add the condition that the function has to contribute to the harmony or homeostasis of the organism. Neither approach works, since the proper function of something could be evil (e.g., a dirty bomb) or destructive to the organism (e.g., a self-destruct gene).

The prospects for a reduction of the claim that cooperation is normal to nonnormative facts about natural selection are bleak. It seems we simply have to accept that the claim that cooperation is normal is not reducible to nonnormative claims within the province of post-Aristotelian science. But nonetheless, we do want an explanation of the normative transition from organisms for which cooperation was not *normal* to ones where it is *normal*. If that explanation is not going to be scientific, it will have to be of some other sort.

Observe that the same question can be asked in the case of other species: what made cooperation normal in social insects? Moreover, the question can be asked in the case of features other than cooperation: what made bipedality normal for us? But the question is more pressing in the case of cooperation in our own species, because for our social practices the normalcy of human cooperation is more important than the normalcy of ant cooperation or of human bipedality. If cooperation is not normal in our species, what right do we have to *expect* it of our fellows, indeed to *compel* it (paying taxes is a form of cooperation, and we compel it)? Ultimately, if a child lacks all musical talent, the parents may reasonably give up on teaching the child to sing, and imposing violin lessons on the child is inappropriate. Not so in the case of a child who lacks a talent for cooperation. Teaching such a child to cooperate is a challenge, but also a duty, and it is no unfair imposition on the child.

Theism can provide two kinds of explanations for the transition from cooperation not being normal to its being normal. Insofar as these are good explanations, their existence provides evidence for theism. Again, how much evidence they provide for theism depends on what alternate hypotheses explaining the same normative transitions can be thought up.

The first kind of theistic explanation, championed, for instance, by Alvin Plantinga (see, for example, Plantinga 1993), is a *reductive* account, but one where the facts reduced to are not scientific ones. On this view, the proper function of something, say a hammer, a car, or a flower, just *is* the function for which its creator intended it. On this view the transition that happened consists simply in the fact that our evolutionary forebears were not intended by God to be cooperative while we *are* intended by God to be cooperative. Thus cooperation was not among these forebears' proper functions, while it is among ours. This is a simple and elegant account, though it is too reductive for my own taste.

The second approach is more metaphysically involved and is based on Aristotelian metaphysics. On this account facts about normalcy or proper function in organisms cannot be reduced further. These facts are grounded in features of the *nature* of the organism. Nonetheless, we may inquire why these features obtain. Aristotle's own answer was that the species were eternal: there have always been human beings on earth with the same normative features that they currently possess. We now know that this was not so, and so we need to seek another explanation.

Now, it seems to be beyond the power of merely physical objects to effect *normative* changes. I can make a sheep have three legs, but I cannot make three-leggedness be *normal* for it. In fact, it seems that, to explain normalcy, a cause capable of producing the *natures* of things would be needed, not just a cause capable of transforming existing things: a cause that can make an animal *with a four-legged—or cooperative—nature*. We need a cause capable of generating new irreducible normative facts that do not simply derive from old ones. God would be such a cause: God is not just a transformative cause, but a creator who can make things that are of a completely new kind.

Thus, we have two ways of theistically explaining how *normative* changes took place, and this is some evidence for the truth of theism. Note, too, that this argument differs from God-of-the-gaps arguments such as those of Paley ([1802] 2006) and Behe (1996) in which positing a designer offers an explanation of something science *currently* cannot explain, because the present argument offers an explanation of a fact that for principled reasons science (of the post-Aristotelian sort) cannot possibly explain.[3]

Radical Moral Altruism

But what if one insists instead that there is, after all, some clever way to reduce normalcy or proper function to selective advantage, even though Wright, Bedau, and Koons have failed? Here I have a final, albeit speculative, argument. It is only *limited* altruism or cooperation that produces a selective advantage. An altruism that embraces genetically distant and nonreciprocating communities, or that reaches out to severely and permanently disabled members of one's own community, or that forgivingly cooperates with the chronically noncooperative, is maladaptive.

I speculate—and here the speculations cross over from philosophy into science—that for altruism to confer a selective advantage, the altruistic individual needs to have some kind of "altruism limiter," a system that limits the outreach of altruism to cases where there is some likelihood of an advantage from the altruistic behavior and where the long-term fitness cost to oneself is not too high. This advantage may accrue through kin- or group-selection, reciprocity, the convenience of simply cooperating rather than bothering to make a careful calculation (if someone asks us what time it is, it is easier to answer than to try to figure out if such cooperation makes sense in terms of fitness), or the like.

If there were an evolutionary account of the normalcy of altruism, it would only be a *limited* altruism that would be normal. Moreover, the evolutionary account would, if I am right, require the presence of an altruism limiter. A failure of the limiter would then be a defect leading the individual to extend her benevolence to genetically distant unhelpful individuals or to severely disabled members of her own community, thereby undercutting the selective advantages conferred by biological altruism.

Consider, for instance, the case of Jean Vanier. Vanier has founded the L'Arche communities, where developmentally disabled people live and pray together with those who do not have developmental disabilities. Is there a fitness advantage to Vanier or his kin from this? Probably not. While the love with which Vanier reaches out to the developmentally disabled is often reciprocated, this reciprocation does not seem to confer a *fitness* advantage on Vanier, and hence is not the kind of reciprocation that is relevant for evolutionary accounts. Nor is it that it was simpler to cooperate than to make some complex calculation to figure out whether there is a fitness advantage, since the costs are just too high: all of Vanier's life is taken up with this activity, and he has opted for celibacy. And of course even if Vanier's case turns out to carry with it some fitness advantage we have not noticed, we can surely find other cases, such as Mother Teresa's ministry to the dying, Albert Schweitzer's work in Africa, or people who rescued Jews from the Holocaust (for a good study of the last case, see Monroe, Barton, and Klingemann 1990).

If an evolutionary account of the normalcy of cooperation could be given in terms of selective advantage, what would be normal would be to have cooperation that is *limited,* and a failure to limit the extent of the cooperation in some way would be a defect, just as it would be a defect to fail to limit one's eating to edibles. Thus, *if* the normalcy of cooperation were to be understood in terms of selective advantage, Jean Vanier would be an unfortunate sufferer from hypotrophy of his altruism-limiter, and it would make sense for him to seek professional help to remedy this defect, perhaps through pharmaceutical means.

And this conclusion is absurd. Quite possibly what Vanier does is *better* than what is normal for our species, but it is not *ab*normal. Here we see that while earlier I was talking of a pair of concepts, the normal and the abnormal, in fact, there are three concepts: the normal, the abnormal (or subnormal) and the supernormal.

For instance, it does not seem normal for humans to have Einstein-level intelligence. But neither is it an abnormality, a defect. Rather, we may call it "supernormal." Note, however, a disanalogy between Einstein-level intelligence and Vanier-level altruism. It is very plausible that any story about the evolutionary origins of cooperation will include the development of an altruism-limiter, since a radical moral altruism that is not limited by anything other than what is morally good (thus, it will be morally bad to steal from Peter to benefit two Pauls) is less effective vis-à-vis fitness. However, it is less plausible that an evolutionary story about the development of intelligence requires a similar intelligence-limiter. While we do have a stereotype that people with Einstein-level intelligence are misfits, it seems not unlikely that they *do* tend to confer a selective advantage to the social groups of which they are members, say by helping to invent new weapons for the defense of these social groups, an advantage that outweighs the disadvantages.

Again, the problem is not whether an account can be given of how radical moral altruism arose in particular cases or in general. It might be quite easy to give such an account, say in terms of an altruism-limiter deficit. Rather, the problem is how one can account for such altruism not being abnormal. Since Wright-style evolutionary accounts of the normalcy of cooperation would seem to make radical moral altruism abnormal rather than normal or supernormal, we have good reason to dismiss such accounts.

However, radical moral altruism's not being abnormal coheres particularly well with *Christian* theistic explanations of the normalcy of cooperation: on such accounts, it is no surprise that radical moral altruism is normal or, more likely, supernormal, rather than abnormal, given that our altruistic activity is not only an image of the activity of a generous God who creates us ex nihilo without deriving any benefit from this, but is the image of a triune God whose nature is the radical mutual self-giving of three divine persons, one of whom died that we might live.

Notes

1. It could perhaps turn out that any adequate account of "fitness" will be through-and-through normative in nature. If that turns out to be true, one would have to either conclude that post-Aristotelian science does include normative concepts, or else selection-based explanations in biology would have to be revised or abandoned. (In this regard, see Fodor (2008) and the essay by Jean Porter in this volume). In the latter case, the problem of explaining the normalcy of cooperation would be supplemented by a problem of giving an explanation of cooperation that makes no reference to selection. If, on the other hand, post-Aristotelian science does include normative concepts, the problem of explaining normalcy would no longer in principle be beyond science. However, the theistic explanation would still, I think, be a quite plausible one. Note too that it does not seem plausible that normalcy reduces to fitness, even if fitness is a normative concept.

2. There could also be a third transition: from a species where it is abnormal to exhibit cooperation to one where it is neither abnormal to exhibit it nor abnormal to fail to exhibit it.
3. In the above, for simplicity, I thought of cooperation as itself selected for. But it could, instead, be the case that what is selected for is a complex of general capacities that make possible the adoption of a cooperative strategy by an individual or group. Yet, it is clear that it is *normal* for members of our species to adopt cooperative strategies in certain contexts. This normalcy, then, needs to be explained, and the same issues come up. Moreover, it seems likely that it is normal for humans to be in contexts where cooperative strategies should be adopted, and hence normal for humans to practice cooperative strategies. Finally, it seems that giving a nontheistic explanation for the normalcy of cooperation *simpliciter*, or of cooperation under appropriate conditions, becomes harder when cooperation is not itself selected for.

References

Bedau, M. 1992. "Where is the Good in Teleology?" *Philosophy and Phenomenological Research* 52: 781–801.

Behe, M. J. 1996. *Darwin's Black Box*. New York: Simon & Schuster.

Fodor, J. 2008. "Against Darwinism." *Mind and Language* 23: 1–24.

Koons, R. C. 2000. *Realism Regained: An Exact Theory of Causation, Teleology, and the Mind*. Oxford / New York: Oxford University Press.

Monroe, K. R., M. C. Barton, and U. Klingemann. 1990. "Altruism and the Theory of Rational Action: Rescuers of Jews in Nazi Europe." *Ethics* 101: 103–22

Paley, W. [1802] 2006. *Natural Theology,* with an introduction and notes by Matthew D. Eddy and David M. Knight. Oxford: Oxford University Press, 2006.

Plantinga, A. 1993. *Warrant and Proper Function*. Oxford: Oxford University Press.

Wright, L. 1976. *Teleological Explanations*. Berkeley, CA: University of California Press.

18

Evolution, Altruism, and God

Why the Levels of Emergent Complexity Matter

PHILIP CLAYTON

I wish to ask how best to connect the various pieces of the rather ambitious narrative of evolution, altruism, and God. Clearly it is a narrative that not all authors in this book wish to affirm; many are writing primarily to help readers understand what game theory in biology, economics, and the social sciences has to tell us about the phenomena of cooperation and altruism. But the fact that this particular group of chapters, on this range of topics, has been collected together does rather unavoidably raise the question of whether there is any unified story that connects the various pieces.

Of course, even to name this question raises doubts about whether "evolution, games, and God" can even be connected. Familiar enough are the voices of those who tell us that the sciences of evolution, properly understood, leave no place whatsoever for religious language. And no less vociferous are those who proclaim that God, properly understood, leaves no place for evolution. Ours is the age of *tertium non datur,* of "there *is* no middle way." Perhaps there is some comfort in this fact, however: nuanced narratives of connection are infrequently enough written, and even less frequently discussed with care, so something novel might actually emerge from the effort.

Before undertaking this task, I should note that much of this volume remains interesting even if the integrative task fails. Although the book's title might suggest that the text is limited to questions of evolution and religion, many of the other chapters offer some of the most sophisticated reflection on cooperation and altruism to be published over the last several decades. Even for those whose interest in religion is strictly limited to its biological functions (or dysfunctions), these treatments of the history, biology, social roles, and ethics of altruistic and cooperative behaviors advance our understanding of these phenomena.

Here, however, I would like to try something more. In what follows I presuppose an interest in the possibility that religious beliefs, behaviors, and values might represent something more than their biological functions. *If* they do, at least some religious phenomena will require explanations that are not given in exclusively biological terms. The question therefore immediately arises: must explanations of that sort negate, or at minimum conflict with, the best biological accounts of human social behavior, or could they supplement and extend the biological accounts? The best way to find out is to construct an account of this supplementing kind and then to see how it fares. In constructing an example of such an account here, I will defend the thesis that *recognizing the distinct features of the various levels of emergent complexity* is an indispensable condition for the success of theological explanations that take the best findings from the sciences and religion seriously. As it turns out, the concepts of cooperation and altruism provide a particularly helpful test case.

How the Definitions Introduce Psychology and Ethics

It is already an achievement, relatively rare in the evolutionary literature, that such a diverse range of authors can even agree on the definitions of terms. From the Introduction on, the various authors agree (by and large) that cooperation is "a form of working together in which one individual pays a cost (in terms of fitness, whether genetic or cultural) and another gains a benefit." Altruism is a logical subset of cooperation, defined as "a form of (costly) cooperation in which an individual is motivated by goodwill or love for another."

Defining terms clearly is one thing. But when confronted with even the *idea* of a theory of cooperation or altruism that runs all the way from unicellular organisms to religiously motivated self-sacrificial behaviors among humans, one may begin to wonder whether any significant commonalities actually extend across such a staggering range of phenomena.

And yet the various chapters of this book, I suggest, have managed to do just that. Already in the introduction, the editors signal that the volume is intended to make the case for the fundamental role of cooperation in evolutionary history, up to and including human culture (see the Introduction to this volume). Although the reader encounters the usual divergences of opinion, caveats, and nuancing along the way, he or she is provided with a wide range of arguments that cooperation is a principle at work across the whole spectrum of evolution (see again Nowak's chapter).

For the moment, let us assume that there are good reasons to think that cooperative behaviors are indeed pervasive across the process of evolution and that they are manifested across vastly different scales, levels of biological complexity, and phases of evolution. This significant result (again, assuming that it holds) immediately turns the spotlight of attention onto the question of altruism. What is it that sets off this particular subset of cooperative behaviors from cooperation in general? What criteria pick out distinctively altruistic actions? What do we assume about the natural world that we can identify such actions within it?

Some biologists will stumble over the definition of altruism given above: "a form of (costly) cooperation in which an individual is motivated by goodwill or love for another." According to this definition, no feature of an overt cooperative behavior is by itself sufficient to make it altruistic. Hence, it appears, no game theoretical studies, no population statistics, indeed no empirical study of behaviors by themselves will suffice to detect altruism. Detecting altruism, the definition suggests, requires one to include reference to the agent's *motivation* for cooperating. But motivations are tricky, even elusive things for empirical natural scientists. How can we test claims about intentions?

Note that the definition seems to imply that only those organisms can be altruistic to whom one can plausibly attribute *motives*. Motives are conscious or semiconscious internal states that serve as grounds for engaging in certain sorts of actions in certain situations and for not so acting in others. On the (plausible) assumption that motives belong only to those organisms of which one can reasonably predicate psychological states, the definition at first seems to run parallel to Sober and Sloan Wilson's famous distinction in *Unto Others* (1998) between evolutionary and psychological altruism. But here the word "cooperation" has been substituted for the set of phenomena that they label as "altruism." In marked contrast, the form of psychological cooperation called altruism in the present book is a *subset* of evolutionary cooperation. Cooperative behaviors count as altruistic only when carried out by an agent when (1) she possesses cognitive, affective, and dispositional states of sufficient complexity that they can count as motivating her actions, and (2) at least some of her actions are motivated by "goodwill" or "love for another."

Of course, one can speak of a given set of environmental stimuli *as if* they "motivated" the subsequent behaviors of an organism. But the language of motives is most naturally at home in (nonreductive) social sciences such as psychology and sociology. If Descartes had been right, such that humans alone possessed "mind" and animals were mere machines, animals would lack all motives. Conversely, when one attributes motives to nonhuman animals, one commits oneself to developing at least a rudimentary psychology in order to (help) explain their actions.

When one reflects more closely on this book's definition of altruism, one unearths a second important entailment. Recall that altruistic behaviors are those that are "motivated by goodwill or love for another." If one is to determine in a given case whether these two conditions are met, one will clearly need the services not only of psychologists but also of ethicists. Indeed, the entailment is stronger; they would have to be ethicists who consider "goodwill" and "love for another" actually to motivate behavior on at least some occasions. Strongly Marxist or Freudian theorists would not affirm that this sort of altruism really exists; nor, one presumes, would representatives of the "strong programs" in sociobiology and evolutionary psychology.

Indeed, some will argue that this definition of altruism has an even more radical consequence—namely, that love in the deepest sense exceeds the boundaries even of ethical theory. Timothy Jackson (Chapter 16), appeals to Anders Nygren (1969) and argues that love is an inner state that eludes ethical analysis because it introduces an irreducibly theological dimension. On this view, referring to "altruistic" or "agapistic" love requires one to include theological possibilities within one's scope of attention. One such possibility is that the existence of persons who "truly love" might turn out to be explainable, at least in part, by the existence of a divine source of love, a God whose nature is agape. That is, the existence of acts of "radical altruism"—acts in which individuals sacrifice their reproductive interests, and perhaps even their lives, without any prospect for even indirect genetic gain—might turn out to be linked in some way to the existence of a religious ultimate. (Again, note that my intent here is not to argue for the existence of such a religious ultimate but rather to explore what biological conclusions might be inconsistent with, or undercut, theological explanations of this sort.)

Now of course, from the mere fact that a definition has theological implications, it does not follow that those implications are true. Still, such implications *are* crucial for understanding the semantic space that is opened up by the definitions in question. These theological possibilities help to structure this book—as the title, *Evolution, Games, and God,* is meant to point out. Just as multiple authors ask whether altruistic phenomena can be fully captured and comprehended by biological explanations—can human altruistic behaviors be selected for and, if not, why do they persist across multiple generations?—so one must also ask whether acts of radical altruism (in the sense just defined) can be accounted for in terms of standard ethical theories. If not, do they involve some sort of "teleological suspension of the ethical," in Kierkegaard's sense of the term? Or could one show, less radically, that they introduce a set of questions that supplement standard biological and ethical theories without negating them? But what would that mean exactly, and how would it work? What would it look like to develop this possibility into a working hypothesis about the relationship between altruism, biology, and theology? It is to this project that we now turn.

Formulating a Theological Hypothesis

Above we considered the possibility that the existence of persons who "truly love" may be explainable, at least in part, by the existence of a divine source of love, a God whose nature is agape. What does it mean to examine this claim as a theological hypothesis? First, it requires some standards or criteria of assessment. For example, it must function in the conceptual space created by the definition of altruism that we have been considering. We should also require that it accepts, rather than undercuts, what the biological sciences have discovered about evolution, which means that it must be compatible with the sorts of conclusions and data presented in other chapters in this book. Finally, if it is to be a theological hypothesis, it must draw in some way on theological traditions. Arguing for the existence of God is not part of our task here; still, the hypothesis will ideally be compatible with the *possible* truth of at least some theological assertions.

For present purposes, we can understand the term "God" to imply that, whatever reality ultimately is, it is not *less* than personal, and that it has at least some of the features predicated of it by (at least some of) the world's theistic religious traditions. Theologians are persons who study and write theories about "God," in one of the many senses allowed for by this definition. Thus, for example, theologians speculate (among other things) on how biological evolution might be compatible with a divine ultimate reality. If some human cooperators are those who are motivated by "goodwill or love for another," theologians will presumably be those who maintain that at least *some* of these agents' altruistic actions—actions that may put them at a selective disadvantage—are viably understood within a theological framework.

One can claim that a theological hypothesis offers the *best* explanation for some set of natural phenomena, or merely that it provides a *viable* explanation (e.g., it is plausible, internally consistent, and does not conflict with scientific results; see Rota, Chapter 19). Claims of the former kind are often called *natural theology*. For example, a natural theologian might argue from (what she takes to be) the fundamental role of cooperation across evolutionary history to the conclusion that one should posit a benevolent God who created and perhaps also guides the evolutionary process. Claims for the *viability* of a theological hypothesis, by contrast, are most often associated with the project of a *theology of nature*. Thus one might argue that, assuming the existence of God, it is plausible to view evolution, in both its process and its results, as compatible with the natural and creative intentions of that divine power.

Note that many of the data and arguments might be shared in common between these two approaches. After all, the main difference between them is that the latter approach claims only *compatibility*, whereas the former claims that the evidential base is strong enough to support an *inference* from the world to God. Since many today find even the claims of a theology of nature to be epistemically suspect (to put it gently), I will limit myself in what follows to claims of this sort. Those who wish to develop my argument further in order to construct a natural theology based on biological evolution will know what additional arguments they need to make to substantiate their more ambitious conclusions.

We are considering the hypothesis that a theological account might provide part of the explanation for acts of radical altruism. How does one go about evaluating such a claim? Here, again, there are two different ways in which one might begin. One might ask what would have to be the case for a "theology of nature" of this sort to be successful. Or one could ask what could *not* be the case if the argument is to succeed. Of the two, the latter approach promises more immediate payoff. After all, if some conclusion *C* is *incompatible* with a theology of nature, and if we then find *C* to be well supported, we know that the theology of nature is in deep trouble. Because of the beauty of this more direct test, and because it is more consonant with the spirit of scientific inquiry, I begin here.

These distinctions allow for a simple, straightforward formulation of my hypothesis: *a theological account of altruism cannot meet the criteria I have formulated if the various stages or levels of evolution do not manifest a certain amount of explanatory independence.* Of course, we expect some general principles or patterns to run across all instances of biological evolution. If they did not, there would be no scientific study of evolution, and the term (used in the singular) would represent sheer equivocation. Still, the hypothesis is plausible only if, above and beyond the commonalities, the various stages or levels of evolution *also* make their own distinct contributions to the overall explanatory task.

Note, for example, that the treatment of cooperation in this volume (whether intentionally or by luck) is divided into distinct areas of study that the editors speak of as "levels." One could even interpret the book (Parts II–VI) as an ascending spiral of different disciplinary perspectives on the phenomenon of cooperation. I am arguing that if these levels turn out *not* to be sufficiently distinct, but instead it turns out that the results and theories formulated on any given level are reducible to the levels below, then the theological hypothesis concerning altruism that we have been considering will to that extent be undermined. Of course, theists can still claim that there are links between human altruistic behavior and a divine influence or model. But it will be much more difficult to conceive how the scientific and theological accounts are connected and how they should be integrated.

Levels of Emergent Complexity

What would it mean to claim that at least some of the levels or stages of evolution are "sufficiently distinct"? I will argue that any two levels in the natural world (call them L_1 and L_2) meet this requirement only if L_2 comes later, is more complex, and exemplifies what in the literature is called *strong emergence* (see Bedau 1997; Silberstein 2002, 2006; Clayton 2004; Clayton and Davies 2006). Before defining strong or "ontological" emergence, a few framing comments are in order.

Although one still occasionally finds authors who claim that all talk of emergent complexity is superficial and antiscientific, the philosophical analysis of the concept of emergence has, in fact, become extremely sophisticated over the last decade or so. In a classic article, el-Hani and Pereira (2000, 133) identify four features generally associated with the concept of emergence:

Ontological physicalism: All that exists in the space-time world are the basic particles recognized by physics and their aggregates.

Property emergence: When aggregates of material particles attain an appropriate level of organizational complexity, genuinely novel properties emerge in these complex systems.

The irreducibility of emergence: Emergent properties are irreducible to, and unpredictable from, the lower-level phenomena from which they emerge.

Downward causation: Higher-level entities causally affect their lower-level constituents.

If classical philosophical reductionism is true, emergence is false (see Howard 2007). There are many examples of this sort of reductionism, but two in particular played an important role in twentieth-century philosophy of science. Perhaps the simplest to grasp is the bridge law theory developed by

Oppenheim and Hempel (1948), Hempel (1965) and Ernst Nagel (1961) and classically summarized in Suppe (1977). Both sides of this debate grant that there are discrete natural sciences that utilize different mathematical (and sometimes nonmathematical) laws, study discrete sets of phenomena, and write down theories about different sorts of individuals. Both sides also agree that natural science is impossible if new "spooky" forces come into existence at the higher[1] levels, which are not composed of matter and energy as understood at the lower levels. Thus the key point of dispute concerns the relations between the levels. If there are bridge laws that can predict all relevant phenomena at L_2 in terms of L_1 laws and initial conditions, then L_2 is reducible to L_1. If, by contrast, descriptive and theoretical terms from L_2 are indispensable for empirically fruitful explanations of the phenomena at L_2, then L_2 is not reducible to L_1. In this case L_2 is *strongly emergent* (compare Chalmers 2006).

One finds a reprise of the bridge laws debate in the extensive discussion of supervenience in philosophy of science and philosophy of mind over the last quarter century (see, e.g., Kim 2002; Kim 2008). According to a well-known definition, "A set of properties A supervenes upon another set B just in case no two things can differ with respect to A-properties without also differing with respect to their B-properties. In slogan form, 'there cannot be an A-difference without a B-difference.'"[2]

Though this is not the place to summarize the supervenience debate, two observations are crucial. First, note that "there cannot be an A-difference without a B-difference" could just happen to be true, or it could be the expression of an underlying natural law that links A-type and B-type phenomena. When supervenience theorists make the latter claim, as many do, they are holding a position very similar to the bridge-law theorists over half a century ago. Philosophers call this view "nomological" supervenience.

If nomological supervenience is true, no actual mental causation takes place. It may seem as though one of your ideas causes you to have other ideas or mental associations. But these are all epiphenomena. In fact, the real causal relations are taking place at the level of biochemistry or perhaps—depending on whether you think biochemistry ultimately reduces to physics—only at the microphysical level (see Kim 1998). By contrast, we saw above that strong emergence affirms real causal effects at emergent levels of organization in the natural world above the level of microphysics. On this view, new structures and forms of organization come to exist over the course of evolutionary history—molecules, cells, organisms, persons—and one can predicate properties and ascribe causal roles to these emergent entities.

Second, one could believe that L_1 and L_2 are linked by nomological supervenience without believing that L_2 phenomena can be predicted on the basis of L_1 phenomena alone (plus the relevant natural laws). Imagine that nomological supervenience holds between two levels, for example brain states (L_1) and specific mental states of an agent (L_2), but imagine that making the actual predictions should prove impossible, even in the long term. One might think that, in such a case, the theological hypothesis is fine, since the realm of what we cannot know (predict) preserves a place for God. But I believe that conclusion is mistaken. If mental experiences are nothing more than manifestations of neurological laws—direct consequences of the brain's anatomy and biochemistry—then there is no room for the sort of theological explanations that we have been considering.

The literature on emergence contains far more nuanced distinctions and criteria than the few I have introduced here, but this less technical description is sufficient for present purposes. The question now arises: should one accept something like strong emergence, or should one build one's arguments on the (explicit or implicit) assumption that philosophical reductionism (say, in the form of nomological supervenience) is true? Put differently, as one does science and interprets the results to the broader public, should one assume—indeed, is one obligated to assume—that sufficient scientific explanations of higher-order phenomena can eventually be given in terms of lower-order data and the relevant natural laws?

Many scientists do, in fact, make this assumption, which means that they are advocates of one or another form of what is known as philosophical reductionism. Most mathematical and evolutionary biologists, for example, assume that one can in principle (and some seem to believe that one can already in fact) formulate biological laws that sufficiently explain the behavior of higher-order organisms, such as human beings. Even human behaviors involving complex cognitive, affective, and moral beliefs and reasoning, they appear to believe, can be (or even have already been) explained using biological laws. Similarly, cognitive psychologists whose work focuses on human mental processing seem to believe that theological beliefs, and indeed the entire range of religious language, can be sufficiently explained in terms of the functions that such ideas and beliefs serve for human organisms and communities. If they are right, there is no need to posit the actual existence of God or any influences associated with deity in order to fully account for the existence of the various human beliefs about deity. On their view, one can just as easily say, following Laplace, "I have no need of that hypothesis."

Note that the affirmation of philosophical reductionism (and thus the denial of strong emergence) is independent of whether or not the individuals in question happen to affirm or deny theism. One can accept strong emergence without holding any theistic beliefs, and one can consistently be a theist while denying that any strongly emergent properties exist in the natural world.

But what would be the *status* of one's theistic belief if one were both a theist and an advocate of philosophical reductionism? After all, there are costs for a philosophical (or theological) position other than outright inconsistency. Recall that we are examining the theological hypothesis that the existence of a deity might play some role in helping to explain acts of radical altruism. We now have enough evidence on the table to see that this theological hypothesis ceases to be credible if philosophical reductionism is true. For, if the reductionists are right, the real explaining of any level of phenomena in the natural world is accomplished by the levels below it. For example, human mental phenomena must be explained at the level of brain anatomy and biochemistry or, ultimately perhaps, at the level of microphysics. If beliefs about deity are sufficiently explained in purely empirical terms (i.e., by evolutionary biology, evolutionary psychology, cognitive science, etc.), it becomes arbitrary to appeal to the *actual* existence of God in order to account for why such beliefs exist. In that case, statements about what God is or does obviously cannot play any role in helping to explain acts of radical altruism.

By contrast, strong emergentists claim that some human actions require explanations that are given, at least in part, in terms of persons: their intentions, thoughts, and self-understanding. By itself, of course, strong emergence does not say anything at all about theology; hence it cannot be sufficient to ground any theological claims. *But it is necessary* if theological assertions are to play any role in explaining any human behaviors. Let me put the point differently: only if reductive explanations are not by themselves sufficient to explain human behaviors (such as acts of radical altruism) does it make sense to endorse theological explanations of such acts.

Emergence and Cooperation

Emergentists accept that there are no nonphysical forces in physics, no *elan vital* that makes living things alive, and no souls, understood as a kind of substance distinct from physical things. More complex objects and organisms are composed out of the same fundamental forces and particles of which all other objects in the natural world are composed. And yet evolutionary processes produce natural systems with diverse sets of new properties. These emergent forms of organization are what we know as the organisms (emergent entities) that are the actors in specific environments. The causal roles played by the various organisms, including their unique sets of properties, are essential to biological explanations at the various levels of emergent complexity in the biosphere.

On this view, it is an empirical fact about this universe that higher-order phenomena cannot be sufficiently explained solely in terms of lower-level laws (philosophical reductionism). In many cases, emergent phenomena must be explained using laws and empirical regularities that are distinctive to a particular level. Thus Stuart Kauffman and I have argued that no amount of knowledge of physics by itself is sufficient to allow one to derive the core principles of biological evolution. For example, the dynamics of Darwinian evolution could in principle run on a different physical base altogether. In (Kauffman and Clayton 2006), we defend five "minimal conditions" for biological agency: autocatalytic reproduction, work cycles, boundaries for reproducing individuals, self-propagating work and constraint construction, and choice and action that have evolved to respond to, for example, food or poison. Unicellular organisms meet these conditions, for example, when they move up a glucose gradient.

What this account offers is a way to conceive the emergence and increasing complexity of agency in the biosphere: "We take agency to be a matter of degree: some minimum conditions must be met for one to ascribe it at all, and then it increases (roughly, as a function of the increase in organizational complexity through evolution) until one encounters full, robust, conscious agency" (Kauffman and Clayton 2006, 517). For explaining the behaviors of some kinds of agents, cell biology is sufficient, whereas for others the complicated dynamics of population biology must be taken into account. But primate behavior is not fully explained without primatology, and for the primate Homo sapiens, social sciences such as psychology, sociology, and cultural anthropology play an indispensable role.

Consider how one would approach the phenomena of cooperation in light of strong emergence. (I can be brief, since other chapters have argued the case in some detail.) The standard interpretation of natural selection focused on competition and "the survival of the fittest." But that, we have learned, is not the full story. Successful biological systems are, in fact, achieved in more complex ways than through a Hobbesian "war of all against all." Ours is not a world in which each organism acts merely as an individual unit to maximize its own fitness. Instead, selection pressures are often better handled by cooperation between organisms of a single species and, indeed, frequently by cooperative behaviors between organisms. The new data reveal the natural world to be far more symbiotic, far more interconnected by webs of interdependent systems, than the standard model had assumed.

These systems, when modeled using mathematical game theory, appear to employ strategies that are analogous to the strategies that cooperating agents use in human social interactions. What is remarkable about the recent results is that they show that virtually all levels of the biosphere, from bacterial behavior to human behavior, are better modeled, in the long run, by giving an account of cooperating strategies alongside selfish or punishing strategies. Since the analogies used to describe human and other animal behavior over the last 150 years were dominated by selfish and antagonistic models, these results require significant changes in how one conceives the process of biological evolution.

This paradigm shift can be read in two distinctly different ways, depending on whether one interprets it in a reductionist or an emergentist manner. For the reductionist, the fact that a single set of equations models the data across such a wide range of systems suggests that fundamentally the same dynamics are occurring at each level. There are regional differences, of course, that are empirically interesting, but on this view the core explanatory work is done by a single overarching theoretical framework.

From an emergentist perspective, by contrast, the game theoretical models do not by themselves complete the explanatory task. Intentional co-operation among human agents is not fully explained by the analogy with simpler biological systems any more than mating rituals among the higher primates are explained by the law of the conservation of energy or the Schrödinger wave equation. On our view, the explanatory task involves discovering not only shared principles of action but also the specific differences that characterize each species and environment. Although one always hopes to discover general laws that hold across a large number of biological systems, accounting for the specific behavioral patterns of a given organism or population remains an equally crucial element of scientific work. The need to be concerned with specific features is a consequence of the fact that the biosphere is organized as a hierarchy of levels of increasing complexity.

In *Closure: Emergent Organizations and their Dynamics*, Cliff Josslyn (2000, 71) argues: "One crucial property entailed by closure is hierarchy, or the recognition of discrete levels in complex systems. Thus, the results of our discussion can be seen in the work of the hierarchy theorists. . . . A number of systems theorists have advanced theories that recognize distinct hierarchical levels over vast ranges of physical space. Each of these levels can, in fact, be related to a level of physical closure . . . that is, circularly-flowing forces among a set of entities, for example among particles, cells, or galaxies." The particular forms of organization at each new level play a crucial role. Biology is not just the static study of forms of organization; it also studies the continually changing forms of newly emergent dynamics. The ways that nature organizes itself, the ways that it produces increasingly complex systems, depend crucially on the form of organization at each particular level. Cooperative behaviors are interestingly different depending on whether one is studying prokaryotes, DNA expression, signal transduction cascades, Hox genes, symbiosis, social signaling, or human language.

This standard for scientific explanation has important ramifications for the study of cooperation at every level at which cooperative behaviors occur in the biosphere. It requires that one name individual chimpanzees and track the idiosyncrasies of individual behavior (Goodall 1986). It focuses attention on the specific features of reconciliation behaviors in higher primates, as in the work of Frans de Waal (1989, 1996, 1998). It requires one to conceive co-operation in complex animal societies as a by-product of the coevolution of culturally learned behaviors and genetically transmitted predispositions (Durham 1991; Deacon 1997). When one comes to the study of humans, it requires special attention to the ways that uniquely human concepts arise (Tomasello 1999) and become the "horizons of meaning" for human existence in the world (Konner 1982). Ultimately, it raises the question of what it means for humans to inhabit worlds constructed, in part, out of their own imagination and projection (Heidegger 1962).

In the past, scientists were often leery of this emphasis on the unique features of human interactions because it was often accompanied by the denial that we are also biological beings, the products of a long and complex evolutionary history. But there is no longer any scientific reason to deny the both/and, the need for both biological *and* cultural/psychological studies of human persons. Studies of emergent systems are important because they explore the dynamics of difference *in* similarity, of discontinuities arising out of deep commonalities that characterize much of evolutionary history, including the emergence of human distinctiveness out of deep genetic similarities with other life forms. We now know that it will be impossible to make sense of human cooperation and altruistic actions without making the fullest possible use of this both/and.

Theological Consequences

There would have been no way to determine in advance whether emergence or reduction better describes the biosphere; nature could have been either way. If I am right, the better evidence today suggests that biological and cultural evolution are characterized by strong emergence. Assume for a moment that this conclusion is correct. One then wants to know: what implications will this result have for religion in general, and for theological accounts of altruism in particular?

Consider, then, the form of a theology of nature that becomes possible if strong emergence is the case. (I use "if" because I do not claim to have established strong emergence in this brief chapter. Readers who doubt the existence of strongly emergent relations in the natural world are welcome to construe this closing section as a thought experiment.)

For a good example of the *wrong* way to approach this question, consider the movement known as intelligent design, or ID. ID defenders, such as

Michael Behe and William Dembski, argue that we should ascribe natural scientific status to certain religious-sounding propositions, such as the proposition that a supernatural intelligence created the universe. They thus introduce the concept of a pre-existent intelligent designer not in the context of psychology or anthropology, not in conjunction with human consciousness and moral valuation but in the domains of physics, astrophysics, and theoretical biology.

Contrast this with an emergence-based approach. We find ourselves compelled to introduce different sorts of explanatory categories across the scale of evolution. Thus, for example, understanding the human psyche and human social interactions requires at least some explanatory categories that do not play a role in explaining the behaviors of other mammals. The human quest for meaning, existential doubt, the role of peak experiences, confidence and sense of self, insecurity and self-doubt—all of these play important roles in making sense of individual and social human behaviors.

Something analogous happens when one turns to religious attitudes and beliefs. Religious persons orient a part of themselves toward an alleged reality that transcends or underlies the empirical world as a whole. Religious persons do not cease to be biological and psychological beings, and explanatory patterns that hold in those disciplines apply to religious persons as well. The difference in the religious case has to do with the object of belief. Of course, there are disanalogies as well. No one doubts that human groups, institutions, concepts, wishes, and intentions exist; the only question about the social sciences is whether they are finally reducible to biological categories. By contrast, very deep disagreements arise concerning the very existence of the alleged object or subject or level of reality with which religious persons believe themselves to be in relation.

"Natural theology" is the project of showing that one or more religious ultimates actually exist. But the framework of a "theology of nature" that has structured this chapter does not claim to be able to demonstrate the reality of a deity or any other religious ultimate. Instead, it aims only to show how and why a given framework of religious belief is plausible. In fact, in these pages I have pursued an even more humble goal. I have only asked what must be presupposed for a given theological hypothesis regarding acts of radical altruism to be viable. Indeed, among what might be a rather large number of necessary conditions, I have focused only on the question of strong emergence.

We found that theological accounts of radical altruism are viable only if one can make a case for the existence of multiple (strongly) emergent levels of organization in the natural world. If all human belief and action is fully explained in the terms of lower-level causal forces, and if human thoughts, values, and intentions are only placeholders for the "real" causal explanations of the future, there is no reason to think that religious motivations and attitudes involve anything more than aesthetic, emotional, and ultimately biological functions. But if, for whatever reasons, the resources of the natural sciences do not fully accomplish the explanatory task, if some human actions are strongly emergent and require irreducibly person-based explanations, it is at least not absurd to consider distinctively theological explanations of human acts of radical altruism.

Notes

1. Throughout this chapter "higher" and "lower" are used without any connotations of value. A level may be "higher" because it manifests more complex structure and functioning.
2. See McLaughlin and Bennett (2008) and the extensive bibliography on supervenience at the end of their article.

References

Bedau, M. 1997. "Weak Emergence." In *Philosophical Perspectives 11: Mind, Causation, and World.* Atascadero, CA: Ridgeview.

Chalmers, D. 2006. "Strong and Weak Emergence." In *The Reemergence of Emergence: The Emergentist Hypothesis from Science to Religion,* ed. Philip Clayton and Paul Davies. Oxford: Oxford University Press, 244–255.

Clayton, P. 2004. *Mind and Emergence.* Oxford: Oxford University Press.

Clayton, P., and P. Davies, eds. 2006. *The Reemergence of Emergence: The Emergentist Hypothesis from Science to Religion.* Oxford: Oxford University Press.

Deacon, T. 1997. *The Symbolic Species: The Co-evolution of Language and the Brain.* New York: W. W. Norton.

Durham, W. H. 1991. *Coevolution: Genes, Culture, and Human Diversity.* Stanford, CA: Stanford University Press.

el-Hani, C. N., and A. M. Pereira. 2000. "Higher-level Descriptions: Why Should We Preserve Them?" In *Downward Causation: Minds, Bodies and Matter,* ed. P. Bøgh Andersen, C. Emmeche, N. O. Finnemann, and P. Voetmann Christiansen. Aarhus, Denmark: Aarhus University Press, 18–42.

Goodall, J. 1986. *The Chimpanzees of Gombe: Patterns of Behavior.* Cambridge, MA: Belknap Press of Harvard University Press.

Heidegger, M. 1962. *Being and Time,* trans. J. Macquarrie. Oxford: Basil Blackwell.

Hempel, C. G. 1965. *Aspects of Scientific Explanation, and Other Essays in the Philosophy of Science.* New York: Free Press.

Howard, D. 2007. "Reduction and Emergence in the Physical Sciences: Some Lessons from the Particle Physics and Condensed Matter Debate." In *Evolution and Emergence: Systems, Organisms, Persons,* ed. Nancey Murphy and William R. Stoeger, SJ. Oxford: Oxford University Press, 141–57.

Josslyn, C. 2000. "Levels of Control and Closure in Complex Semiotic Systems." In *Closure: Emergent Organizations and their Dynamics,* ed. Jerry Chandler and Gertrudis van de Vijver. Annals of the New York Academy of Science Series. New York: New York Academy of Sciences. Vol. 901, 67–74.

Kauffman, S., and P. Clayton. 2006. "On Emergence, Agency, and Organization." *Biology and Philosophy* 21: 501–21.

Kim, J. 1998. *Mind in a Physical World: An Essay on the Mind-Body Problem and Mental Causation.* Cambridge, MA: The MIT Press.

———. 2008. *Being Reduced: New Essays on Reduction, Explanation, and Causation,* ed. Jakob Hohwy and Jesper Kallestrup. New York: Oxford University Press.

———. ed. 2002. *Supervenience.* Brookfield, VT: Ashgate.

Konner, M. 1982. *The Tangled Wing: Biological Constraints on the Human Spirit.* New York: Holt, Rinehart, and Winston.

McLaughlin, B., and K. Bennett. 2008. "Supervenience." In *The Stanford Encyclopedia of Philosophy,* ed. Edward N. Zalta. Fall 2008 edition, http://plato.stanford.edu/archives/fall2008/entries/supervenience/. Last accessed December 17, 2009.

Nagel, E. 1961. *The Structure of Science: Problems in the Logic of Scientific Explanation.* New York: Harcourt, Brace & World.

Nygren, A. 1969. *Agape and Eros,* trans. Philip S. Watson. New York: Harper & Row.

Oppenheim, P., and C. G. Hempel. 1948. "Studies in the Logic of Explanation." *Philosophy of Science* 15: 135–75.

Silberstein, M. 2002. "Reduction, Emergence, and Explanation." In *The Blackwell Guide to the Philosophy of Science,* ed. Peter Machamer and Michael Silberstein. Oxford: Blackwell Press, 182–223.

———. 2006. "In Defense of Ontological Emergence and Mental Causation." In *The Reemergence of Emergence: The Emergentist Hypothesis from Science to Religion,* ed. Philip Clayton and Paul Davies. Oxford: Oxford University Press, 203–26.

Sober, E., and D. S. Wilson. 1998. *Unto Others: The Evolution and Psychology of Unselfish Behavior.* Cambridge, MA: Harvard University Press.

Suppe, F., ed. 1977. *The Structure of Scientific Theories,* 2nd ed. Urbana, IL: University of Illinois Press.

Tomasello, M. 1999. *The Cultural Origins of Human Cognition*. Cambridge, MA: Harvard University Press.

de Waal, F. 1989. *Peacemaking among Primates*. Cambridge, MA: Harvard University Press.

———. 1996. *Good Natured: The Origins of Right and Wrong in Humans and Other Animals*. Cambridge, MA: Harvard University Press.

———. 1998. *Chimpanzee Politics: Power and Sex among Apes*. Baltimore, MD: Johns Hopkins University Press.

19

The Problem of Evil and Cooperation

Michael Rota

> Anyone who considers the course of nature, without the usual pre-determination to find all excellent, must see that it has been made, if made at all, by an extremely imperfect being. . . . Mankind can scarcely chuse to themselves a worse model of conduct than the author of nature.
>
> John Stuart Mill, Unpublished letter to the *Reasoner*

> If we are not obliged to believe the animal creation to be the work of a demon, it is because we need not suppose it to have been made by a Being of infinite power. But if imitation of the Creator's will as revealed in nature, were applied as a rule of action in this case, the most atrocious enormities of the worst men would be more than justified by the apparent intention of Providence that throughout all animated nature the strong should prey upon the weak.
>
> John Stuart Mill, "Nature"

John Stuart Mill wrote these words before Darwin's work on natural selection had become widely known. With the subsequent ascendancy of evolutionary theory, the picture of nature as "red in tooth and claw" seems to have become still more pronounced. Yet as Curtis (Chapter 3) discusses in this volume, the writing of Henry Drummond (1894, 19) suggests that, by the close of the nineteenth century, the Darwinian notion that life is characterized by incessant struggle was often unduly emphasized: "The final result is a picture of Nature wholly painted in shadow—a picture so dark as to be a challenge to its Maker, an unanswered problem to philosophy, an abiding offence to the moral nature of Man. The world has been held up to us as one great battle-field heaped with the slain, an Inferno of infinite suffering, a slaughter-house resounding with the cries of a ceaseless agony." In the minds of some of Drummond's contemporaries, certain facts about evolutionary history, facts relating to animal suffering, raise the problem of evil[1] in a particularly serious way. We might develop this line of thought as follows:

> If an all-powerful, all-knowing, perfectly loving God exists, he would not have made a world in which the strong survive by preying on the weak, and in which the way for one animal to thrive is by devouring or displacing other animals. A loving God would not have set up a world in which the life of animals is an antagonistic struggle of one (or a few together) against all others. But the life of animals is such a struggle, and so there must be no such God.

An immediate response to such an argument is that it misdescribes nature. While the life of animals is indeed characterized by struggle, it cannot accurately be described as an antagonistic struggle of one against all. The preceding line of argument focuses exclusively on pain, suffering, and competition, but this is to focus on what is in fact just one side of the story evolutionary biology has to tell us. Pleasure, flourishing, and cooperation are the other side of the story. Indeed, certain examples of cooperation and helping behavior among the higher mammals are quite striking. Robert Trivers (1985, 385), for example, describes a case in which a dying adult false killer whale *(Pseudorca crassidens)* had rolled over on his side and was "bleeding from the right ear, heavily infested with parasitic worms." Whales and dolphins in this situation will sometimes support a dying or wounded animal (holding it up near the surface), and this dying whale was surrounded by twenty-nine other adult false killer whales. This species, Trivers (1985, 385) continues, "[I]s normally an open-water species, but the adults remained with this stricken male [close to shore] for three days until he died. Prior to his death, the whales resisted being pushed out to sea, but after his death many left on their own and the others no longer resisted being pushed out to sea." Nature is indeed "red in tooth and claw,"[2] but it is not *just* red in tooth and claw. As the work of Martin Nowak, Karl Sigmund, and others has indicated (Nowak 2006 and this volume; Nowak and Sigmund 1998 and 2005; Dugatkin 1997; Sober and Wilson 1998), cooperation has been an indispensable factor throughout the evolutionary process. It even appears that the world we are in was poised from the beginning to develop forms of life that exhibit highly advanced types of cooperative, almost self-sacrificial, behavior (see also Conway Morris 1998 and 2003). Such advanced forms of cooperation among higher animals in turn provide a basis for truly altruistic behavior in any rational animals inheriting instincts and behavioral traits from those animals. Taking into account this fuller picture of the natural world, it is less difficult to believe that God might have chosen to create a world like this one.

In this way, an understanding of the role of cooperation in the evolutionary process can give a different tone to discussions concerning evolution and the problem of evil. It is just not the case that nature is nothing more than "a slaughter-house resounding with the cries of a ceaseless agony," to use Drummond's words. Still, it remains quite natural to wonder, if an omnipotent God exists, why did he allow the vast amount of animal suffering that the evolutionary process undeniably involves? Let me distinguish two related questions here. First, if God exists, why did he bother to use an evolutionary process at all—why not just create the various creatures specially and directly, and so avoid the millions upon millions of years of animal suffering that the evolutionary process involved? Second, given that God wanted to use an evolutionary process, why did he allow the animal suffering it involved, when he could have miraculously prevented it?

I turn to the first question first: why would God bother to use an evolutionary process at all? It could be said, in response, that we should not expect to be able to discover the answer to this question. If God exists, then God's intellect is immeasurably greater than ours, and there is little a priori reason to expect that we would be able to grasp his purposes. (As the heavens are higher than the earth, so are God's thoughts higher than our thoughts, to paraphrase Isaiah 55:8–9.) An inability on our part to think of a justifying reason for God to use an evolutionary process does not imply that there is no such reason.[3]

While not disagreeing with this initial point, I think we can say more. In what follows, I consider two possible reasons God might have had to employ an evolutionary process, rather than create specially and directly.

The Dignity of Causality

The first reason is suggested by a comment made by Thomas Aquinas. In a discussion of providence (*Summa theologiae* I.23.8), Aquinas considers the sense in which human beings can be said to help God. Characteristically, he begins by making a distinction:

> There are two ways in which someone is said to be assisted by another. In the first way, he is assisted in the sense that he receives power from the other, and this is the way in which someone who is weak is assisted. Hence, God cannot be assisted in this way. . . . In the second way, someone is said to be assisted by another in the sense that his action is carried out by the other, in the way that a lord is assisted by his ministers. It is in this way that God is assisted by us to the extent that we carry out what He has ordained. . . . This is not because of any defect in God's power, but rather because He makes use of mediating causes in order to preserve the beauty of the order of things and in order to communicate even to His creatures the dignity of being causes. (Aquinas, I.23.8 ad 2)

Setting aside Aquinas's point about the beauty of order, consider his idea about the dignity of being a cause, about the value that attaches to having the ability to exercise causality. Let us take this idea and apply it not just to human beings, but to created things in general (as Aquinas himself appears to do at *Summa theologiae* I.22.3). On this suggestion, it is a good thing for created things to share in this attribute of God—namely, that they can cause other things to be. In general and considered in itself, there is a value in being able to exercise causality. A being that can exercise causality is more perfect than is a causally inert being; it is higher up the chain of possible beings, so to speak.

Applying this general principle to the case of the production of new species, we reach the conclusion that it is a good thing for creatures to have the ability to produce new species. Perhaps God has chosen to bring about the development of material creation by means of an evolutionary process as a way of letting created things "do more"—as a way of giving creatures an important role in the way reality develops. Paraphrasing the words of philosopher Peter Kreeft (2002), God does nothing by himself that he can possibly delegate to his creatures. On this suggestion, God has selected an evolutionary process (rather than special creation ex nihilo) in order to confer upon his material creatures an important causal role in the development of material life.

Was there, perhaps, a different way to allow living material beings to produce the diversity of species we see today (or a similarly impressive range of species), a way that involved significantly less animal suffering than the amount that has actually occurred? It is possible that there just is no such way. And if there is no such way, not even an omnipotent God could bring about the good in question without allowing the level of animal suffering we see in our world (namely, the good that consists in living material beings playing a similarly active role in the production of future species). So it is possible that a level of prehuman animal suffering similar to the level that has actually occurred might be the necessary cost of giving to creatures the project of producing the flora and fauna we see around us.[4]

What are we to say about this proposal? Is the good of conferring the dignity of causality really worth all the animal suffering that has occurred? In assessing this question, we should consider the possibility that animal life is richer and more worthwhile for the struggle it involves. Some comments by Richard Swinburne are relevant here. Swinburne (1998, 160–92) provides an account of animal suffering within the context of a broader discussion of natural evil. He first argues that the natural evil that humans suffer makes possible the existence of other valuable goods, such as the courage of the person who does something good in the face of danger, or the sympathy of someone expressing compassion for the pain of another. Because Swinburne thinks that higher-level animals can have simple sorts of beliefs and desires and can perform simple sorts of intentional actions, he can then apply these same sorts of points to higher-level animals. He writes (1998, 171): "It is good that the intentional actions of serious response to natural evil which I have been describing should be available to simple creatures lacking free will. As we saw earlier, good actions may be good without being freely chosen. It is good that there be animals who show courage in the face of pain, to secure food and to find and rescue their mates and their young, and sympathetic concern for other animals. An animal life is of so much greater value for the heroism it shows." After developing these considerations, he asks (1998, 173),

> [I]s the opponent of theodicy really right to insist that the world would be better without the challenges to and courage shown by animals? I do not think so. The world would be much the poorer without the courage of a wounded lion continuing to struggle despite its wound, the courage of the deer in escaping from the lion, the courage of the deer in decoying the lion to chase her instead of her offspring, the mourning of the bird for the lost mate. God could have made a world in which animals got nothing but thrills out of life; but their life is richer for the complexity and difficulty of the tasks they face and the hardships to which they react appropriately. The redness of nature "in tooth and claw" is the red badge of courage.

With these thoughts, I leave it to the reader to evaluate the "dignity of causality" approach, and turn to a second possible reason for God to employ an evolutionary process, despite the animal suffering it involves.

Divine Hiddenness

The problem of divine hiddenness refers to the problem of why God does not make his existence completely evident to all people. If God greatly desires that human beings believe in him, why does God not give all of us undeniable proof of his existence? Why does God not make his existence obvious or completely evident, in the way that the existence of one's chair is obvious, or in the way that the roundness of the earth is now completely evident to us? Surely an omnipotent God *could* do this.

In order to discuss the hiddenness of God, I want to adapt a parable told by the Danish philosopher Søren Kierkegaard in his *Philosophical Fragments*. Suppose that there is a king who falls in love with a peasant woman. The king wants to marry her, but he does not want her to love him just because he is the king. He wants a real relationship with her, and wants her choice to be completely free. He fears that if he comes to her accompanied by his retinue, in all his kingly majesty and splendor, he will make it difficult for her choice to be one of complete freedom, and he will make it difficult for her to freely choose to marry him for the right reasons. So what can he do? He can disguise himself.

Analogously, God wants to bring us into personal communion with himself. He does not just want our belief, he wants our friendship, our love. But if love is to be love, it must be freely given. And if love is to be the best sort of love, it must be freely given *for the right reasons*—reasons having to do with our awareness of the goodness of the other, and not just a focus on our own satisfaction and self-interest. If God made his existence indubitably apparent to everyone, he might very well frustrate his goals in two ways. First, he might make it difficult for some people *freely* to choose to enter into a relationship with him. Second, even in a case where the indubitable awareness of God's existence did not take away someone's freedom (with respect to the choice for God), that indubitable awareness might make it hard for a person not to let the choice for God turn into a matter of mere Machiavellian calculation.

By hiding himself, God gives us a particular sort of freedom. This freedom in turn makes possible a much greater good than mere belief—namely, the good that consists in a person's choosing, freely and for the right reasons, to give his or her life back into God's hands.

In addition to these considerations about freedom and proper motivation, there is also a point to be made about deep moral change, about conversion of the heart. According to more than one religious tradition, a genuine response to God often involves a redirection of one's life and a renunciation of various cherished desires, including the desire to run one's own life in just the way one pleases. Faith can require a radical change of heart, and this fact has implications for human receptivity to the divine initiative. In the words of philosopher Laura Garcia (2002, 89), "not everyone will be eager to know the truth about God, since it calls for a kind of radical conversion to which most of us will find some level of internal resistance."

There is a certain frame of mind that results in one's seeing the existence of God not as good news but as a threat to one's own plans and one's own autonomy.[5] The news that God exists, if I am in this mindset, would strike me as bad news, because it would mean that there is a person who has the right to tell me what to do, who knows all of my actions, and who will judge me and punish me if I do not do what he wants rather than what I want. Suppose God revealed his existence to a person entrenched in this frame of mind. It is not at all clear that the person would immediately be moved closer to a genuine love of God and moral goodness. He might simply reject God's authority. Or he might resolve to toe the line but only because of a new application of self-concerned, egoistic reasoning. And even this resolution might be fleeting, given the human capacity for self-deception. It seems unlikely that God would facilitate a real conversion of heart by clearly revealing his existence to a person of this mindset.

Indeed, it is possible that a belief in God that was forced on someone (say, in virtue of signs and wonders) would only raise emotional barriers and produce resentment. Conclusive external evidence might push some people away from a relationship with God, and actually decrease the chance of their slowly opening up to God of their own accord.

So there are reasons to think that God would want to remain at least partially hidden from human beings. With this in mind, we can see a possible reason God might have had for bringing about the existence of life by means of an evolutionary process rather than through special creation. If God had created the various species directly and immediately, humans would very probably, at a certain point in the development of their scientific knowledge, come into possession of an almost unassailable theistic argument from biological design.[6] And perhaps this would be more evidence of his existence than God wishes us to have, or evidence of the wrong sort.

So perhaps one reason God had for using an evolutionary process (despite the animal suffering it involves) was to preserve divine hiddenness. If God had created all the species directly, he would have provided a level of external evidence for his existence that might well have hindered many people from freely choosing to enter into a relationship with him for the right reasons, and might even have hindered genuine conversion of heart, at least for some.

The foregoing also suggests an answer to our second question: given that God wanted to use an evolutionary process, why did he allow the animal suffering it involved, when he could have miraculously prevented that suffering? While length constraints prevent me from addressing this question in much detail, I shall try to sketch what I see as the broad outlines of an answer. Very briefly: an evolutionary process with no selection pressure will go nowhere (see van Inwagen 2006, 115). Suppose God's goal was to use an evolutionary process to bring about the existence of rational creatures, and yet do so without there being serious animal suffering in the process. If God had intervened to prevent all pain and bodily harm visited upon animals, he would have had to bring about the development of new species in miraculous ways.[7] Further, he would have had to deal with the effects of eventual overpopulation and competition over limited resources in a way that did not involve large-scale animal suffering, and this would surely require miraculous activity on a large scale. Now either this frequent miraculous action would be done in such a way as to be one day discernible by natural science, or not. If so, it would jeopardize divine hiddenness, and if not, it would involve some sort of systematic deception on the part of God.[8] But God would not want to do either of these two things (jeopardize divine hiddenness or systematically deceive humans).[9] So, given that God wants to use an evolutionary process in the first place, God would have a reason to refrain from miraculously preventing all the animal suffering that a normal evolutionary process would involve.

Conclusion

The view expressed by Mill, in the quotations with which I began, is still present today. Consider, for example, the assertions of philosopher David Hull (1991, 486), writing in *Nature:*

> What kind of God can one infer from the sort of phenomena epitomized by the species on Darwin's Galapagos Islands? The evolutionary process is rife with happenstance, contingency, incredible waste, death, pain and horror. . . . Whatever the God implied by evolutionary theory and the data of natural history may be like, He is not the Protestant God of waste not, want not. He is also not a loving God who cares about His productions. He is not even the awful God portrayed in the book of Job. The God of the Galapagos is careless, wasteful, indifferent, almost

> diabolical. He is certainly not the sort of God to whom anyone would be inclined to pray.

But perhaps things are not so clear as Professor Hull would have us suppose. When we take a fuller view of the natural world, observing its beauty as well as its brutality, the view that an omnipotent and loving God might have created a natural world like ours becomes considerably more credible. Add to this the observation that, if an infinite God does indeed exist, there is a good chance that God could have a reason for using an evolutionary process of which we are unaware. Add, further, that the use of an evolutionary process allows for the emergence of rational beings in a world where nature retains great causal integrity and allows for creatures to have an important causal role in the development of life. Consider, finally, that the use of an evolutionary process would preserve divine hiddenness. With all this in mind, it seems to me that the proposition, "If an omnipotent, omniscient, perfectly loving God exists, then there would not have been so much animal suffering" is dubious. And this is enough to undercut the claim that what we now know about evolutionary history, and the suffering it has involved, gives us good reason to think that God does not exist.[10]

Notes

1. In contemporary English usage, "evil" applies most naturally to that which is morally wrong. In discussions of the problem of evil, though, the term "evil" has a broader meaning, including (a) suffering of any sort (physical, emotional, or psychological), and (b) the physical defects or personal situations that cause such suffering, as well as (c) moral wrongdoing or moral fault. "Evil", in the present context, just refers to bad situations or events.

   "The problem of evil" typically refers to a problem for those who believe in God, understood as an all-powerful, all-knowing, and morally perfect being. Why, if God exists, does he permit the vast amount of pain and suffering that we observe in our world? This question suggests an argument against the existence of God, the argument from evil, which can be summed up in three sentences: We find vast amounts of horrendous evil in the world. If there were a God, we would not find vast amounts of horrendous evil in the world. There is, therefore, no God. (This formulation is quoted, with a few changes, from van Inwagen 2006, 56.)

2. To use Tennyson's phrase (from Canto 56 of his poem *In Memoriam A. H. H.*)
3. See Wykstra 1984 and Tracy 2007.
4. More precisely: it is possible that a level of prehuman animal suffering similar to the level that has actually occurred might be the necessary cost of giving to creatures the project of producing the flora and fauna we see around us, conditional on God's preserving other important goals that might be compromised by miraculous intervention to preclude all animal suffering.
5. The comments in this paragraph are drawn almost entirely from van Inwagen 2006, Lecture 8, especially 146–51. For more on divine hiddenness, see that lecture and Howard-Snyder and Moser 2002.
6. Perhaps I should say, "If God had created the various species directly and immediately, and not also 'covered up' the evidence of his doing so, humans would very probably, at a certain point in the development of their scientific knowledge, come into possession of an almost unassailable theistic argument from biological design." But because God is not a deceiver, we can rule out the scenario in which God creates species directly and then leaves evidence of an evolutionary process to deceive us.
7. When an evolutionary process reaches a point in which species capable of feeling pain appear, the further development of those species into new, more advanced species will almost certainly require either pain (to allow sentient creatures to navigate the world) or nearly continuous miraculous intervention. See van Inwagen 2006, 119.
8. Humans would be subject to systematic divine deception if, for example, God had created fossils in the ground 6,000 years ago, ex nihilo, with all the observable properties that real fossils actually have. I claim that such systematic deception would be incompatible with God's nature and purposes. For a defense of the claim that God has an obligation not to engage in such systematic deception, see Swinburne's discussion of the principle of honesty, in Swinburne 1998, 138–40, and 165.
9. What is more, the level of miraculous intervention required to sustain an evolutionary process without animal suffering would require or entail a massive irregularity in nature—the laws of nature would have to be suspended in massive ways. This might itself be a great defect in the world. On this point, see van Inwagen's approach to the problem of animal suffering, in van Inwagen 2006, 113–34.

   With some risk of oversimplification, van Inwagen's defense with respect to animal suffering can be summed up as follows. The existence of higher-level sentient creatures (such as the higher mammals) is a great good, but the only possible worlds in which they exist but do not suffer (more precisely: do not suffer in a way morally equivalent to the way they suffer in the actual world) are worlds that are massively irregular. A massively irregular world is a world in which the laws of nature fail to hold in some massive way. The worlds in which the higher-level animals exist but do not suffer (better: do not suffer in a way morally equivalent to the way they suffer in the actual world) either require God to intervene in miraculous ways, almost ubiquitously, to protect animals from suffering, or are systematically deceptive (as would be a world where animals behaved just as they actually do but do not, in fact, feel pain). But massive irregularity is a defect in a world, a defect at least as grave as the defect of containing animal suffering of the sort that has occurred in our world. So, for all we know, God has permitted animal suffering because the alternatives (a world without animals at all, or a massively irregular world with animals) were worse, or at least not better.

10. For their helpful comments on this material, I am grateful to the members of the Evolution and Theology of Cooperation project at Harvard University, especially Sarah Coakley, Justin Fischer, Maurice Lee, Philip Clayton, and Martin Nowak, as well as Jon Jacobs and Jon Stoltz, my colleagues at the University of St. Thomas. I am also thankful to the John Templeton Foundation for its generous support during the time in which this paper was written.

References

Aquinas, T. *Summa theologiae,* trans. A. J. Freddoso, http://www.nd.edu/~afreddos/summa-translation/TOC.htm. Last accessed August 7, 2012.

Conway Morris, S. 1998. *The Crucible of Creation: The Burgess Shale and the Rise of Animals.* Oxford: Oxford University Press.

———. 2003. *Life's Solution: Inevitable Humans in a Lonely Universe.* Cambridge: Cambridge University Press.

Drummond, H. 1894. *The Ascent of Man.* New York: James Pott and Co.

Dugatkin, L. A. 1997. *Cooperation among Animals: An Evolutionary Perspective.* New York: Oxford University Press.

Garcia, L. L. 2002. "St. John of the Cross and the Necessity of Divine Hiddenness." In *Divine Hiddenness: New Essays,* ed. Daniel Howard-Snyder and Paul K. Moser. Cambridge: Cambridge University Press.

Howard-Snyder, D., and P. K. Moser, eds. 2002. *Divine Hiddenness: New Essays.* Cambridge: Cambridge University Press.

Hull, D. L. 1991. "The God of the Galapagos." *Nature* 352: 485–86.

Kreeft, P. 2002. "Time and Eternity." Lecture given for the C. S. Lewis Institute at Oxford, http://www.peterkreeft.com/audio/20_cslewis_time-eternity.htm. Last accessed December 15, 2009.

Mill, J. S. [1847] 1986. "Unpublished Letter to the *Reasoner.*" In *Collected Works of John Stuart Mill,* ed. A. P. Robson and J. M. Robson. Toronto: University of Toronto Press, v. XXIV, 1082–84.

———. [1854] 1969. "Nature." In *Collected Works of John Stuart Mill,* ed. J. M. Robson. Toronto: University of Toronto Press, v. X, 373–402.

Nowak, M. 2006. "Five Rules for the Evolution of Cooperation." *Science* 314: 1560–63.

Nowak, M. A., and K. Sigmund. 1998. "Evolution of Indirect Reciprocity by Image Scoring." *Nature* 393: 573–77.

———. 2005. "Evolution of Indirect Reciprocity." *Nature* 437: 1291–98.

Sober, E., and D. S. Wilson. 1998. *Unto Others: The Evolution and Psychology of Unselfish Behavior.* Cambridge, MA: Harvard University Press.

Swinburne, R. 1998. *Providence and the Problem of Evil.* Oxford: Clarendon Press.

Tracy, T. 2007. "The Lawfulness of Nature and the Problem of Evil." In *Physics and Cosmology: Scientific Perspectives on the Problem of Natural Evil,* ed. R. J. Russell, N. Murphy, and W. R. Stoeger. Vatican City State: Vatican Observatory Publications and Berkeley, CA: Center for Theology and the Natural Sciences.

Trivers, R. 1985. *Social Evolution.* Menlo Park, CA: The Benjamin/Cummings Publishing Co.

van Inwagen, P. 2006. *The Problem of Evil: The Gifford Lectures Delivered in the University of St. Andrews in 2003.* Oxford: Clarendon Press.

Wykstra, S. 1984. "The Humean Obstacle to Evidential Arguments from Suffering: On Avoiding the Evils of 'Appearance.'" *International Journal for Philosophy of Religion* 16: 73–93.

20

Evolution, Cooperation, and Divine Providence

Sarah Coakley

This final chapter in our interdisciplinary volume is about the seemingly problematic relation of evolutionary processes, as scientifically understood, and the classic Christian doctrine of divine providence. Let me divide what I have to say into two main parts, with a hinge passage between them that will explain why I see most (popularly known) *current* options in response to the problems of evolution and divine providence as unsatisfactory. In the first part, I shall examine what I see as the *three* most profound problems for Christian theism since the advent of Darwinism, so profound as to cause many to see Darwinism as a defeater of Christian belief. These problems were certainly not absent before the discovery of evolution; indeed they are classic inheritances from Christian philosophical theology and apologetics, which have exercised Christian thinkers at least since the third century CE. But evolutionary theory has certainly sharpened them in particular ways that—I would insist—responsible contemporary Christians cannot now avoid confronting. In the second section, I shall highlight the *particular* insights and nuances that our research team at Harvard ("Evolution and the Theology of Cooperation," 2005–2008) brought to the solution of these questions. In short, I want to suggest not only that we can effect a convincing response to the problems in the first section, but that, in addition, we can provide a distinctive and novel approach resulting specifically from our new understandings of the significance of cooperation as an evolutionary phenomenon.

Three Fundamental Problems in the Relation between the Created Realm and Divine Providence, Given Evolutionary Theory

In this section I am assuming a classical understanding of the Christian God—that is, a God who is Being itself, creator and sustainer of all that is, eternal (i.e., atemporal, omnipresent), omniscient, omnipotent, all-loving, indeed the source of all perfection. One solution to the problems we confront in this section is to give up on one, or more, of these classical attributes for God. But for the meantime I shall not entertain that systematic option—since I suspect it results from a failure to think through the full logical implications of divine atemporality (see Leftow 1991)—even though it cannot, a priori, be ruled out. One of the deficiencies in many previous accounts of the problems addressed in this section, however, has been artificially to extrapolate the debates from the trinitarian and incarnational dimensions of this classical Christian theism—that is, covertly to assume that it is deism, rather than Christian theism, that is a stake in the attempt to construe the relation between God and the created process. In what follows I shall attempt to avoid this mistake, anticipating some of the themes that the consideration of cooperation will also bring to bear in the second section.

What, then, are the three problems, mentioned above, that confront us when we try to see a coherent relation between a good, providential deity, and the unfolding created process? First, there is the issue of how we should understand the relation of God's providence to prehuman dimensions of creation and their development. Second, there is the issue of how God's providence can relate to the specific arena of human freedom and creativity. Then third, there is the problem of evil, the question of why what happens in the first two realms manifests so much destructiveness, suffering and outright evil, if God is indeed omnipotent, omniscient, and omnibenevolent.

Why does modern evolutionary theory intensify these problems? They were, after all, already confronted and tackled with some sophistication in classical Greek philosophy and in early Christian thought, and refined further in the much-ramified discussions of high scholastic medieval theology. But modern Darwinian evolutionary theory appears: (a) to underscore the contingency or *randomness* of evolutionary mutation and selection, and thus to render newly problematic the possibility of a coherent divine guidance of precultural evolution; (b) to bring further into question the compatibility of divine providence with the human freedom of the cultural evolution stage, given the deterministic and reductive assumptions of much evolutionary theory—bolstered more recently by genetic accompaniments to the original Darwinian vision. ("Freedom", for many secular scientists and philosophers, now looks like little more than an elbow room within a causally predetermined nexus [so Dennett 1984]; yet—paradoxically—one represented in much modern thought as straining toward an *autonomous* will to power that would precisely compete with, and cancel, an undergirding divine impetus); and thus (c) modern evolutionary theory appears to intensify thereby the problem of evil intolerably. If God is, after all, the author and sustainer of the destructive mess and detritus of both precultural and cultural evolutionary processes, why is s/he so incompetent and/or sadistic as not to prevent such tragic accompaniments to her master plan? If intervention is an option for God, why has s/he not exercised it?

A complete and detailed answer to these conundrums cannot be essayed here (for instance, I am not going into all the complexities of divine "middle knowledge" as one possible solution to the second problem); but some broad strokes and intuitions will help lead the way through to a preliminary solution. In the case of each problem there is a common contemporary misapprehension to be avoided, on the one hand, and some important enrichment and coloring from the Christian doctrines of Trinity and Incarnation, on the other hand, to add crucially to our reflection.

First, then, it is vital to avoid, in the case of precultural evolution, the presumption that God *competes* with the evolutionary process as a (very big) bit player in the temporal unfolding of natural selection. Once we are released from that false presumption, God is no longer—and idolatrously—construed as problematically interventionist (or feebly failing in such) along the same temporal plane as the process itself.[1] Rather, God is that without which there would be no evolution at all; God is the atemporal undergirder and sustainer of the whole process of apparent contingency or randomness, yet—we can say in the spirit of Augustine—simultaneously closer to its inner workings than it is to itself. And as such, God is *both* within the process and without it. To put this in more richly trinitarian terms: God, the Holy Spirit, is the perpetual invitation and lure of the creation to return to its source in the Father, yet never without the full—and suffering—implications of incarnate sonship. Now once we see the possibility of understanding the contingency of precultural evolution in this way, we need not—as so much science and religion dialogue has done in recent years—declare the evolutionary process as necessarily deistically *distanced* in some sense from God (see, eg., Polkinghorne 1989, 45: God *gets out of the way* so that evolution can happen contingently, and Polkinghorne calls this kenosis). Rather, let me propose in contrast—in a rather different understanding of divine "self-emptying"—that God is kenotically infused (not by divine loss or withdrawal, but by effusive pouring out[2]) into every causal joint of the creative process, yet precisely without overt derangement of apparent randomness. How can this be? First, it can be so because God's providential impinging on the evolutionary process, on this view, is not a miraculous or external *additum,* an occasional tinkering with evolutionary developments from the outside but, rather, the intimate, undergirding secret of the whole maintenance of the created order in being. And secondly, it can be so because we now know with ever greater precision, given the aid of the mathematical calculus of game theory, that evolutionary processes do occur within certain particular patterns of development. Even epistemically, then, we can now chart processes of remarkable evolutionary regularity, and, ontologically, there seems no *irrationality* in positing the existence of a transcendent (and immanent) divine providence, albeit one that kenotically "self-hides" in the spirit of incarnational presence.

But how, the skeptic might object, is evolutionary contingency, and (in a minute when we get to it) genuine human freedom, to be seen as *logically* compatible with this secret divine guidance? The intuition pump I want to propose here is what Peter Geach once called the chess-master model (Geach 1977). The basic idea is this: God is like a chess master playing an eight-year-old chess novice. There is a game with regularities and rules, and although there are a huge number of different moves that the child can make, each of these can be successfully responded to by the chess master—they are all already familiar to him. And we have no overall doubt that he is going to win. The analogy with God and the evolutionary process, or with human freedom, admittedly involves some stretching: for a start, God is the One here who has created the whole game. Also God timelessly knows what will happen on any different scenario depending on what moves occur. But there is a crucial difference here between God *knowing* what will occur and God directly *causing* what occurs. For, on this model the contingent variables and choices occur at the level of secondary causation (albeit undergirdingly sustained and thus primarily caused by God). What this means is that, in the case of atemporal divine providence, God's primary causation is not in any way in competition with secondary causations in the created realm.[3]

So now let us apply this same model to the problem of divine providence and human cultural evolution, including the evolution of genuine (indeterministic) freedom. The modernistic danger here is a slightly different—but closely related—one from the danger that we saw in the first problem (that is, the danger of assuming that God is a mere item, albeit big, in the temporal universe itself). For the problem here, secondly, is to think falsely of God as making human *autonomy* competitively constrained by divine action, rather than thinking of true human freedom as precisely right *submission* to the graced will and action of God (see Coakley 2002). In other words, once again we can think not deistically but trinitarianly and incarnationally of God. We can make Christ's agony in the garden, or his submission to divine will on the cross, as the hallmark and pattern of achieved human freedom rather than its supersession. Once we see human freedom, in its truest and best sense, as freedom *for* God, rather than freedom *against* God, then much of the force of the second problem falls away.[4] Not that suffering and sin do not remain—and that in apparently gross quantities. And that brings us immediately to our third problem.

And here, once more, there is an equally seductive modern misapprehension to avert, and that is the presumption that dying, or indeed evolutionary extinction, is the worst thing that can happen to anyone (or thing). But that, again, I would contest on theological grounds. This point is *not* to be misheard, note, as a seeming justification for avoidable suffering, victimization, or abuse, but it is to be heard *christologically,* as an insistence that the deepest agony, loss, and apparent wastefulness in God's creation may, from the perspective of atemporal divinity (and yet also *in the Son's* agony and "wasted" human death), be spanned by the Spirit's announcement of resurrection hope. Evil, from this perspective, is mere absence of good; death is the prelude to resurrection.[5] To be sure, the risk God takes in human "freedom" is the terrible risk that humans announce their *false* autonomy in cruelty and destructiveness; yet the risk is the only risk out of which the worthiest, and—again—most incarnational, forms of participation in God can arise. To underscore my profound difference here from some of the forms of late nineteenth-century optimistic meliorism that flourished in liberal theologies that seized incautiously on the recently discovered evolutionary phenomenon of altruism (see Curtis, Chapter 3): cooperation, as an evolutionary development, implies no facile moral optimism. Cooperation can ultimately lead, in its transformed human state, to very great good *or* very great evil: even dictators can at times be splendid cooperators for their own nefarious ends.

To sum up my argument so far: it is not that God has *not* intervened in the history of the evolutionary process to put right the ills of randomness and freedom. For, in one sense God is intervening constantly—if by that we mean that God is perpetually sustaining us, loving us into existence, pouring God's self into every secret crack and joint of the created process, and inviting the human will, in the lure of the Spirit, into an ever-deepening engagement with the implications of the Incarnation—its "groanings" (Romans 8. 23) for the sake of redemption. God, in this sense, is *always* intervening, but only rarely do we see this when the veil becomes thin, and the alignment between divine, providential will and evolutionary or human cooperation momentarily becomes complete. Such, indeed, we might hypothesize speculatively, was Christ's resurrection, which we call a miracle (or even an ahistorical event) because it seems, from a natural and scientific perspective, both unaccountable and random. Yet, from a robustly theological perspective, it might be *entirely* natural, the summation indeed of the entire trinitarian evolutionary process and thus its secret key.

Transition: Unsatisfactory Options in the Current Debates about Evolution and Divine Providence

These thoughts, now briefly enunciated, help us to see why the particular range of options currently popularized in the press in response to the evolution/God debate, seem curiously inept alternatives. Dogmatic scientific atheism, first, constantly goes well beyond the empirical evidences of evolution itself, and can give no convincing account of its own pessimistic reductionism; it thus falls on its own methodological sword. A suitably "apophatic"[6] Christian doctrine of creation is ironically far less "ideological, and thus—dare we say it—more *scientific*" than this sort of reductive neo-Darwinism (so Hanby 2006, 482). Intelligent Design, in inverse contrast, tends to assume a God who only occasionally bestirs himself to action; even if this were not already unacceptable theistically, its "solutions" prove deeply problematic and vulnerable scientifically as well. (This is not to say that there are not remaining areas of uncertainty, even nescience, in current evolutionary theory; nor is it to deny a priori—as we have stressed already—the possibility of ostensibly interventionist miracles. The issue is one of divine coherence and consistency.) The third option, however, which we may here call the no-contest position (as evidenced in much fine Roman Catholic theology of the twentieth century), also has its problems. Since our own view most closely approximates to this third option (of the three popularly discussed), it is worth clarifying what is deficient about a certain sort of no-contest position (let us dub it the lazy no-contest stance) before we pass into our last section and attempt to indicate how these problems might be rectified by a particular attention to the evidences of cooperation.

Lazy no-contesters, I would suggest, threaten to undermine their own intellectual credibility in at least three overlapping ways. First, by hermetically sealing the boundaries between science and theology, they merely invite the (obvious) scientific response of Laplace that "I have no need of this hypothesis." God, in other words, is so effaced from possible evidential discovery as to render her invisible, and thus fully *dispensable,* on scientific grounds. Secondly, such a divide tends to reinforce the—admittedly often smudged—separation between church and state that, at least in North America, keeps religious commitment in a subjective realm of preference rather than in a public realm of rational negotiation (witness the impassioned attack in 2007 by secular science professors on the Reason and Religion option proposed for the Harvard "core" undergraduate curriculum). Thirdly, and correlatively, the lazy noncontestation view therefore implicitly encourages the presumption that religious belief is *irrational,* or "personal/affective," rather than rational, accountable and arguable (albeit within a realm that also embraces significant mystery).

In short, the no-contest position is to be affirmed for its right insistence that God and the evolutionary process are not, so to speak, on the same level, whether temporally or in substance. But the same position is to be resisted if it attempts to avoid *all* critical vulnerability to the creative discoveries of empirical science. So we now need to consider, in our second and last section, how the discovery of "natural cooperation"—as what Martin Nowak in Chapter 4 calls the "third fundamental principle of evolution" (alongside mutation and natural selection)—might modify, or nuance, the no-contest position. Only thus, I shall suggest, can one avoid the dangers and pitfalls we have just outlined, and to which the position is so often subject.

How Cooperation Makes a Theological Difference

I shall confine myself to two basic points in this last section. The two points form a pincer movement in that they enunciate, both from a scientific and from a theological perspective, a necessarily *dialectical* pattern in the relationships between evolutionary and providential understandings of the world's processes, and one therefore that cannot leave the no-contest position unaffected, or in its "lazy" stand-off posture. It causes me to think of a model of science and theology as disciplines that mutually inspire, but also mutually chasten, each other. In short, if my intuition is correct here, the cooperative tendencies of evolution themselves *suggest* a natural *praeparatio* in the processes of selection for the potential later heights of saintly human self-sacrifice (only ultimately comprehensible as a response to divine grace), whereas the "eyes of faith," on the religious side, discern the phenomena of cooperation as already indications precisely of trinitarian and incarnational effects. What we have here, in other words, is a manifestation of the two-sided bridging model of the relation between evolutionary biology and philosophical theology in which empirical science acknowledges its explicative strengths *and* its philosophical limitations, and theology and metaphysics together strive to complete the vision toward which evolutionary cooperation seemingly gestures.[7]

On the scientific side first, then, the phenomenon of cooperation, seen now to be as deeply incalcated in the propulsion of evolution—from the bacterial level upward—as Darwin's celebrated principles of mutation and selection, provides a significant modification of the nature-red-in-tooth-and-claw image that Darwinism early accrued to itself. There is no less suffering or "wastage" on this model of evolution, to be sure, but what there *is* is an ever-present tendency *against* individualism or isolationism that only the application of the game theoretical calculus has been able successfully to explicate. The fear, then, often expressed in the twentieth century by the Vatican, that the embracing of Darwinism somehow encourages hostile competitiveness or individualism has to be severely modified. At the very least, and in advance of any ascription of religious meaning to the phenomenon of cooperation, evolution at significant junctures *favors* cooperation, costly self-sacrifice and even forgiveness; it favors in due course a rudimentary human ethical sensibility (so Marc Hauser, Chapter 13), and thus delivers—already in the realm of the higher pre-human mammals—tendencies toward empathy, toward a desire to protect others close to one at the cost of personal risk. At the very least, then, this is the seed bed for higher, intentional forms of ethical virtue, although these latter developments (with their complex forms of human intentionality and freedom of choice) are of a distinctively different sort from the prehuman varieties of cooperation and cannot, in my view, be *reductively* subsumed under mathematical prediction.

From the philosophical or theological side, on the other hand, and secondly, these same phenomena may seem to suggest the possibility of a new form of "moral/teleological" argument for God's existence (so Alexander Pruss, Chapter 17). Not that such an argument could ever amount to a "proof" in the deductive sense but rather be a constituent in a *cumulative* set of considerations that would together mount a case precisely for an incarnational God, a God of intimate involvement in empathy, risk, and suffering.[8] In this sense, then, not only would the no-contest view be modified and enriched, but both sides of the evolution and science divide significantly transformed in their understanding of their relation. To be sure, the agnostic or atheistical evolutionary biologist would continue to question (if not actively resist), the necessity of any such metaphysical speculation about the existence of divine providence. But the difference from an older perception of the two discipline's relations would be the explication of at least a *theoretical* capacity for bridging (not merging) the two discourses by discussion of particular evidences and their potential meanings and undergirding metaphysics. On the theological side, the great advance that this development would bespeak would lie in the intrinsic and immediate attention given to the doctrines of Incarnation and Trinity, rather than to the covertly deistic God who has—to great spiritual detriment and imaginative constriction—so dominated the science/religion debates since the Enlightenment.

Notes

1. Davies 1992 supplies an excellent account of this fundamental theme, and its implications, in one notable classical exponent: Thomas Aquinas. Many contemporary discussions of the (supposed) clash between evolutionary theory and theological metaphysics founder on a failure to understand this fundamental point about the absolute ontological difference between Creator and creation.
2. See Coakley 2002 for an exhaustive account of different possible theological renditions of "kenosis" in Christian tradition, and Coakley 2006 for my preferred understanding of Philippians 2 as a "suffusion" of divine love in the Incarnation, as classically expressed by Gregory of Nyssa in the fourth century.
3. There is a huge and complex literature in contemporary philosophy of religion about whether divine "foreknowledge" necessitates direct divine control of a sort that undermines human freedom. In the early-modern and modern period many lost confidence in Thomas Aquinas's own sophisticated response to this problem, in which he carefully distinguishes divine knowledge and divine causation and then subdivides causation into primary and secondary forms. For one of the best recent expositions of Thomas's position on this point, see Goris 2005. For a longer discussion of this issue in relation to the Christian doctrine of providence, see Coakley 2009.
4. Stump 2003 presents one of the most ingenious recent analytic renditions of Thomas Aquinas's understanding of human freedom's sustenance by divine primary causation. The exposition is, however, not without contention, given that Thomas himself appears to have changed his mind more than once on how best to express this paradox of the relation of divine providence and human freedom.
5. Davies 2011 provides one of the richest accounts of Aquinas's "solution" to the problem of evil, one which must be understood against the backdrop of his trinitarianism and his entire religious metaphysic, along with his important notion of evil as *privatio boni*.
6. In other words, one aware of what *cannot* be said about God (on account of his/her divine mysteriousness), as well as what can.
7. For a much more developed account of this line of approach, including a new rendition of "natural theology", see Coakley 2012.
8. Swinburne 2004 is the most celebrated contemporary account of a "cumulative case" for God's existence along probabilistic lines. Swinburne, however, does not

deal in his volume with the doctrinal complications of Trinity and Christology (which he treats only in subsequent monographs), and his "case" relies on a Bayesian account of probability, which can only deliver a result *just* sufficient for rational plausibility. For a different rendition of how arguments for God's existence might work in relation to the phenomena of cooperation and altruism, see Coakley 2012.

References

Coakley, S. 2002. "*Kenosis* and Subversion: On the Repression of 'Vulnerability' in Christian Feminist Writing." In *Powers and Submisions: Spirituality, Philosophy and Gender.* Oxford: Blackwell. 1–39.

———. 2006. "Does Kenosis Rest on a Mistake? Three Kenotic Models in Patristic Exegesis." In *Exploring Kenotic Christology: The Self-Emptying God,* ed. C. Stephen Evans. Oxford: Oxford University Press, 246–64.

———. 2009. "Providence and the Evolutionary Phenomenon of 'Cooperation': A Systematic Proposal." In *The Providence of God: Deus Habet Consilium,* ed. Francesca Aran Murphy and Philip G. Ziegler. Edinburgh: T&T Clark, 181–95.

———. 2012. "Sacrifice Regained: Evolution, Cooperation and God." The 2012 Gifford Lectures at Aberdeen University, http://www.abdn.ac.uk/gifford/about/.

Davies, Brian. 1992. *The Thought of Thomas Aquinas.* Oxford: Oxford University Press.

———. 2011. *Thomas Aquinas on God and Evil.* Oxford: Oxford University Press.

Dennett, D. 1984. *Elbow Room.* Cambridge, MA: The MIT Press.

Geach, P. 1977. *Providence and Evil.* Cambridge: Cambridge University Press.

Goris, H. 2005. "Divine Foreknowledge, Providence, Predestination, and Human Freedom." In *The Theology of Thomas Aquinas,* ed. Rik van Nieuwenhove and Joseph Wawrykow. Notre Dame, IN: Notre Dame University Press, Chap. 5.

Hanby, M. 2006. "Reclaiming Creation in a Darwinian World." *Theology Today* 62: 476–83.

Leftow, B. 1991. *Time and Eternity.* Ithaca, NY: Cornell University Press.

Polkinghorne, J. 1989. *Science and Providence: God's Interaction with the World.* London: SPCK.

Stump, E. 2003. *Aquinas.* London: Routledge.

Swinburne, R. 2004. *The Existence of God.* 3rd ed. Oxford: Clarendon.

Contributors

Johan Almenberg is Political Adviser to the Swedish Ministry of Finance. He holds a PhD in Economics from the Stockholm School of Economics.

John Hedley Brooke is an Emeritus Fellow of Harris Manchester College at the University of Oxford, where he was, before his retirement, Andreas Idreos Professor of Science and Religion. He is also a Distinguished Fellow of the Institute for Advanced Study at the University of Durham.

Philip Clayton is Ingraham Professor and Dean at Claremont School of Theology, and Provost at Claremont Lincoln University.

Sarah Coakley is Norris-Hulse Professor of Divinity at the University of Cambridge, and Deputy Chair of Arts and Humanities.

Heather D. Curtis is Assistant Professor in the Department of Religion at Tufts University.

Thomas Dixon is Senior Lecturer in History at Queen Mary at the University of London.

Anna Dreber is Assistant Professor in the Department of Economics at the Stockholm School of Economics.

Justin C. Fisher is Assistant Professor in the Department of Philosophy at Southern Methodist University.

Ned Hall is Professor in the Department of Philosophy at Harvard University.

Christoph Hauert is Associate Professor in the Department of Mathematics at the University of British Columbia.

Marc D. Hauser is an American evolutionary biologist and researcher in primate behavior, animal cognition, and human behavior.

Timothy P. Jackson is Professor of Christian Ethics at the Candler School of Theology and Senior Fellow at the Center for the Study of Law and Religion at Emory University.

Dominic D. P. Johnson is Alastair Buchan Professor of International Relations and Fellow of St Antony's College, Oxford.

Stephen M. Kosslyn is Director of the Center for Advanced Study in the Behavioral Sciences at Stanford University.

Maurice Lee is Assistant Professor in the Department of Religious Studies at Westmont College, Santa Barbara.

Friedrich Lohmann is Professor of Protestant Theology and Ethics at the Bundeswehr University of Munich.

Martin A. Nowak is Director of the Program for Evolutionary Dynamics and Professor of Mathematics and Biology at Harvard University.

Jean Porter is John A. O'Brien Professor of Theology at the University of Notre Dame.

Alexander Pruss is Associate Professor in the Department of Philosophy at Baylor University.

Michael Rota is Associate Professor of Philosophy at the University of St. Thomas.

Jeffrey P. Schloss is Distinguished Professor of Biology, T. B. Walker Chair of Natural and Behavioral Sciences, and Director of the Center for Faith, Ethics, and Life Sciences at Westmont College, Santa Barbara.

Index